STEM CELLS
Scientific Facts and Fiction

SECOND EDITION

STEM CELLS

Scientific Facts and Fiction

SECOND EDITION

Edited by

CHRISTINE MUMMERY
Leiden University Medical Center, Leiden, The Netherlands

ANJA VAN DE STOLPE
Formerly Hubrecht Institute, now Philips Research, The Netherlands

BERNARD A.J. ROELEN
Utrecht University, Utrecht, The Netherlands

HANS CLEVERS
Hubrecht Institute, KNAW, and University Medical Center Utrecht, The Netherlands

AMSTERDAM • BOSTON • HEIDELBERG • LONDON
NEW YORK • OXFORD • PARIS • SAN DIEGO
SAN FRANCISCO • SINGAPORE • SYDNEY • TOKYO

Academic Press is an imprint of Elsevier

Academic Press is an imprint of Elsevier
32 Jamestown Road, London NW1 7BY, UK
225 Wyman Street, Waltham, MA 02451, USA
525 B Street, Suite 1800, San Diego, CA 92101-4495, USA

First edition, 2010
Second edition, 2014

British Library Cataloguing-in-Publication Data
A catalogue record for this book is available from the British Library

Library of Congress Cataloging-in-Publication Data
A catalog record for this book is available from the Library of Congress

ISBN: 978-0-12-411551-4

For information on all Academic Press publications
visit our website at elsevierdirect.com

Typeset by MPS Limited, Chennai, India
www.adi-mps.com

Printed and bound in China

13 14 15 16 17 10 9 8 7 6 5 4 3 2 1

Contents

Preface

Throughout history, mankind has looked for the fountain of youth; the elixir for eternal life, keeping us forever young in body and mind. In the discovery of stem cells, has the modern biomedical scientist finally succeeded where wizards and alchemists failed? This book was written to address this and other questions on what stem cells can mean for health and society, now and in the near future. And because the field of stem cell research is moving so rapidly, just four years after its first edition, this book was in need of an update. This second edition includes new sections on biotechnology, "organs-on-chips," and the facts and fiction of stem cell therapies.

The organs of our body are built of a mixture of specialized cells that determine how each organ functions and what it does. These specialized cells have much shorter lives than our own: brain cells live for many years, but white blood cells, skin cells, and cells of the intestine survive sometimes for just a matter of days. This means that these cells continually need replacement. Stem cells hidden away in our body are the source of these new specialized cells. They have the lifelong task of waiting quietly until they are needed, and then they divide and replace a specialist cell that has in some way been destroyed.

Stem cells in bone marrow, which make all of our blood cells, have been used for decades to restore the blood and immune system after patients have been treated for leukemia, a cancer of the blood. These stem cells turned out to be remarkably simple to transplant from a healthy donor to a patient once their basic biology was understood. Following in these footsteps, many different types of stem cell have been discovered over the last 20 years, even in tissues where it was thought that none existed, such as the brain and the heart. By understanding their biology better, scientists hope to use them in the clinic one day, much like bone marrow stem cells now. The most amazing, yet controversial stem cells are undoubtedly those found in embryos just a few days old: embryonic stem cells. Until recently, these were the only stem cells able to make all of the 200 or so specialized cells of the body. Yet just when we thought this was an intractable problem, with ethical objections countering the potential benefit to very sick patients, stem cell research took a new and completely unexpected turn. This was the discovery of induced pluripotent stem cells, based on reprogramming of adult cells, for which Shinya Yamanaka and Sir John Gurdon were jointly awarded the Nobel Prize in 2012. This discovery has in less than 10 years transformed the field and allowed scientists throughout the world to contribute to pluripotent stem cell research. We are just beginning to see the impact of this in biomedical research, and the new edition of this book anticipates what this will mean in the near future.

There is growing excitement surrounding stem cells as their power is harnessed and they make inroads into clinical and academic research. Scientists see an amazing technology emerging, the boundaries of which are unknown. Doctors see new opportunities to treat the diseases of old age. Entrepreneurs in biotechnology expect new marketing opportunities to develop for commercial products based on stem cells. Among political and religious leaders, questions are asked: do we want this technology, is it safe, and where will it lead? Where do we set limits for science and society?

As is often the case when science opens a new door, things can go wrong. In good faith, we may move forward too fast, and spectacular scientific claims may turn out not to be as robust as we thought. They might not even be true. Just as in many other professions, personal integrity and the attraction of fame and fortune can all influence individual behavior, and stem cell research has already been among those areas with serious cases of fraud. *Mala fide* stem cell practices continue to grow at epidemic rates and this growth of what has become known as "stem cell tourism" shows no signs of stopping, however much the experts warn patients of their lack of effect or even risks.

This book places the stormy developments in the stem cell field clearly in context. It provides an understandable and comprehensive overview of the history and state of the art, and distinguishes truth from fiction and empty promises from fact for nonexperts in the field. Whether you are a student or patient, politician or lawyer, entrepreneur or just plain interested, we hope you enjoy this book as a much as we enjoyed researching and writing it!

Christine Mummery
Leiden University Medical Center, Leiden, The Netherlands

Anja van de Stolpe
Formerly Hubrecht Institute, now Philips Research, Eindhoven, The Netherlands

Bernard A.J. Roelen
Utrecht University, Utrecht, The Netherlands

Hans Clevers
Hubrecht Institute, KNAW, and University Medical Center Utrecht, The Netherlands

Acknowledgements

We would like to express our gratitude to all the people that made this book possible, including all the scientists, past and present, who contributed to the stem cell field but to whom we were unable to give personal credit and acknowledgement because of this book's informal nature.

We thank the following people for their excellent contributions to some of the specialized chapters—often including unique illustrations:

Reinier Raymakers (University Medical Center Utrecht) for his contribution on hematology in Chapter 7; Pieter Doevendans, Joost Sluijter, and Linda van Laake (University Medical Center Utrecht), and Marie Jose Goumans (Leiden University Medical Center) for their contribution on stem cell applications in clinical cardiology (Chapter 9); Ricardo Fodde (Josephine Nefkens Institute of the Erasmus Medical Center, Rotterdam) for fruitful discussions and supplying essential information on the chapter on circulating tumor stem cells (Chapter 12); Stefan Braam and Robert Passier (Leiden University Medical Center) for their contribution on stem cell applications for Pharma companies, and Galapagos-Biofocus BV (Leiden) for information and illustrations on their drug target discovery technology; Mark Einerhand (Vereenigde, The Hague) for contributing his expertise to the information on intellectual property (Chapter 15).

James Thomson (University of Wisconsin), Shinya Yamanaka (University of Kyoto), Susan Solomon (New York Stem Cell Foundation), Alan Trounson (California Institute for Regenerative Medicine), Nancy Witty (International Society for Stem Cell Research), Alex Damaschun, Anna Veiga, Joeri Borstlap (hESreg), Michele de Luca (University of Modena and Reggio Emilia, Modena), Anthony Hollander (University of Bristol), Susana Chuva de Sousa Lopes (Leiden University Medical Center), Rui Monteiro (University of Oxford), Mark Peschanski (I-Stem), Megan Munsie (Australian Stem Cell Centre), Jan Barfoot (EuroStemCell), and Elena Cattaneo (Senator of Italian government and University of Milan) and Kastuhiko Hayashi (Kyoto University/Kyushu University, Fukuoka), are all warmly thanked for their rapid responses to requests for input to information boxes within the text. In addition, the following are thanked for illustrations and photographs: Sebastiaan Blankevoort (Leiden University Medical Center) (Box 2.1). Martin Evans (Cardiff University), Kelly Hosman (Utrecht University), Ian Wilmut (Centre for Regenerative Medicine, Edinburgh, UK), Leon Tertoolen (Box 9.12), Uwe Marx (Berlin) (Box 13.2), Oliver Brüstle (Box 15.1).

CHAPTER

1

The Biology of the Cell

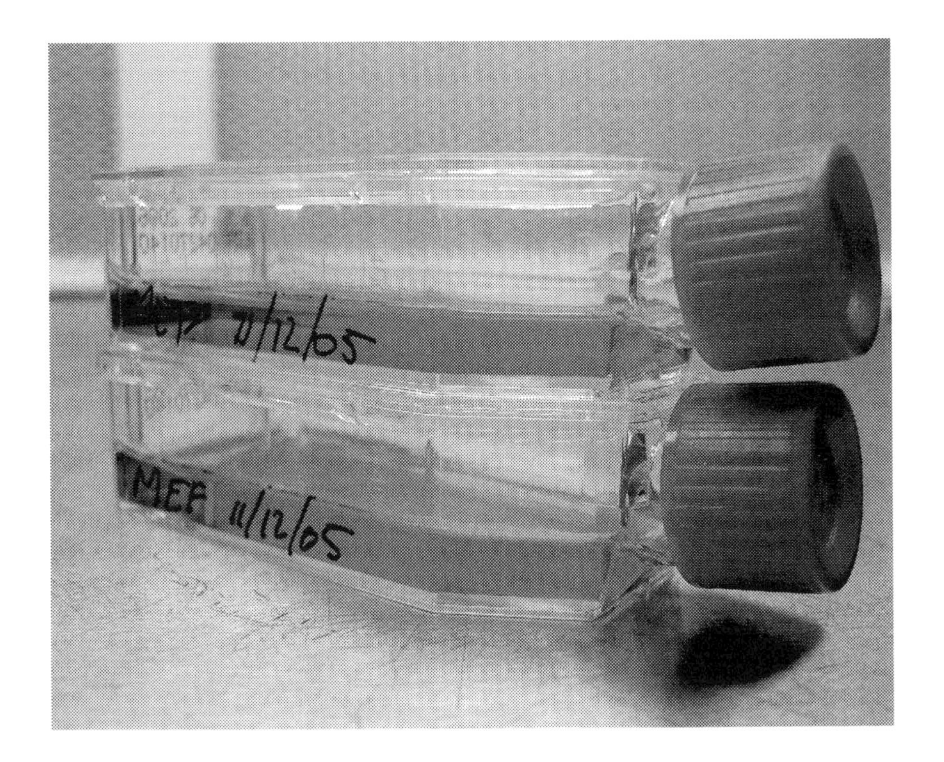

OUTLINE

This book is about stem cells. Stem cells and their applications in clinical medicine, biotechnology, and drug development for pharmaceutical companies involve many facets of biology, from genetics, epigenetics, and biochemistry to synthetic scaffolds and three-dimensional architecture for tissue engineering. For this reason, the most important molecular and cell biological principles needed to understand stem cells will be introduced to the reader in this chapter.

1.1 ORGANISMS' COMPOSITION

Humans and animals, as well as plants and trees, contain many different functional organs and tissues. These, in turn, are composed of a large variety of cells. Cells are therefore the basic building blocks that make up the organism. All animal cells have a similar structure: (1) an outer layer called the plasma membrane, which is made up of a double layer of lipid molecules, and (2) an inner fluid known as cytoplasm. The cytoplasm contains a variety of small structures called organelles, each of which has a specific and essential function within the cell. Most cell organelles are themselves separated from the cytoplasm by their own membrane. The shape of the cell is determined and supported by the cytoskeleton, a flexible scaffolding composed of polymers of protein molecules which form a network

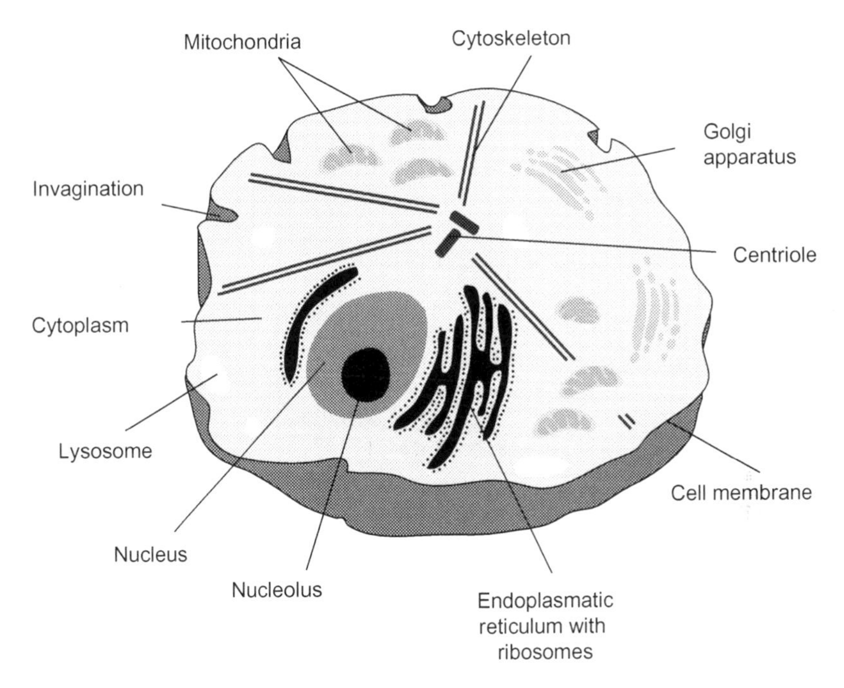

FIGURE 1.1 Schematic representation of an animal cell. The cell contains a fluid called the cytoplasm, enclosed by a cell (or plasma) membrane. The nucleus contains the genetic information; the DNA. The shape of a cell is determined by its cytoskeleton. Proteins and lipids are generated and assembled in the endoplasmic reticulum. The Golgi apparatus is then responsible for further transport within the cell. Lysosomes are small vesicles with enzymes that can break down cellular structures and proteins that are no longer required. The energy necessary for the cell is generated by the mitochondria. *Source: Stamcellen Veen Magazines.*

that shapes the cell and allows it to move and "walk." Inside the cell, countless proteins facilitate the chemical and physical reactions and transport of other molecules required for carrying out specific cellular functions (Figure 1.1).

The most prominent organelle when a cell is viewed under the microscope is the nucleus. This contains the chromosomes, which are in part made up of deoxyribonucleic acid (DNA)—one long molecule of DNA per chromosome—representing the organism's "blueprint" for what it is and what it does. Although cells can have different shapes and functions, the DNA sequence in all cells of one individual is, in principle, identical (with the exception of certain blood cells). Other prominent structures in the cell are the mitochondria. These organelles are present in large numbers and generate the energy required by the cell. Cells with very large energy requirements, such as heart cells, contain correspondingly higher numbers of mitochondria. Energy is also required for the formation of proteins using the genetic code as a template. Rudimental proteins made in this way are delivered to the tubular structures of the endoplasmic reticulum, where they are processed into actual working proteins in yet another organelle, the Golgi apparatus. They are then transported in small vesicles, termed *vacuoles*, to the site in the cell where they are required for their own specific function. Each cell is, thus, a highly dynamic structure with its own powerhouse, factories, and transport systems.

1.2 DEOXYRIBONUCLEIC ACID, GENES, AND CHROMOSOMES

The deoxyribonucleic acid (DNA) in each cell of our body contains all of the information needed to create a complete individual. In humans, DNA is divided into around 23,000 different genes, each of which encodes the blueprint for one or more proteins. What does the information in the DNA look like, and how is it translated into the production of proteins? How does the cell decide which proteins to make? This together determines what stem cells can (or cannot) do and is important for understanding what stem cells can mean for medical research and biotechnology (Figure 1.2).

A DNA strand is composed of a long series of nucleotides. Each nucleotide consists of a deoxyribose molecule, which forms the backbone of the DNA molecule, and is linked to one of four bases: adenine (A), guanine (G), thymine (T), or cytosine (C). Nucleotides are connected by *phosphate groups* (molecules containing the element phosphor) and, as a result, form a long chain. The specific order (or sequence) of the different bases represents the core of the *DNA code* that contains the blueprint of an organism. The DNA sequence is therefore usually written as a series of the letters A, G, C, and T, just as letters in a book. Two single strands of DNA combine to form a double-stranded DNA molecule as complementary bases form base pairs held together through hydrogen *bridges* (or links): adenine binds to thymine, while guanine always binds to cytosine (Figure 1.3). The two strands are therefore complementary; when the sequence of one strand is known, the sequence of the other strand is also known. If, for example, part of one strand is AGTATTC, the other strand would read TCATAAG. The information in the nucleus can be compared to a library, with the genes represented as books and the nuclear code represented by letters in the books.

Normally, the DNA in a cell nucleus is double stranded and forms a long chain (Figure 1.4). These long, double-stranded DNA molecules (one DNA molecule can be up to 10 cm long) form the famous Watson

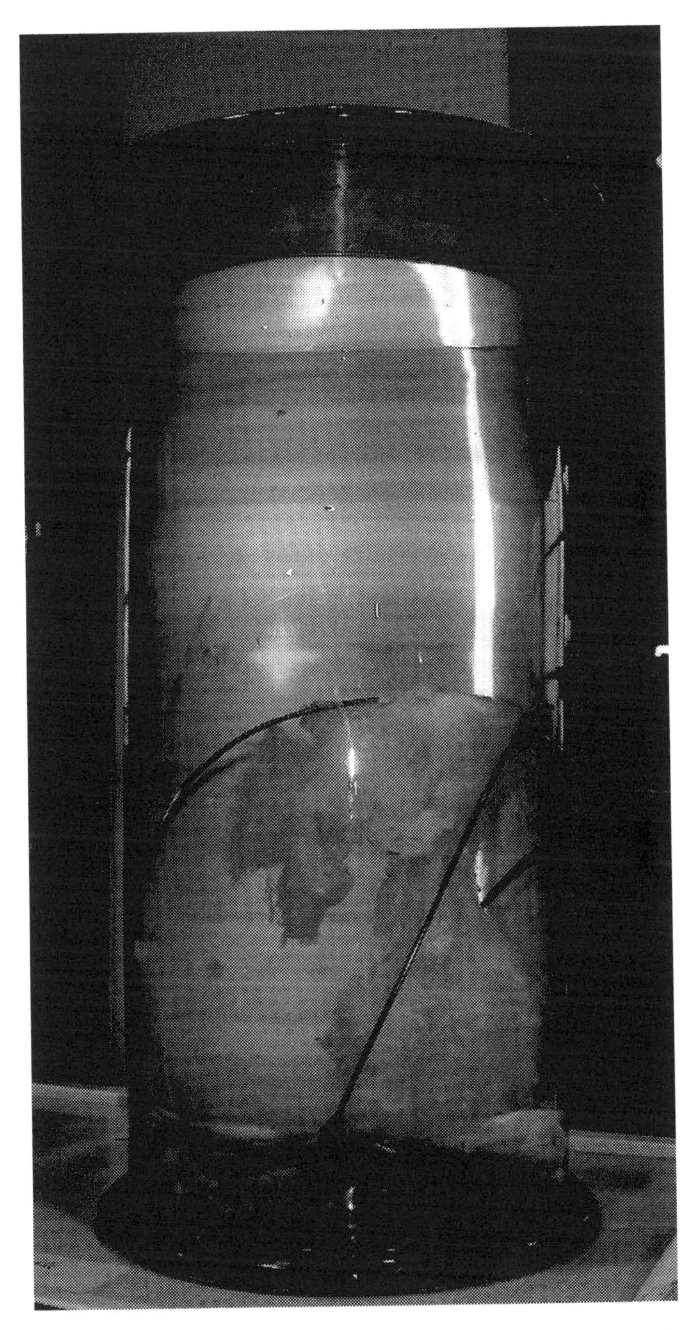

FIGURE 1.2 DNA can be isolated from cells and precipitated (separated from liquid). It then appears as a white, glue-like substance. *Source: Stamcellen Veen Magazines.*

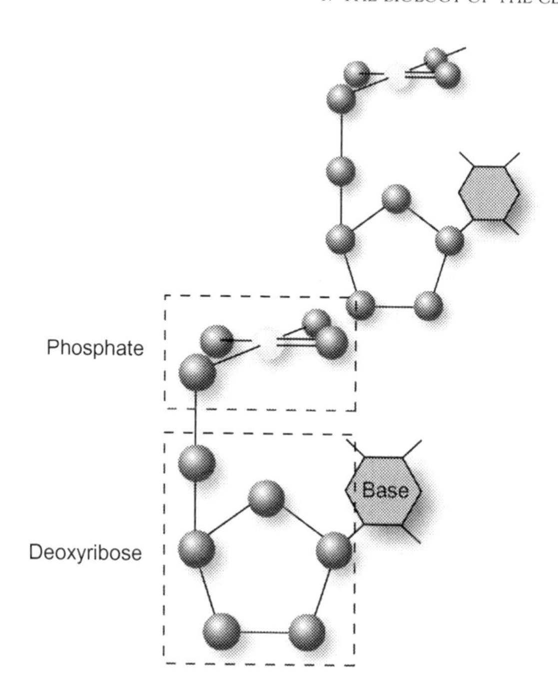

FIGURE 1.3 A single DNA strand is a long polymer composed of sugar (deoxyribose) and phosphate groups that together form the backbone of the DNA. Any one of the possible four bases—adenine, thymine, cytosine, or guanine—is coupled to the deoxyribose strand. *Source: Stamcellen Veen Magazines.*

and Crick DNA "double helix" structure, itself wrapped around a core of a special family of proteins called histones (Figure 1.5). If all of the DNA in one cell was unrolled it would be about 2 m long. This means that in an adult person made up of $\sim 10^{13}$ cells, the total length of the DNA is an astonishing 2×10^{13} m. For comparison, this is equivalent to going ~500,000,000 times around the world (Figure 1.6).

Histone proteins provide an extremely long DNA molecule with the support and guidance to fold into a complicated three-dimensional form that fits into the nucleus. This intricate combination of DNA and proteins is called chromatin, and each long DNA strand folded around proteins called a chromosome. The ends of the DNA molecules that form the *caps* of the chromosomes are called telomeres. These telomeres protect the chromosome ends from DNA damage but are shortened after each cell division (Figure 1.7). Cells can continue dividing until the telomeres are "used up;" this is the process of cellular aging or senescence. Cells that can divide indefinitely, such as some stem cells and cancer cells, have an enzyme called telomerase that builds the telomeres back on after each division. They therefore never get shorter in these cells, which are then immortal.

The number of chromosomes is specific for each animal species. Human cells have 46 chromosomes. These chromosomes occur in pairs: one chromosome of each pair is inherited from the mother, the other chromosome is inherited from the father. The 46 chromosomes are thus

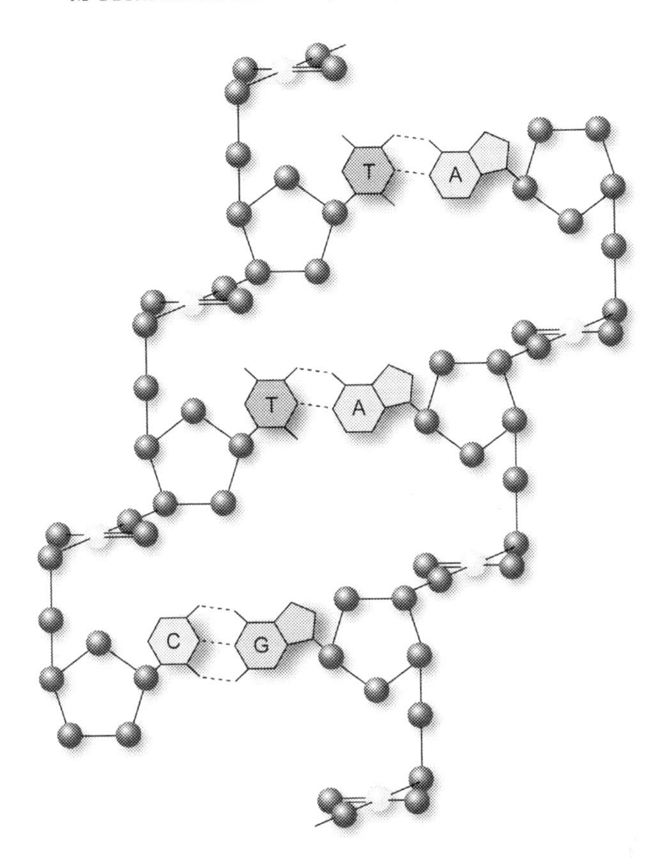

FIGURE 1.4 A complete DNA molecule is composed of two strands that are coupled by hydrogen bonds. Adenine is always coupled to thymine; guanine is always coupled to cytosine. *Source: Stamcellen Veen Magazines.*

present in the cell as 23 pairs. The chromosomes that are numbered from 1 to 22 are called autosomal chromosomes or autosomes. In addition, there is one pair of sex chromosomes, X and Y, which are not equal. Females have two X chromosomes in each cell, while males have one X and one Y chromosome. Chromosomes in a cell can be visualized using a special staining technique; this reveals a *karyotype* image of the chromosomes. This technique is used in the clinic to investigate whether cells from a patient have a normal number of chromosomes, whether the chromosomes are intact, and whether an XY (male) or XX (female) pattern is present (Figure 1.8).

Functional units of DNA are called genes, and these are more or less equally distributed over each of the single strands of the chromosomal double-stranded DNA molecule. Each gene consists of a long stretch, or sequence, of nucleotides. This sequence is divided into different

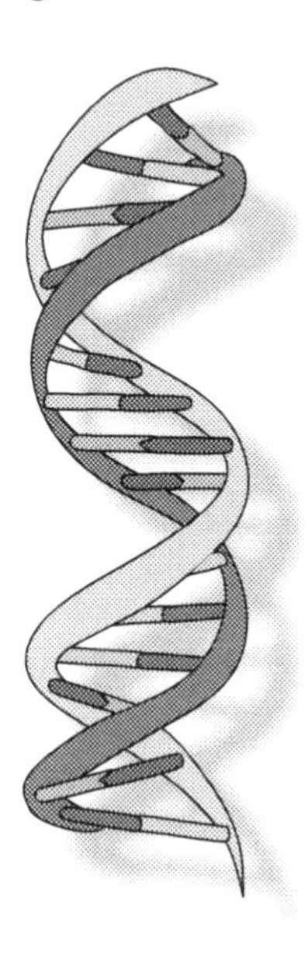

FIGURE 1.5 When two DNA strands are bound together they form a double helix structure, with bases (orange, red, green, purple) on the inside of the helix and the sugar-phosphate backbone (blue) on the outside.

FIGURE 1.6 James Watson (here with Anja van de Stolpe, one of the authors), who, with Francis Crick, unraveled the molecular structure of DNA. The pair discovered that DNA forms a double helix, in which complementary bases are coupled through hydrogen bonds. *Source: Anja van de Stolpe.*

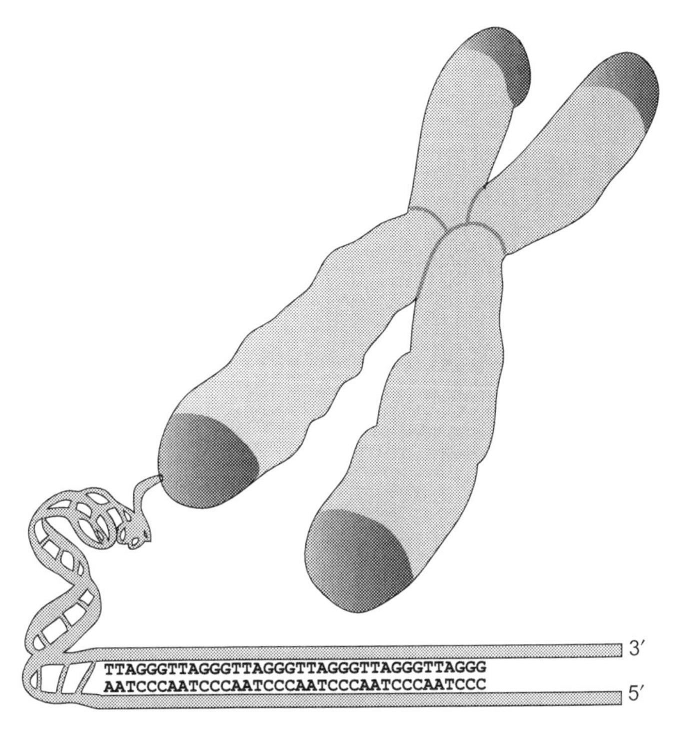

FIGURE 1.7 The ends of chromosomes (brown) contain thousands of copies of the sequence TTAGGG. These form the telomeres that protect the ends of the chromosomes from damage. With each cell division, some of the telomeres break off until they are completely gone. When the telomeres of chromosomes are entirely used up, the cell will stop dividing or die.

functional regions. The core region is called the coding sequence, and it is this sequence that will be translated into functional molecules: proteins. Somehow, the four-base code of the DNA needs to be translated to make specific proteins. How does this work? First, the information carried by the gene sequence needs to be transferred out of the nucleus of the cell. To do this, the genomic DNA is copied (or *transcribed*) to a single strand of molecules called messenger ribonucleic acid, usually referred to as messenger RNA or simply mRNA. A new mRNA molecule is built by coupling nucleotides complementary to the nucleotides in the coding DNA strand. For example, where the DNA nucleotide contains a cytosine (C) base, the mRNA molecule will incorporate a corresponding nucleotide with a guanine (G) base. RNA molecules are therefore complementary to the coding DNA strand. An RNA molecule is, in essence, similar to single-stranded DNA, except for two important differences: (1) RNA contains ribose molecules instead of deoxyribose, and (2) RNA always incorporates the RNA-base uracil (U) instead of

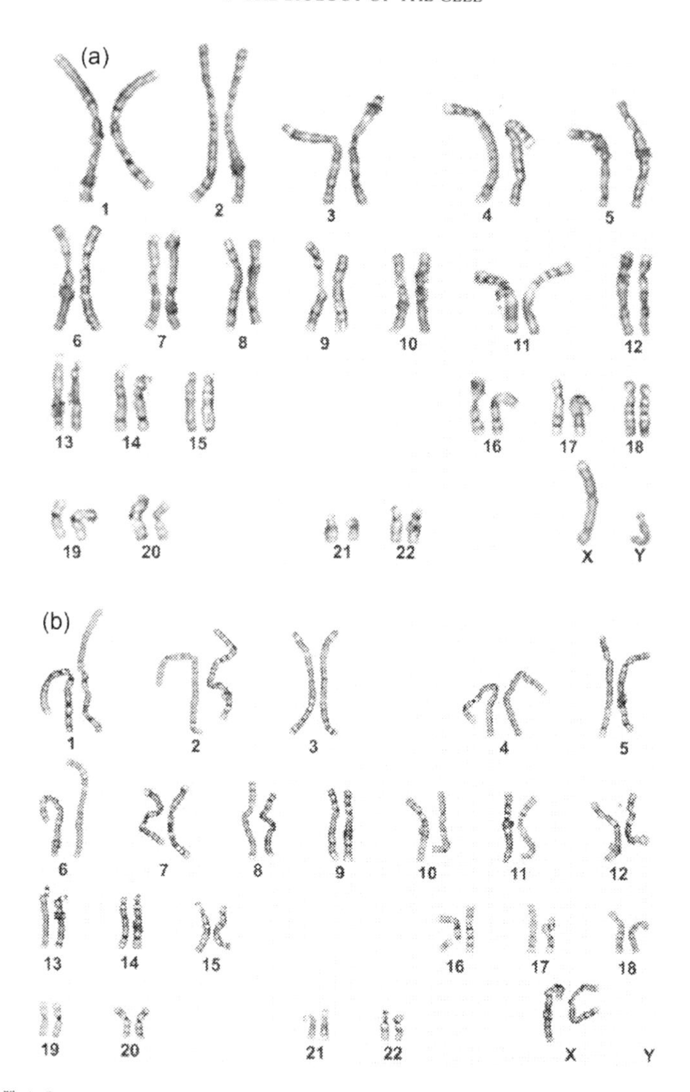

FIGURE 1.8 Human chromosomes, shown here in a karyogram of a man (a) and a woman (b). Human cells contain 23 pairs of chromosomes, of which 22 pairs are similar in males and females. The 23rd chromosome pair is different in males and females: cells of men contain an X and Y chromosome, cells of women have two X chromosomes. *Source: Hans Kristian Ploos van Amstel, University Medical Center Utrecht, Netherlands/Stamcellen Veen Magazines.*

thymine (T). Furthermore, RNA molecules are single stranded; in general they do not form double-stranded structures like DNA. Instead, they can form complex looped structures with base pairing between complementary regions along the sequence (Figure 1.9).

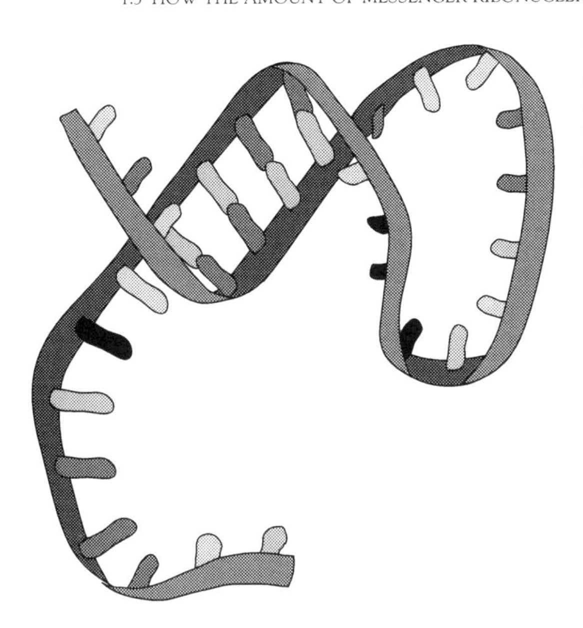

FIGURE 1.9 RNA molecules are single stranded but can form complex looped structures with pairing between complementary bases along the sequence.

1.3 HOW THE AMOUNT OF MESSENGER RIBONUCLEIC ACID IS REGULATED

The coding section of the gene requires a special regulator called a promoter to enable the transcription of a series of RNA molecules. This process is controlled by a specific protein, the RNA-polymerase enzyme. In addition to this core enzyme, multiple proteins and protein complexes recognize and bind in a highly specific manner to certain nucleotide sequences spread all along the promoter region of a gene. These regulatory proteins are called transcription factors. Together, interactions between these proteins and the promoter DNA regulate how many RNA molecules will be produced from which gene.

1.1

ANTONI VAN LEEUWENHOEK

Although plants and animals have been studied for centuries, the realization that organisms are composed of cells is of relatively recent origin. This is simply due to the fact that in the past technologies to visualize cells were lacking. It was only after the invention of the microscope (around 1595) that cells could be made visible for the first time.

1.1 *(cont'd)*

The Dutchman Antoni van Leeuwenhoek was one of the first microscopists dedicated to the discovery and description of the hitherto invisible world of biology. Van Leeuwenhoek was born October 24, 1632 in Delft, The Netherlands. Quite unlike other great scientists of his day, he did not receive a university education, and was entirely self-taught. His naïve approach, disregarding any form of scientific dogma, allowed him to think freely and be guided only by his own enthusiasm and interest (Figure B1.1.1).

Antoni van Leeuwenhoek was by trade a salesman of household linen and used magnifying glasses to examine the quality of cloth. He

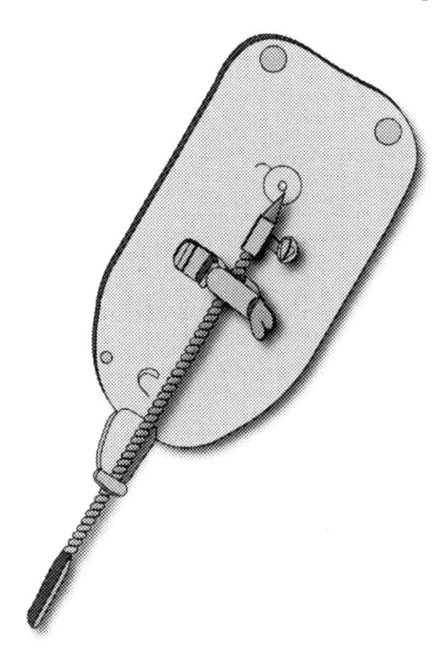

FIGURE B1.1.1 The microscopes made by van Leeuwenhoek were, in fact, magnifying glasses of outstanding quality. *Source: Stamcellen Veen Magazines.*

ground his own lenses using diamond shavings, which he obtained from Delft diamond cutters. He also built his own microscopes, which were basically simple instruments containing a single lens but ground with high precision, sufficient to achieve magnifications of around 300 ×. Van Leeuwenhoek's microscopes consisted of two metal plates fixed to each other with a lens between them. The lens was fixed and the object to be examined was placed on top of a metal holder that could be moved using a set-screw; focusing occurred through a screw at the back of the instrument. The entire construction was less than 10 cm in size. Van Leeuwenhoek's microscopes were, in essence, only very strong magnifying glasses, quite different from the composite microscopes that also existed at the time. However, it was his curiosity and insight, combined with the quality of the lenses and his ability to illuminate the objects properly, that allowed him to discover the microscopic world. He examined water from ditches, tooth plaque, bakers' yeast, stone dust, blood, and sperm. (Figure B1.1.2)

The Delft physician Reinier de Graaf introduced van Leeuwenhoek to the Royal Society in London, after which he published his findings in a total of 200 letters (in Dutch), which he sent to the Society. The

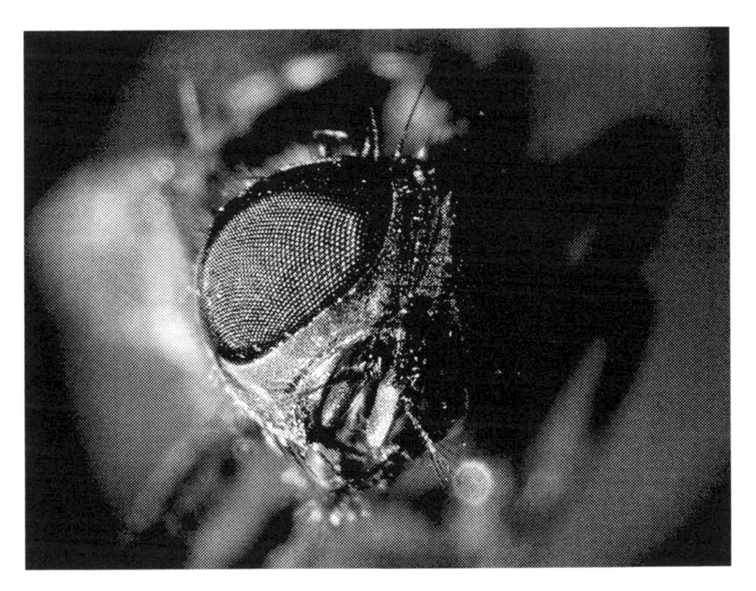

FIGURE B1.1.2 Antoni van Leeuwenhoek studied many small objects with his microscope, among them the composite eyes of insects: he tried to determine the number of facets on each eye. *Source: Stamcellen Veen Magazines.*

letters, which needed to be translated to English or Latin for publication, are anecdotal, containing a plethora of random observations, but uniquely detailed in their descriptions. Van Leeuwenhoek achieved international fame with his observations, but in a letter written in 1716 he said that he "did not strive for fame, but (was) driven by an inner craving for knowledge." He believed this drive was stronger in him than in most other people. Van Leeuwenhoek died August 16, 1723.

1.4 FROM MESSENGER RIBONUCLEIC ACID TO A FUNCTIONAL PROTEIN

The newly transcribed RNA molecule is processed within the cell nucleus into a genuine messenger molecule, mRNA, and transported via specialized shuttling proteins from the nucleus to the ribosomes in the cytoplasm. A protein itself consists of a series of amino acids linked together like beads on a string. The nucleotide sequence of a gene determines the amino acid sequence of the protein, so that it is the DNA that determines which proteins are actually made by a cell. The genetic code of the DNA and mRNA, therefore, needs to be translated into the correct order of amino acids so that they link up properly for each protein. The genetic code itself consists of "triplet" units of three nucleotides, called codons. Different triplets code for the different amino acids. With 4 different nucleotides, 64 different triplets (e.g., AUG, AUU, AUC, AUA, etc.) can be made, but there are "only" 20 amino acids to choose from (Figure 1.10). This means that the code is *degenerate*, and that most amino acids are designated by more than one triplet. The triplet that codes for the amino acid methionine also functions as the *start codon*, telling the ribosome where to start making a protein. In addition, there are 3 triplets that do not code for an amino acid but instead terminate translation. These are referred to as *stop codons*. Most interestingly, the genetic code is universal. DNA in all organisms has the same structure and is composed of the same nucleotides; only the sequence of nucleotides is different. Moreover, triplets of all organisms code for the same amino acids. For instance ACG codes for the amino acid threonine, whether it is in a human cell, a cell from a fruit fly, or baking yeast. The combination of amino acids that make up a protein can be different, however, between different organisms. Human actin, for instance, has a slightly different amino acid composition compared with that of a mouse (Figure 1.11).

Translation of the mRNA sequence to a chain of amino acids requires a cell organelle known as the ribosome. This ribosome binds the mRNA and subsequently recruits molecules that recognize both the nucleotide sequence and the accompanying amino acid. These adaptor molecules are small RNA molecules known as transfer RNA (tRNA). The tRNA molecules carry amino acids that are transferred to the growing amino acid chain like beads on a string.

To make a functional protein from the amino acid chain requires it to be correctly folded, usually followed by chemical adaptations such as *cross-linking* between specific amino acids in the protein molecule. Sometimes proteins only function in a complex with one or more of the same or different protein molecules, and these protein complexes need to be organized. Finally, the protein is transferred to the correct location inside or outside

1st position	2nd position U	C	A	G	3rd position
U	Phe	Ser	Tyr	Cys	U
	Phe	Ser	Tyr	Cys	C
	Leu	Ser	STOP	STOP	A
	Leu	Ser	STOP	Trp	G
C	Leu	Pro	His	Arg	U
	Leu	Pro	His	Arg	C
	Leu	Pro	Gln	Arg	A
	Leu	Pro	Gln	Arg	G
A	Ile	Thr	Asn	Ser	U
	Ile	Thr	Asn	Ser	C
	Ile	Thr	Lys	Arg	A
	Met	Thr	Lys	Arg	G
G	Val	Ala	Asp	Gly	U
	Val	Ala	Asp	Gly	C
	Val	Ala	Glu	Gly	A
	Val	Ala	Glu	Gly	G

FIGURE 1.10 Although it was discovered in 1953 that DNA forms a double helix structure with complementary bases, the genetic code was only deciphered in 1966. A set of three nucleotides (triplet) codes for an amino acid, with different triplets able to encode the same amino acid. The triplet that codes for methionine also functions as a start signal (green). There are three triplets that do not code for an amino acid but function as a signal to stop translation (red). *Source: Stamcellen Veen Magazines.*

the cell. Reversible chemical changes are frequently made to the protein later in its life cycle, to change its function or activity. For example, phosphorylation (or addition of a phosphor molecule) of a specific single amino acid in a transcription factor protein can cause it to become activated or deactivated, or determine its intracellular localization. Another common example is a hormone, present outside the cell, which binds to a membrane receptor on a cell and induces phosphorylation of the intracellular part of the receptor, causing it to transmit a message into the cell.

1.5 FROM DEOXYRIBONUCLEIC ACID AND PROTEINS TO A CELL WITH A SPECIFIC FUNCTION

DNA is not simply present in the cell as an independent entity. Genes can only function within the context of a cell, tissue, or organ, and finally, organism. To understand the role of DNA and genes in a cell, we need to take a deeper look inside the cell, to see where proteins are at work.

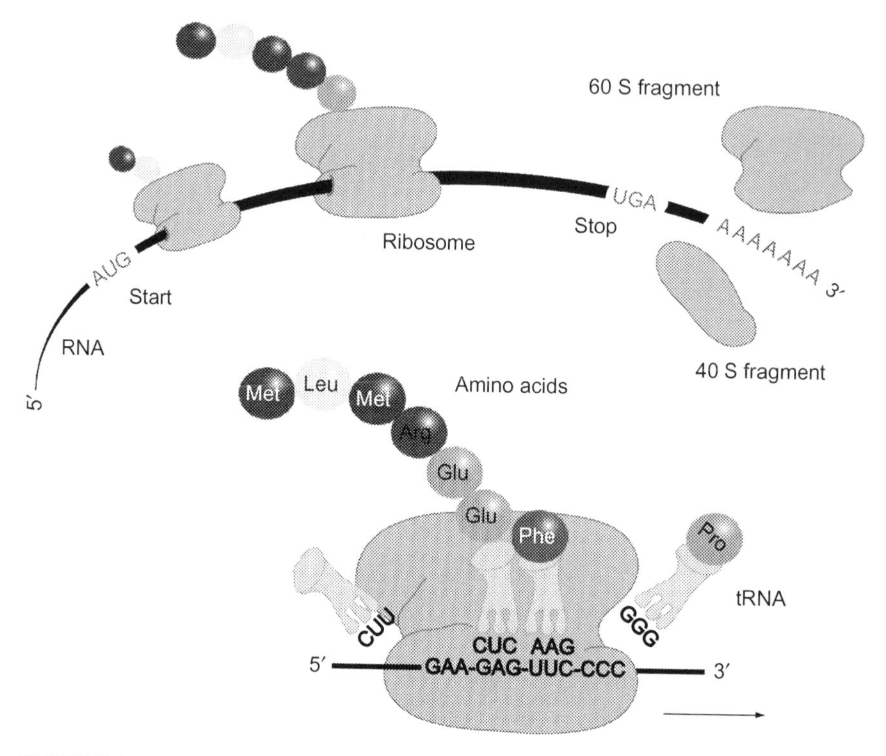

FIGURE 1.11 Messenger RNA (mRNA) is "read" by a large complex composed of proteins and RNA, called the ribosome. A triplet of three nucleotides (a codon) in the mRNA codes for an amino acid. The amino acids are independently coupled to transfer RNA (tRNA), which contains a sequence complementary to that of the codon. Because amino acids are specifically coupled to different tRNA molecules, the mRNA code is translated to a strand of amino acids, a protein. The AUG sequence provides the start signal as well as the signal for the amino acid methionine. Three codons encode a stop signal, which causes the ribosome to disintegrate into a large (60 S) and a small (40 S) fragment. The protein is then released into the cell for use. *Source: Stamcellen Veen Magazines.*

Consider that every healthy cell in an individual contains in principle the same genetic material, or DNA. The only difference between cells (say brain cells or heart cells), is how they use the DNA. Transcription of genes and turning them into mRNA results in the production of thousands of proteins that determine which type of cell is created and which characteristic set of proteins, with their associated functions, each cell will have. However, some types of protein products are present in every cell. Examples include proteins coded by "housekeeping" genes, which are needed for the survival of the cell itself and that most cells have in common. These housekeeping genes or proteins include, for example, actin, which is essential for the scaffold structure of a cell that maintains its shape and form.

Proteins also play a crucial role in communication between cells, and cellular communication is essential for the function of the cell within its environment, for cell division, and for differentiation to a cell type with a specific assignment. Cells communicate with each other through direct contact or via secreted signaling molecules, which recognize a specific receptor on (or in) the target cell. Signaling molecules may be present in the immediate environment of the cell, such as a growth factor, or originate from further afield, for example, the hormone estrogen. Direct contact between cells as well as signaling molecules can transfer a signal into a cell with the message to change a certain aspect of cell function. This is called *signal transduction*. The change in function of the target cell can be caused by altering the transcriptional activity of certain genes in the cell nucleus, leading to synthesis of new proteins, or by rapid chemical modification of molecules already present in the cell.

1.5.1 Epigenetic Regulation

To make things even more complex, RNA synthesis and degradation, and, consequently, protein synthesis, are also controlled at other levels. The genomic DNA in almost all cells of an organism is complete, which means that in every cell all information is available to make a complete organism. However, the *accessibility* of the information differs per cell type. Going back to the metaphor of the cell nucleus represented by a library, the library stays intact in every cell (all books are there) but not all books within the library can be accessed (some books are in closed rooms). Indeed the genomic DNA is not freely accessible (Figure 1.12). The state of the histone proteins, around which the DNA is wrapped to form the chromatin, can make genes more or less accessible for transcription. Additionally, cytosine nucleotides in the DNA sequence of a promoter area of a gene can be methylated—meaning a *methyl group* is added to the nucleotide—which is associated with reduced transcription of mRNA molecules. This methylation pattern is normally considered permanent (Figure 1.13). Together with the histones, it determines which genes in a cell are no longer accessible by transcription factors, and thus their information is not available to make proteins.

The chromatin structure and the state of DNA methylation are copied to daughter cells during cell division. This regulation of gene use is called epigenetic regulation. Such mechanisms are important during development of the embryo and the differentiation of cells into a specific cell type, a process that permanently inactivates a number of genes not necessary for the function of the mature cell. In general, the differentiation process of cells and the epigenetic changes that the cells undergo are one-way traffic. The rooms in the library that are closed by epigenetic

FIGURE 1.12 A cell nucleus can be thought of as like a library with the different genes represented as books. Some books/genes are more accessible than others.

FIGURE 1.13 In DNA, a cytosine can be methylated when it is positioned next to a guanine. A methyl group (CH_3) is added at the 5′-position of the molecule. This changes the accessibility of the DNA to DNA binding proteins such as transcription factors.

changes will normally remain closed. Once a cell is differentiated, for example, it has become a nerve cell, it will not change to a different cell type, say a liver cell, under normal circumstances in the body. In some cases, however, this *transdifferentiation* can be induced in the laboratory by forced expression of transcription factors that are not normally expressed in that particular cell. These very dominant cell type specific transcription factors are called *master regulators*.

FIGURE 1.14 (a) In addition to our genome structure and composition, the types of food we consume can play important roles in the development of diseases. (b) Cells can be grown in specially designed sterile plastic dishes. The cells are cultured in a liquid (or medium) containing sugars, amino acids, salts, and minerals as nutrients for the cells. The color of the medium is due to a pH-sensitive coloring agent: when the liquid is too acid (low pH) the medium turns to orange; when the medium is too basic (high pH) the liquid changes to purple. *Source: Bernard Roelen/Stamcellen Veen Magazines.*

Of increasing interest are abnormalities in epigenetic regulation, because these appear to play a role in various forms of cancer and are caused by certain genes being abnormally methylated. Consequently, the corresponding proteins, which are important for proper control of cell division, are no longer produced and tumor growth is the result (Figure 1.14).

1.5.2 Ribonucleic Acid Interference

There is a third layer of mRNA regulation, which was discovered much later than all the rest. This is mediated by a family of small RNA molecules that do not encode amino acids. For this discovery, Andrew Fire and Craig Mello were awarded the Nobel Prize in Physiology or Medicine in 2006. They discovered that the DNA also codes for small double-stranded RNA molecules, which do not contain a code for proteins although they do travel to the cytoplasm. Instead, these RNA molecules appear to be processed into single-stranded small (~20–30 nucleotides) RNA molecules inside a special protein complex, which facilitates binding to a partly complementary nucleotide sequence in mRNA molecules. These bound mRNA molecules are then rapidly degraded, eliminating any further protein production. This whole process is called RNA interference. Since these so-called microRNA (miRNA) molecules are not very selective in choosing which mRNA to stick to, one type of miRNA can usually cause degradation of a whole bunch of different mRNAs. It seems as though they may act as a buffer against too much stress in a cell, by reducing the production of many related proteins.

These three levels of gene regulation play essential roles in both the maintenance of stem cells and their differentiation to specific cell types, such as heart muscle cells.

1.6 DEOXYRIBONUCLEIC ACID DIFFERENCES BETWEEN GENOMES

A human genome sequence referred to as the "reference" human genome sequence can now be found on the internet (*www.ncbi.nlm.nih.gov*). However, no two genomes are exactly the same, and even between identical twins, differences in epigenetic patterns exist. If a nucleotide in the genomic DNA is switched to a nucleotide with another base attached, for example, C is replaced by T, this is called either a mutation or a variation. The latter is often called a single nucleotide polymorphism (SNP). Put simply, a mutation is a change that hardly ever occurs but can be associated with a (serious) disease; in contrast, a variation or polymorphism is a more common change that has few, if any, consequences for the individual.

Mutations can be either confined to a single nucleotide switch or involve a multiple nucleotide sequence. A single nucleotide change in the DNA, for example, a C changed to a T in one of the strands, means that a G changes to an A in the complementary strand after the next cell division. Alternatively, larger DNA fragments can be either deleted from the DNA or amplified, which means that, for example, a certain nucleotide sequence now occurs three times on a stretch in the DNA molecule instead of the usual once. Also, a stretch of DNA can be inverted (turned around), so that the tail of the sequence now comes first on the DNA molecule, or exchanged with a sequence from another part of the genome—even from another chromosome.

Depending on where in the genome a DNA change takes place, it may either affect the function of the gene and the formation of the corresponding protein, or have no effect at all. Even mutations in the part of the gene that codes for amino acids do not necessarily lead to protein dysfunction; it all depends on the particular location and the consequences for the amino acid sequence. Since several triplets of nucleotides can code for the same amino acid, a single nucleotide change may not change the protein for which the sequence is the blueprint. A change, for instance, in the coding sequence of a gene from AAA to AAG will still code for the amino acid lysine and there will be no effect on the protein that is formed. On the other hand, a single nucleotide change in the sequence from AAA to AAC will lead to the formation of the amino acid asparagine instead of lysine, which can have serious consequences for the function of the protein. Outside the coding area of a gene, the chances that a change of nucleotide has no effect, or the effect is minimal, are even larger. This is because part of the DNA surrounding the coding regions is relatively unimportant for the gene's function and protein synthesis.

1.7 DISEASES DUE TO VARIATIONS AND GENOME MUTATIONS

Recently it has become clear that many diseases have a genetic basis. This means they can be passed from parents to children and grandchildren. Genetic *predisposition* (an increased chance of getting the disease in certain families) to common conditions such as cancer, cardiovascular diseases, and diabetes seems to be determined by a complex combination of nucleotide variations, or polymorphisms, in a number of genes, each of these variations being fairly common and inconsequential on their own. In combination, however, these variations can be associated with an increased disease risk; although predisposition for a disease does not mean that the person will actually develop it. Considerable research has been invested into trying to find out what makes those with an increased risk actually develop the disease. Nevertheless, for many genes and their nucleotide variants, it is still unclear whether they actually contribute to the risk of acquiring a specific disease: the connection has not been proven. For several common conditions, a clear pattern of heredity has not even been identified, and external factors such as nutrition are thought to play an important role in whether or not the disease becomes manifest, although this is still far from certain. In contrast, in some rare diseases heredity is often evident and can usually be pinpointed to a specific gene, such as in families with hemophilia and with certain relatively rare forms of diabetes. In these cases, a specific DNA abnormality in a gene sequence plays a direct causal role in the disease.

1.8 DOMINANT OR RECESSIVE

The terms *dominant* and *recessive* are commonly used in genetics particularly when discussing whether a gene mutation will actually cause a physical condition, whether it will be inherited, and what the chance is of children being afflicted by the same complaint. It is also often talked about in a more trivial context: predicting the color of a baby's eyes when those of its parents are, for example, brown and blue. We will therefore explain what saying genes are dominant or recessive actually means. As discussed previously, almost all genes in the genome are present in duplicate: one copy is inherited from the father, the other copy is inherited from the mother. In a dominant mutation, only one of the two gene copies needs to be defective to cause disease. Even when a patient still has an intact gene on the second chromosome (the patient is *heterozygous* for the mutation), this second gene is unable to prevent occurrence of the disease. In the next generation, on average half of all the children inherit the defective gene and also suffer from the disease or affliction; the other half does not. For recessive mutations, individuals who are heterozygous for the mutation are not affected by the disease and are only "carriers" of the mutation. This means the disease will only manifest itself if the gene is defective on both chromosomes, and no normal copy of the gene is left. If both parents are heterozygous and, thus, carriers of the mutation but do not suffer from disease themselves, one in four of their children runs the risk of being *homozygous* for the disease, which means that they will have the mutation on both gene copies and therefore develop the disease (Figure 1.15).

1.9 DEOXYRIBONUCLEIC ACID OUTSIDE THE NUCLEUS: BACTERIAL REMAINS

The genomic DNA that is part of chromosomes is not the only DNA present in the cell. Mitochondria, the powerhouses of cells, also contain DNA, in this case a circular strand, which codes for a number of proteins used mainly for energy metabolism of the mitochondrion. This DNA probably has its evolutionary origins, oddly enough, as bacterial DNA that somehow ended up in a cell. This proved to be a win–win situation where the bacteria evolved into the cell organelles that we now know as mitochondria, while their DNA has been

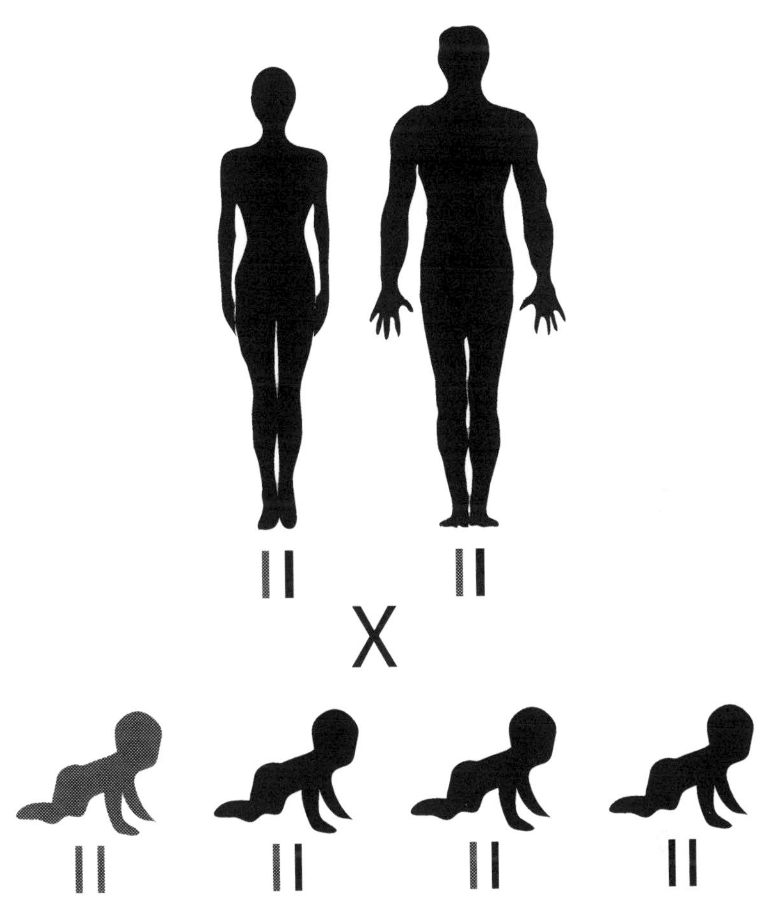

FIGURE 1.15 When a patient is heterozygous for a recessive disease-causing mutation (red), the disease will not develop because almost all genes are present in the duplicate DNA strand, and the second "normal" gene (black) can compensate and prevent disease. The patient is only a carrier of the mutation. If both parents are heterozygous, there is a 25% chance that their children will carry the mutation on both genes and suffer from the disease, 50% of their children will be a carrier of the disease just like the parents and have the mutation on one of the two genes, and 25% of their children will not have the disease or the mutation on either of the two genes. *Source: Stamcellen Veen Magazines.*

maintained alongside the chromosomal DNA in the nucleus. It is quite amazing that in many respects this mitochondrial DNA still closely resembles the DNA in bacteria. After fertilization, the mitochondria of the sperm cells are degraded by enzymes in the egg's cytoplasm, while the mitochondria from the egg remain intact. Since a fertilized egg contains the cytoplasm with mitochondria from the mother, mutations in this mitochondrial DNA are inherited down the mother's side.

1.2

HELA

The Legacy of Henrietta Lacks

On February 1, 1951 Henrietta Lacks, a 30-year-old African-American woman and mother of five children, visited Johns Hopkins Hospital in Baltimore, Maryland with abnormal vaginal blood loss. Nobody could have predicted that this would mark the beginning of a remarkable series of medical discoveries, including the development of a vaccine against polio.

Doctors diagnosed Henrietta Lacks with cervical cancer, and while chances of recovery were minimal at the time, radiotherapy was started a week later. However, prior to the start of radiation therapy, a tissue biopsy was taken from the cancer and transferred to George Gey (1899–1970), a scientist working at Johns Hopkins. Together with his wife Margaret, he had been performing pioneering work on cell culture and for years had tried to grow human cells in the laboratory, to no avail. Cell culture, or attempts at it, was performed very differently in those days. No special sterile cabinets with filtered air were available; cultures were grown on open laboratory tables using a Bunsen burner to create a small, semisterile environment. No commercial culture media were available. Instead, a mixture of chicken blood plasma, cow embryo plasma, placental blood, and salts was used (Figure B1.2.1).

The cells, derived from the cancer of Henrietta Lacks, were added to a culture bottle with this homemade culture fluid, and something very surprising happened. The cells started dividing faster than ever, and could be transferred from one bottle to the next. Unfortunately,the tumor cells in Henrietta's body grew just as fast as those in the culture bottles, and she died on October 4, 1951, only 8 months after the initial diagnosis of cervical cancer. After her death, her cells continued to grow in culture bottles at incredible speed. The first human cell line in the laboratory had been created.

The cells were named HeLa cells, the first two letters of Henrietta's first and last name. To maintain some privacy for her family, the cells were said to have come from Helen Larson or Helen Lane. At the time, it was not usual to ask patients or family members for written permission to use tissue for scientific research. David Lacks, Henrietta's husband, had no idea Henrietta's tumor cells continued to grow in the laboratory. HeLa cells became crucial for cancer research, but also for fundamental research on cell function, protein synthesis, and effects of medication and radiation. Moreover, they were used to develop and

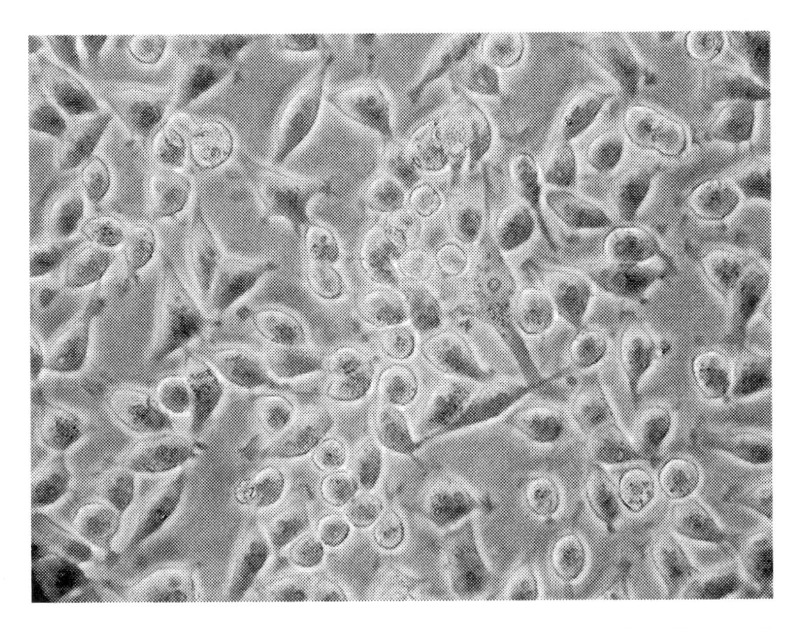

FIGURE B1.2.1 Micrograph of HeLa cells. Even today, HeLa cells are cultured in many laboratories to study the biology of cells. *Source: Stamcellen Veen Magazines.*

refine culture methods and identify which components are essential constituents in a culture medium.

The cells proved easy to infect with polio virus, which at the time was a frequent cause of child paralysis; this enabled use of the cells for testing vaccines against the disease, eliminating animal experiments for this purpose. Carefully packaged bottles containing HeLa cells were sent across the globe to Gey's colleagues for research purposes. Within 2 years of the initial successful culture, 600,000 bottles of cells had been shipped. The cells simply continued to grow. They were even used in space shuttle experiments, 300 km above Earth, at zero gravity (Figure B1.2.2).

HeLa cells are still being used in laboratories across the world, and are, in fact, the most commonly used human cancer cell line. They are in part responsible for our ability to culture stem cells. Henrietta Lacks' family was completely unaware of the existence of the cells. Only in 1975, 24 years after Henrietta's death, did the children accidentally discover that their mother's cells were still alive. Henrietta herself never knew what happened to her tumor cells, or the important role that they played in biology and medicine. A book was even published about the cells and their origin (Figure B1.2.3). The story did not stop

1.2 (*cont'd*)

FIGURE B1.2.2 HeLa cells have even been sent into space in space shuttle missions to study the effects of weightlessness on cells. *Source: Alfred Roelen/ Stamcellen Veen Magazines.*

there, but continued into 2013, when researchers from different laboratories determined the complete genome DNA sequence of the cultured HeLa cells. They made this information publically available by publishing in a scientific journal and posting the sequence on a public database. Although originally derived from Henrietta Lacks, the HeLa cells had been cultured for over 60 years, and culturing for so long inadvertently leads to structural changes in the genome. That in itself is interesting. In addition, for medical purposes, sequencing the genome of HeLa cells may lead to clues as to why these tumor cells behaved so very aggressively, even compared to other tumor cells, and how to treat such tumors more effectively. It also meant, however, that very private information about Henrietta Lacks had become publically available, again without the family's knowledge. Although by itself not illegal, it did upset the Lacks family because it also revealed aspects of their own DNA.

Although the genome currently cannot say much about an individual person or family, public databases are being studied using search

FIGURE B1.2.3 Award-winning science writer Rebecca Skloot published a book describing the origin of HeLa cells. Because neither Henrietta Lacks nor her family were properly informed about what would happen to the cells if the experiments were successful, the case led to changes in the ethical considerations and law around collecting biological samples from humans.

engines and from this information the chances of developing particular diseases or character traits can be estimated. After conversations with the Lacks family, led by Francis Collins, former director of the National Institutes of Health (NIH), it was decided that the genome sequence be entered in a controlled-access database that can only be reached after agreeing to specific terms. Because of the question that arose on HeLa cells, the U.S. Department of Health and Human Services is preparing a new proposal regarding the protection of participants in research.

1.10 CELL LINES AND CELL CULTURE

Cell *lines* refers to cells that can be kept growing in culture for long periods of time. To grow in the laboratory, cells are placed in a plastic dish (or culture bottle) containing a liquid culture medium. This fluid contains the necessary nutrients for the cells such as sugars, amino acids, and minerals. Cells maintained in warm culture medium in an

FIGURE 1.16 Cells are cultured in special "incubators" that maintain the correct conditions of temperature (37°C or body temperature), humidity, and concentrations of oxygen and carbon dioxide.

incubator at 37°C divide every 10 to 24 hours, depending on whether they are normal or cancer cells. Once they reach a certain number and density, they can be used for experiments. For certain cells that need to attach to the bottom of the plastic culture dish to grow, such as skin fibroblasts, growth density is expressed as the degree of confluence. So 50% confluence means that cells have covered about half of the surface of the dish or flask. If there is no space between cells, they form a closed (confluent) cell layer with no space between, which is referred to as 100% confluent. The characteristics of the cultured cells may depend on their degree of confluence; many normal cell types, for instance, stop dividing when they have reached 100% confluence. Tumor cells often do not stop dividing and will continue to pile up until all nutrients in the culture medium are used up (Figure 1.16).

Because the ingredients of the (liquid) culture medium feed the cells, the medium will become exhausted after several days in the presence of metabolically active dividing cells, even if they are normal. Cells also secrete degradation products into the medium. This means that the culture medium must be replaced at regular time intervals, for instance, every day or three times per week depending on how fast the cells are growing. Simultaneously, because the cells in the dish are dividing, the growing cell mass must be diluted at regular intervals. The cells are therefore removed from the dish using enzymes that dislodge them, and a small proportion of the cells is transferred to a new dish with fresh culture medium, while the rest of the cells can be used for an

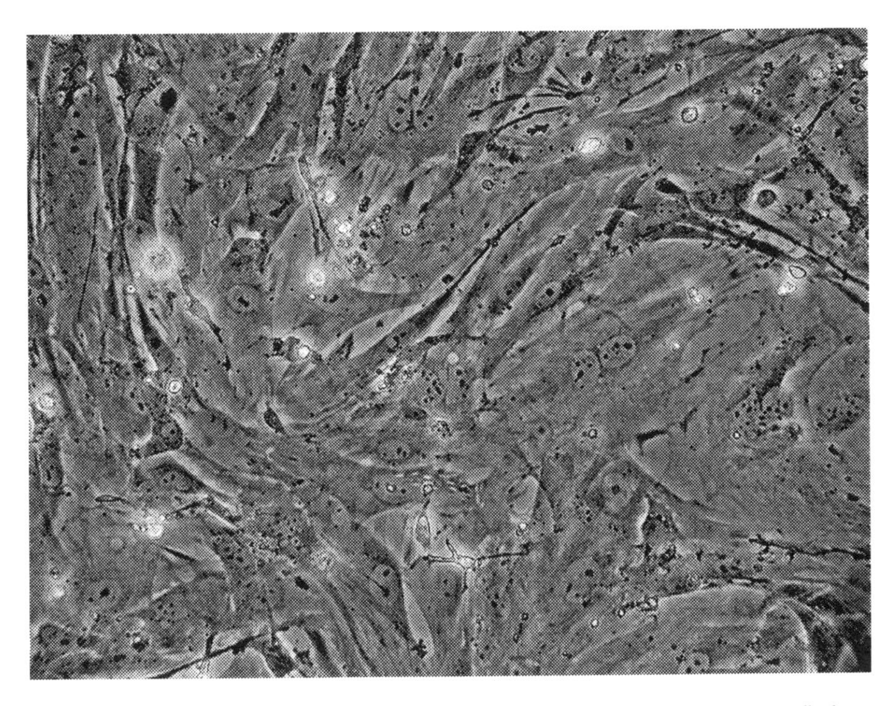

FIGURE 1.17 When the cultured cells cover the entire surface of the dish or flask in which they are cultured, the layer of cells is said to be confluent. *Source: Stamcellen Veen Magazines.*

experiment. Every time a cell line is transferred in this manner, the cells are referred to as being "passaged." Cells are often only used for experiments over a few "passages," because their properties can change over time with increasing number of passages. They can, for example, acquire mutations in their DNA (Figure 1.17).

For many types of biomedical research, no suitable human cell lines are available. Sometimes animal cell lines, usually obtained from mice or rats, can be used as substitutes for this. Most cell types do not proliferate indefinitely in culture; rare exceptions are some types of stem cells and cell lines obtained from cancer tissue. These cells have unique characteristics that provide them with immortality. They specifically have telomerase, which build on new telomeres at each division, as discussed earlier. For cell types that do not have the capacity to divide indefinitely, immortalized behavior can be induced by artificially introducing specific growth promoting genes into the cells. These procedures may, however, change the nature of the cell, such that they no longer resemble the original cell type. They may then actually behave like tumor cells.

1.3

HOOKE

Hooke, a Cork, and a Cell

The invention of the microscope and the work of Englishman Robert Hooke are responsible for introduction of the word "cell" to indicate the structures we now consider the building blocks of tissues. Hooke studied thin layers of cork under his microscope and noted that these were built of blocks he called "cellulae," Latin for "small spaces." The cells in the cork were already dead; what Hooke described were the remains of cell walls. However, the term has been used ever since as a structural unit for all living organisms.

Hooke was born July 18, 1635 (O.S., i.e., Julian calendar) in Freshwater, England. Even as a child he expressed a strong interest in nature and the creation of mechanical toys. He was also a skilled artist and, following the death of his father, he was sent to London in 1648 (at the age of 13) to work in the studio of the Dutch portrait artist Peter Lely. However, Hooke realized he lacked a proper education, and went on to attend the Westminster School that same year. There he proved to be a veritable polymath, excelling in mechanics and mathematics, as well as being a skilled musician. In 1653, he moved to Oxford to continue his studies, and was surrounded by other great scientists, including Christopher Wren and Robert Boyle (Figure B1.3.1).

Hooke specialized in astronomy and mechanics and studied the skies using a primitive telescope. He also used a microscope to examine both living and dead objects. His artistic skills served him well; he accurately drew everything he saw under the microscope. These drawings and descriptions were published in 1665 in the book *Micrographia*, to general acclaim. It is in this book that Hooke described the structure of cork and coined the term *cell*.

Hooke also did a great deal more: he was the first to describe Jupiter's rotation around its own axis; he observed planets, stars, the sun, and comets; and discovered a star. He experimented with integrating springs into a clock to make it portable, and studied light, air pressure, and fossils. However, he never gave himself time to investigate any of these things in real depth. Once an idea had been roughly described, he would be distracted by another train of thought and move on to something else. He became involved in a conflict with fellow scientist Isaac Newton; Hooke felt that Newton had stolen his findings on light and color, which were described in his famous work *Principia*. In part due to his tendency never to fully develop his ideas

FIGURE B1.3.1 Hooke discovered that cork is made up of many small compartments that he called cellulae. This is the origin of the word cell that we use today to describe the basic building block of all living tissues. *Source: Stamcellen Veen Magazines.*

into theories, Hooke got into more conflicts with scientists, such as Christiaan Huygens and Johannes Hevelius. Heated arguments earned him the reputation of a grumpy cynic. However, his talents are undeniable: he was a member of the Royal Society in London, defined Hooke's law of elasticity, and even worked as an architect when a large part of London was destroyed by the Great Fire of 1666. Hooke died on March 3, 1703 in London.

1.4

SERENDIPITY AND THE DISCOVERY OF RIBONUCLEIC ACID INTERFERENCE

Craig Mello and Andrew Fire brilliantly unraveled the mysteries of RNA interference (RNAi) at the molecular level, for which they were awarded the Nobel Prize in Physiology or Medicine in 2006. The actual discovery of RNAi had, however, occurred much earlier, simply by accident (or, as we call it, "serendipity"). Serendipity means the accidental discovery of something that one is not in search of, even sometimes an experiment gone awry; the discovery of penicillin by Alexander Fleming is probably the best-known example. Alexander Fleming was culturing bacteria on plates and when he went on holiday he forgot to clean the

1.4 (*cont'd*)

plates. He discovered on his return that on some plates fungi were growing but no bacteria, thereby discovering that the products of these fungi could kill bacteria. Similarly, plant scientists were striving to improve the color intensity of a blue petunia flower, when to their surprise the petunia turned white instead of blue (Figure B1.4.1). As it turned out, small interfering RNA (siRNA) had removed all the mRNA for blue color. Much later, microRNA was discovered as the normally occurring human correlate of siRNA, acting through a largely similar mechanism.

FIGURE B1.4.1 Petunias led to the accidental discovery of small interfering RNAs.

CHAPTER

2

Embryonic Development

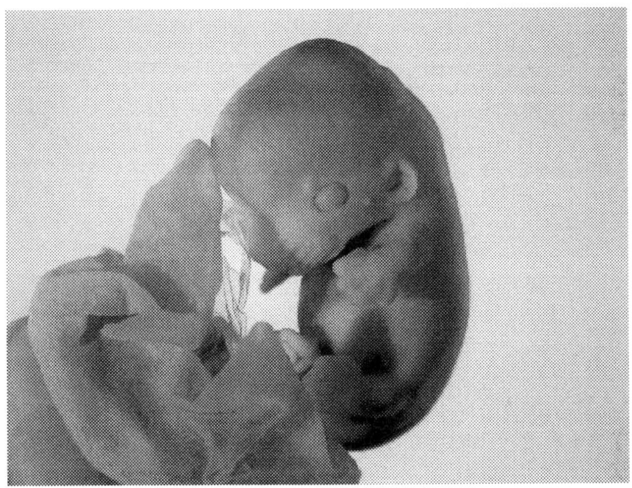
Stamcellen *Veen Magazines*

OUTLINE

Embryonic development has fascinated scientists and philosophers from ancient culture to the present day. The Greek philosopher Aristotle (384–322 B.C.) was among the first to describe the process of fertilization and embryonic development in detail; in his Περι ζωιων γενεσεωσ or *De generatione animalium* (*On the Generation of Animals*) he discussed how a

FIGURE 2.1 (a) The Dutch professor of zoology Ambrosius Hubrecht (1853–1915) compared the development of various animal embryos to gain insight into the processes that underlie evolution. (b) Charles Darwin (1809–1882) showed that evolution is driven by natural selection. *Source: (a) Hubrecht Institute, Netherlands/Stamcellen Veen Magazines; (b) Archive NWT, Netherlands/Stamcellen Veen Magazines.*

living animal and its internal organs are formed. He studied native animals, such as snakes, turtles, fish, and insects, and attempted to create a systematic overview of nature, making use of his observations and anatomical studies. This attempt, understandably, was not without error, an example being his conviction that, while some animals originated from the fertilization of an egg, many others—such as flies and eels—grew spontaneously from rotting meat. Only centuries later was this principle of *generatio spontanea* (spontaneous generation) finally disproved through careful experiments by the Frenchman Louis Pasteur (1822–1895).

In attempts to unravel the mysteries of embryonic development, a few specific animal species have been used repeatedly, most often for straightforward practical reasons such as whether the animals of interest were readily available, or easy to handle and manipulate experimentally. In the early twentieth century, for example, the Dutch professor Ambrosius Hubrecht (1853–1915) became interested in embryonic development of mammals (Figure 2.1). He chose the hedgehog as his preferred animal model and recruited the inhabitants of the city of Utrecht to help gather his research material: hedgehog embryos. Hubrecht paid a "kwartje," a Dutch quarter or 25 cents, for every live hedgehog that was delivered to his door, hoping for some pregnant females. In this way he built up a large collection of embryos at almost every developmental stage; this was wonderful for gathering study objects but it decimated the province of hedgehogs for many years, and such a method would certainly not be

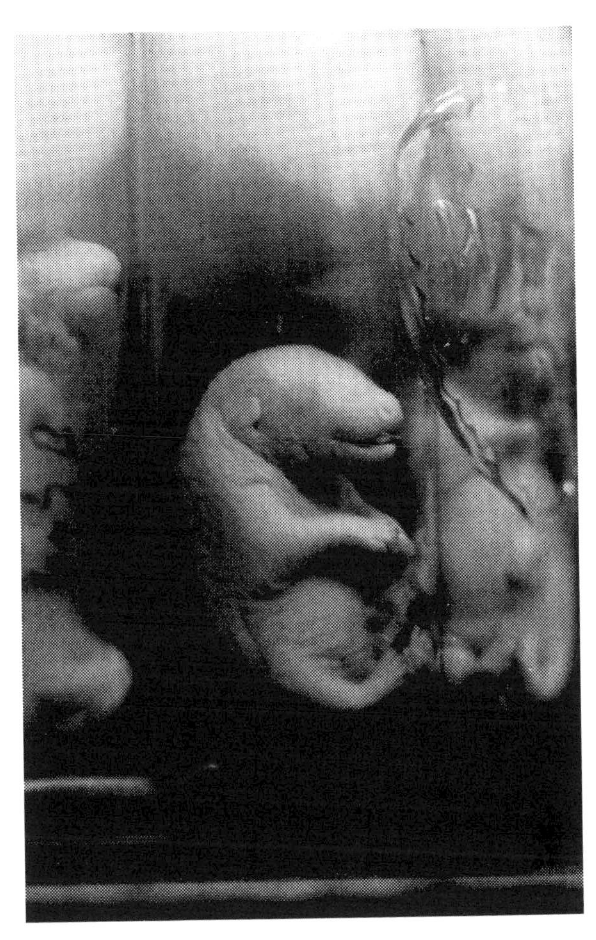

FIGURE 2.2 In the early days of experimental embryology, the availability of embryos was important in the choice of experimental models for scientific research. The Dutch embryologist Hubrecht studied hedgehog embryology and paid citizens of the city of Utrecht 25 cents for each hedgehog they delivered to him. Hubrecht kept the hedgehog fetuses that he collected from pregnant females in preserving jars until further research. The hedgehog population in the province of Utrecht was, however, severely reduced. *Source: Stamcellen Veen Magazines.*

allowed today. His collection was preserved, however, and is still accessible in a Berlin museum for study by those interested in comparative embryology.[1]

Over time, other species have gained popularity for embryo research, again initially for reasons of availability, but also for simplicity and short generation (or gestation) times (Figures 2.2 and 2.3). These include

[1]Hubrecht Collection of the Institute of Systematic Zoology, Museum for Natural Sciences, Humboldt University of Berlin, Germany.

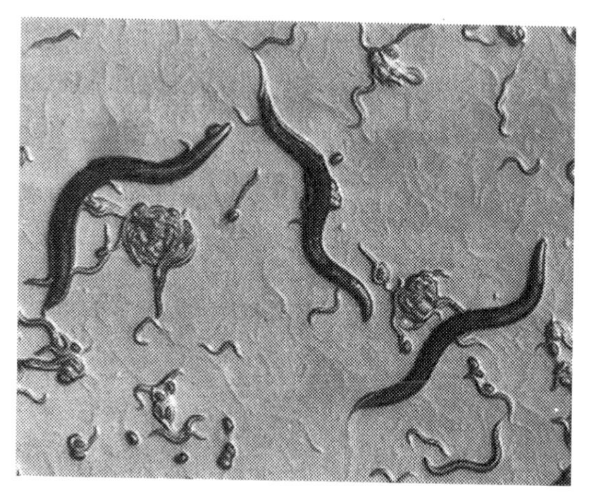

FIGURE 2.3 *Caenorhabditis elegans*, a nematode (or roundworm), is so small that it is barely visible to the naked eye. Each animal has exactly the same number of cells in its body (1090) and the function of each cell is exactly the same. A total of 131 cells die during development of the embryonic worm as a result of programmed cell death (apoptosis). Experiments with these worms actually led to the discovery of programmed cell death, essential for development of all animal species and involved in some aspects of disease. The Nobel Prize in Physiology or Medicine was awarded to Sydney Brenner, H. Robert Horvitz, and John E. Sulston in 2002 for the discovery of apoptosis. *Source: Henri van Luenen, Netherlands Cancer Institute, Amsterdam, Netherlands/Stamcellen Veen Magazines.*

the roundworm *Caenorhabditis elegans*, the fruit fly (*Drosophila melanogaster*), the zebrafish (*Danio rerio)*, the African clawed frog (*Xenopus laevis*), the chicken (*Gallus domesticus)*, and the house mouse (*Mus musculus*) (Figure 2.4 through 2.7). It has become increasingly clear that the key features of embryonic development have for the most part remained unaltered by evolution. While animals show obvious differences in appearance, the majority of their genes have been well preserved, demonstrating roughly similar structure and function. For this reason, the fruit fly, for example, can be used almost as the equivalent of an animal model for research into human embryonic development, physiology, and disease. This is particularly useful because its embryos develop so quickly, and it is convenient for genetic studies to determine the effects of errors in particular genes on offspring.

Despite years of dedicated research, much still remains to be discovered about reproduction and development in almost all species, from the formation of gametes (the "sex cells;" eggs, and sperm) and fertilization, to the subsequent development of the embryo, and formation of a new individual. Both ethical concerns and limited availability have always restricted research using human embryos, so that most knowledge on human embryonic development is based on assuming that the process is

FIGURE 2.4 The fruit fly *Drosophila melanogaster* has been extremely useful for embryological and genetic research. The German professor Christiane Nusslein-Volhard received the Nobel Prize in Physiology or Medicine in 1995 for her pioneering research on the function of *Drosophila* genes. *Source: Stamcellen Veen Magazines.*

much the same in experimental animals such as laboratory mice. While there are many similarities between embryonic development in mice and humans, there are also significant differences, most notably in how long development takes. For example, a mouse fetus develops in the uterus in 18 to 20 days, compared to a human pregnancy of 9 months. In addition, immediately following implantation in the womb, the mouse embryo forms a hollow cylinder, consisting of two different cell types, while the human embryo, like most other mammals, becomes shaped like a flat disk (Figure 2.8). A human baby can already be born alive around the 26th week of pregnancy, but, by contrast, a mouse pup needs to be carried almost to term to survive outside the womb. Many more differences have been discovered, as human embryo research has revealed more and more detail, especially about when and how tissue function begins. It is clear that extrapolating knowledge from animals to humans should be done with few assumptions. Nevertheless, basic principles have been conserved between species, so that it is the same genes that, for example, determine

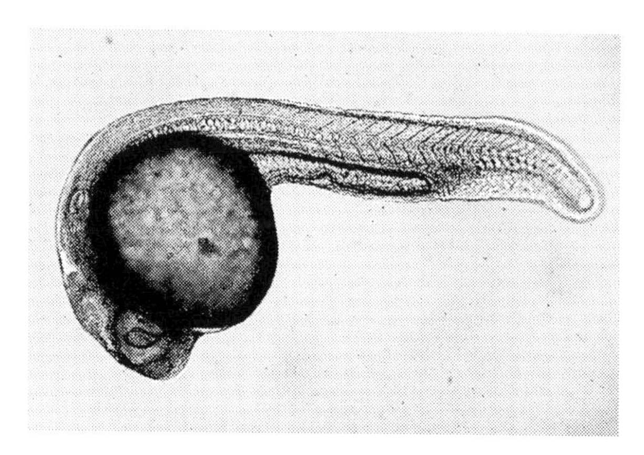

FIGURE 2.5 A zebrafish (*Danio rerio*) embryo 24 hours after fertilization of the egg. This fish species is used extensively for genetic research on vertebrates. The genome of the zebrafish can be altered relatively easily, for example, by the random inactivation of genes through chemical mutagenesis of sperm in male fish. The effects of the resulting mutations on morphology, physiology, or behavior of their offspring can then be studied. Once an interesting mutant has been found, the gene responsible can be identified by DNA sequencing. Alternatively, morpholinos (short DNA sequences) can be used to reduce expression of particular genes to determine their effect on development. The advantage of zebrafish for these mutagenesis and gene knockdown screens is that they are transparent, which enables good morphological observation to be made under the microscope. Many genes identified in this way have later been shown to cause congenital defects in humans. *Source: Astrid van der Sar/Jeroen den Hertog, Hubrecht Institute, Netherlands/Stamcellen Veen Magazines.*

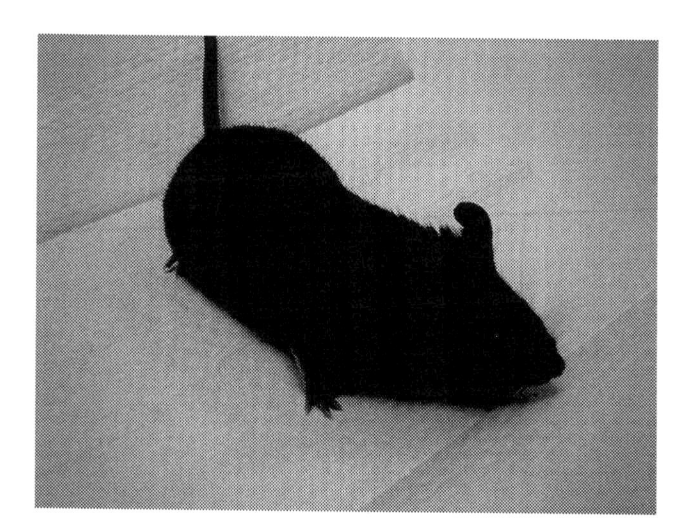

FIGURE 2.6 The laboratory mouse (*Mus musculus*) is one of the most frequently used animals for many kinds of biomedical research. *Source: Susana Chuva de Sousa Lopes/Stamcellen Veen Magazines.*

FIGURE 2.7 The African clawed frog *Xenopus laevis* has "earned its stripes" in embryological research. The large size of its eggs and early embryos made it one of the easiest vertebrates in which to study and manipulate development. *Source: Hubrecht Institute, Netherlands/Stamcellen Veen Magazines.*

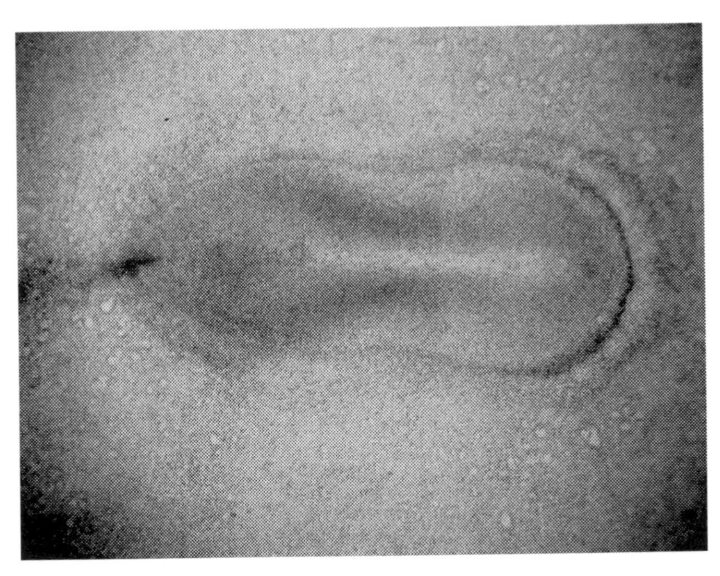

FIGURE 2.8 As in many other mammalian species, the human embryo adopts a planar (or flat) morphology after it reaches the womb. In this image, a 16-day-old horse embryo is depicted, which has a similar planar morphology. The part of the embryo that will become the head is on the right, the future tail is on the left.

the head end and the tail in fruit flies, mice, and humans. Likewise, the same genes or gene combinations determine the rest of the body plan: where the limbs form, their length and size, and where the organs are placed. This is known as the *homeobox*, or *Hox*, gene code of the body plan.

2.1

HOX GENES: HEADS OR TAILS?

All mammals, including man, have a *body plan* that is more or less similar, independent of the species: a head with eyes, a nose, ears and a mouth; neck; chest and belly; arms and hands with fingers; and legs and feet with toes. This incredible organization of the body, in which cells know what to become just because of their position is not present at the beginning of development, but gradually appears over time. The fertilized egg is a simple spherical cell with no distinguishable top or bottom, or left or right. During the first divisions of the embryo, all of the cells can form a complete fetus and not just, say, only the head or only the legs. How then are the different structures formed in the correct place? Why are our heads always at the top of our body, or at the front in animals? And how are our bodies configured such that our arms stick out of the shoulders and legs stick out of the pelvis but never the other way round?

The regulation of the head-to-tail axis is surprisingly similar in many animals and trying to find out why has fascinated scientists for centuries. The first indications that specific genes are, in fact, responsible for correct positioning of the body came from experiments with fruit flies. In the first half of the 1900s, fruit flies, or *Drosophila melanogaster*, were extremely popular as experimental animals for studying genetics because of their very short generation time. Geneticists found out that exposing these animals to X-rays to damage their DNA occasionally led to flies with slightly different shapes: mutant flies. Once in a while, fruit flies were found that looked very strange indeed. Some flies, instead of having one pair of wings and one pair of halters (organs that stabilize the fly during flight), had two pairs of wings. The body part that in a normal fly formed the halters was transformed into a segment that was normally destined to form wings (Figure B2.1.1). These kinds of mutations are called *homeotic transformations* from the Greek *homoiosis* "becoming like": one body part (or segment) is replaced by another. Because insects are segmented in a head-to-tail fashion, such

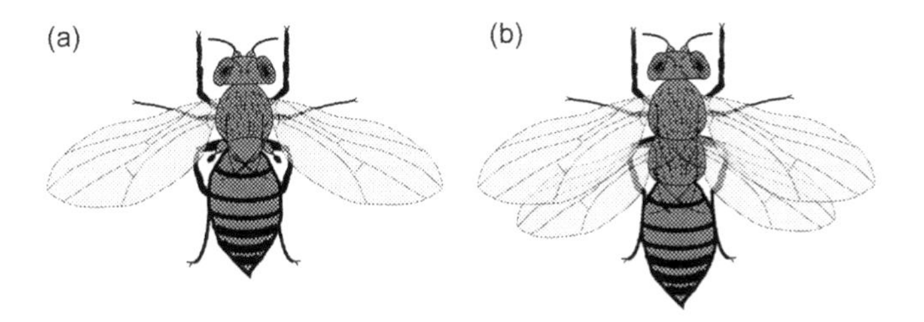

FIGURE B2.1.1 A normal fruit fly (a) has one pair of wings and one pair of halters for stability. In a mutant four-winged fly (b) the segment containing the halters has been transformed into a segment that contains wings.

replacement means that a part of the animal has been transformed to a more head-like or more tail-like segment.

The experiments with X-rays were not performed to create flies with superpowers, as popular comic books might have us believe. Rather, X-rays or radiation can cause changes in the sequence of the DNA of exposed flies, and by studying their DNA, the gene responsible for causing the homeotic transformations can be identified.

It was discovered that the gene responsible for changing halters into wings was part of a group of very similar genes that are close together in the genome, in two different complexes. Incorrect functioning of these genes led to homeotic transformations, and as a result the genes were called *homeobox* genes or *Hox* genes. Clearly, the genes responsible for these very striking transformations must be important for establishing a correct body plan.

It was a complete surprise when it was discovered that Hox genes are present in mammalian genomes, not just in the fly. All mammals, including humans, have Hox genes with similar functions as those in flies: determination of the body plan. This was so amazing and groundbreaking that it resulted in Edward Lewis, Christiane Nüsslein-Volhard, and Eric Wieschaus being awarded the Nobel Prize in Physiology or Medicine in 1995 for "their discoveries concerning the genetic control of early embryonic development," work that included studies of Hox genes.

Hox genes code for transcription factors: proteins that can bind sequence-specific regions of DNA and regulate the transcription of genes. The part of the DNA in Hox genes that is responsible for the DNA-binding part of the protein is called the homeobox. Hox genes regulate the formation of the body plan in animals throughout the animal kingdom and they are united by several remarkable features:

2.1 (cont'd)

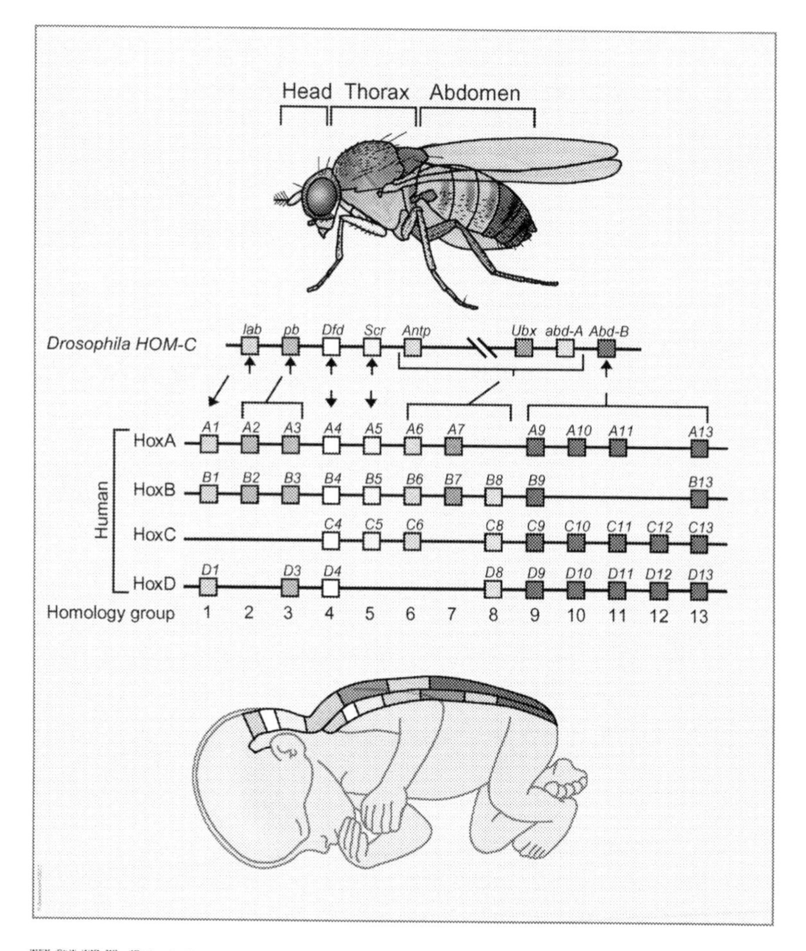

FIGURE B2.1.2 Hox genes are responsible for defining what will become head and what will become tail in embryos. Hox genes in flies are present in two clusters of DNA, while in humans four different clusters are located on four different chromosomes. Genes on one side of the clusters will be expressed in the tail and genes along the chromosome will be progressively expressed further toward the head region, as indicated by the different colors in the scheme. Similar Hox genes also determine the body pattern in humans, for example, that our head is on our shoulders, our shoulders are on our rump, and our legs form at the base of our spine. *Source: Sebastiaan Blankevoort, Leiden University Medical Center, Netherlands.*

they are present in "clusters" or groups in the genome, and their order on the genome faithfully represents expression of their proteins along the head-to-tail (antero–posterior) body axis. Hox genes in flies are present in two clusters of DNA, while in humans there are 39 Hox genes that are found in four different clusters located on four different chromosomes. Genes on one side of the clusters will be expressed in the tail and genes along the chromosome will be progressively expressed further toward the head region. This intriguing order of expression is known as *colinearity*. In this way, the expression of a specific set of Hox genes (the Hox code) introduces polarity to the embryos and determines identity along the head-to-tail axis (Figure B2.1.2). Although homeotic transformations in flies can lead to wings on their heads, humans never have legs attached to their shoulders, even though homeotic mutations are known in humans. At most they cause the identity of vertebrae in the spinal column to be altered or other relatively minor skeletal changes to the body plan.

2.1 FERTILIZATION AND EARLY EMBRYO DEVELOPMENT

Before the male and female gametes, the sperm and the egg, can combine to start the formation of an embryo, they need to be prepared or "ripened" for fertilization. A normal nucleus in the human body cell contains 23 pairs of chromosomes, or a total of 46: the cell is then described as *diploid*. One chromosome of each pair is derived from the father, the paternal chromosome; the other is from the mother, the maternal chromosome. The number of chromosomes in the gametes needs to be halved prior to fertilization to prevent unwanted doubling of the 23 chromosome pairs, and this reduction occurs during a process called meiosis (Figure 2.9). In addition, the gametes need to be made fully functional for fertilization. Mature human gametes, both sperm and egg cells, thus contain half the normal number of chromosomes. The cells are then described as *haploid*. When the two fuse at fertilization, the 23 chromosomes of the sperm enter the egg, and in the resulting fertilized egg the normal number of chromosomes found in a normal cell is thus restored, creating a diploid embryo.

Maturation of the egg cell largely occurs in the ovaries, where it lies in a fluid-filled cavity, called a follicle, and is surrounded by other cells, the cumulus and granulosa cells (Figure 2.10). During maturation of the egg (or ovum), it is separated from the surrounding cumulus cells by a thick

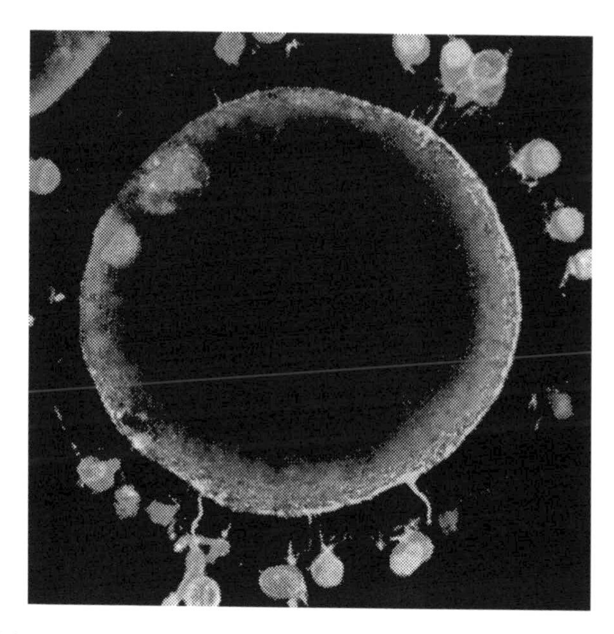

FIGURE 2.9 When mammalian oocytes (or eggs) mature they undergo reduction divisions to become "haploid" instead of the usual "diploid" cells that have two chromosome sets. This is called meiosis. Here, a pig oocyte is shown in which meiosis is almost complete and which is ready to be fertilized. The chromosomes are visible as blue areas. The egg is surrounded by cumulus cells, which aid in the maturation process. *Source: Jurriaan Hölzenspies.*

FIGURE 2.10 A pig ovary after removal from the sow. The vesicle-like structures (follicles) that are clearly visible each contain one egg. *Source: Stamcellen Veen Magazines.*

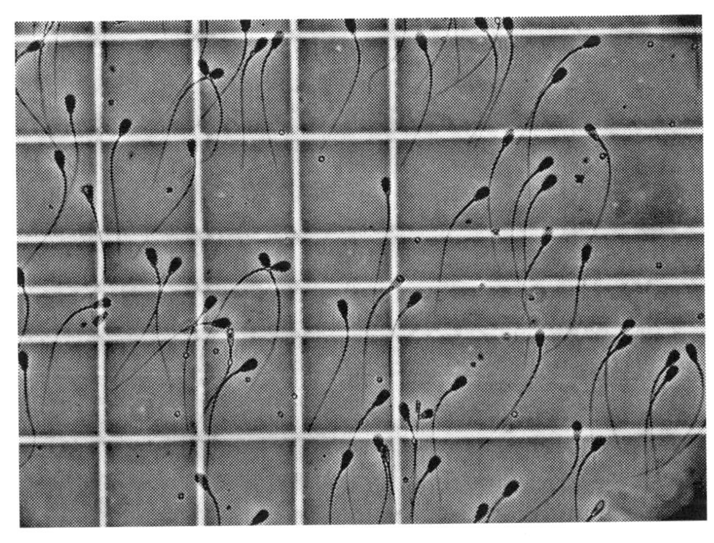

FIGURE 2.11 Sperm cells are composed of a head that contains the DNA and a tail for movement.

shell called the zona pellucida, which forms around the egg. This shell is not composed of cells but of proteins secreted from the egg. The zona pellucida is very important in protecting the egg and, later, the early embryo as it moves from the ovary toward the womb, but it also functions as "receptor" for the sperm cell, allowing the sperm to bind to the egg as a first step in fertilization. The mature egg, surrounded by the zona pellucida and cumulus cells, is released into the Fallopian tube during ovulation, where it is fertilized by a sperm cell (Figure 2.11). The sperm cell penetrates the zona pellucida and fuses with the egg cell, bringing together male and female chromosomes (DNA) into a single nucleus in the cytoplasm of the fertilized egg cell, now called a *zygote*. The egg cell is many times larger than the sperm cell. As a result, the zygote's cytoplasm is derived almost entirely from the egg; it contains all the necessary components to support combining of the chromosomes and the first cell divisions. It is also the source of all of the mitochondria, which supply cells with energy. This means that most of the mitochondria in our body are inherited from our mothers and not our fathers, and is the reason why transferring a healthy nucleus into a new egg may be a way to rescue embryos with some mitochondrial diseases in the future (see Box 5.4 "Three-Parent Embryo and the Law" in Chapter 5, "Origins and Types of Stem Cells: What's in a Name?").

The special nature of egg cytoplasm is also illustrated by the fact that it can reprogram the nucleus of an adult cell that is already differentiated (specialized) and initiate the growth of a new individual.

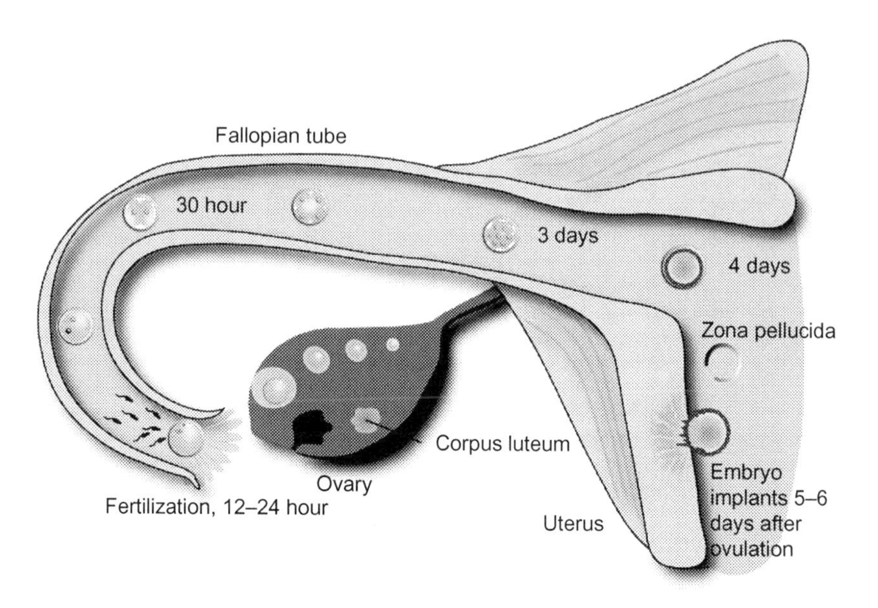

FIGURE 2.12 Schematic representation of the first week of human embryonic development. The egg is fertilized 12–24 hours after ovulation (release from the ovary). Within the egg, a male and female pro-nucleus are formed and these subsequently fuse. About 30 hours after fertilization the zygote "cleaves" (or divides into two). The newly formed embryo is transported through the oviduct to the womb (uterus) as it continues to develop. The zona pellucida, which has surrounded the embryo during its journey, is shed approximately 4.5 days after fertilization, after which the embryo is ready for implantation in the wall of the uterus. If the oocyte has been successfully fertilized, the remnants of the ruptured follicle that have become a corpus luteum remain active in the ovary and produce the hormone progesterone, essential for the preparation of the mother's body for the pregnancy. *Source: Bernard Roelen/Stamcellen Veen Magazines.*

A complete living animal may result from placing the (diploid) nucleus of a differentiated cell in an egg cell from which its own genetic material has been removed. This procedure is called cloning, and the sheep Dolly was the first, and still the most well-known, mammal that was created this way (see Chapter 6, "Cloning: History and Current Applications").

The zygote divides into a two-cell embryo, which is transported to the uterus (womb), pushed along by ciliae (fine, finger-like structures) in the Fallopian tubes. A human embryo implants into the wall of the uterus five to seven days after fertilization (Figure 2.12). During its journey through the Fallopian tubes to the uterus, the embryo divides a few more times. These initial cell divisions (called *cleavages*) are special because they are not accompanied by cell growth, so the cytoplasm of the egg is divided among the newly formed cells, which, of course, become progressively smaller. The total volume of the early embryo remains

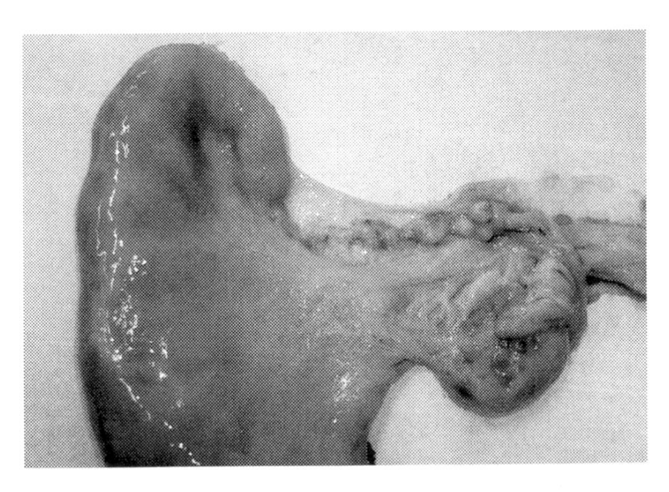

FIGURE 2.13 A cow uterus (left) is attached to the oviduct (coiled structure in the middle). The ovary (bottom left) is positioned close to the oviduct by ligament structures. *Source: Bernard Roelen.*

essentially unchanged compared to the fertilized egg. Two cells become four, four cells become eight, and so forth (Figure 2.13).

At this point, each of the cells making up the embryo, called blastomeres, are capable of forming a complete individual if they are separated from the embryo. The blastomeres, just like the zygote, are still *totipotent*. This also means that if a single cell is carefully removed from the embryo, for example, for preimplantation diagnosis, this will not interfere with the normal development of the embryo. An embryo can sometimes spontaneously split into two, or very rarely into three, at this early stage and in humans this can give rise to a small number of pregnancies with identical twins (or triplets!).

During early embryonic development, before implantation in the uterus, the totipotent character of the blastomeres quickly fades. Successive cell divisions lead to a morula, a solid lump of cells no larger than the period at the end of this sentence. From this stage onward, visible differences appear in the embryo and different cell types can be clearly seen. The outer cell layer differentiates into a thin epithelial layer, the trophectoderm, and the cells inside the embryo form a compact clump, the inner cell mass. A fluid-filled cavity, the blastocoel, develops in the middle of the embryo as fluid is transported in from the outside through the trophectoderm. This cavity grows in size, giving the embryo—now called a blastocyst—the appearance of a balloon covered on the outside by the trophectoderm, with the inner cell mass squashed to one side. The inner cell mass will ultimately form all the fetal tissue and is also the source of embryonic stem cells; the trophectoderm later contributes to the placenta (Figure 2.14).

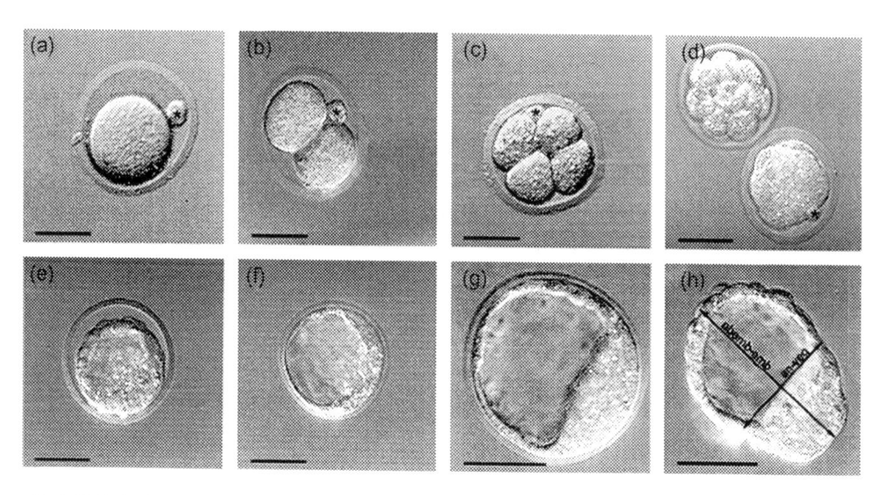

FIGURE 2.14 The development of a mouse embryo from a fertilized egg (zygote) (a) to a two-cell stage embryo (b), four cell-stage embryo (c), morula (d), and blastocyst (e) through (h). Three cell types can be distinguished in the blastocyst: trophectoderm (green), primitive endoderm (yellow), and the inner cell mass (orange) that gives rise to the embryo proper. *Source: Susana Chuva de Sousa Lopes. Reprinted with permission from Lanza (Ed.), Handbook of Stem Cells, Elsevier, 2005/Stamcellen Veen Magazines.*

The embryo reaches the uterus around the blastocyst stage. Here, it expands and breaks out of the zona pellucida, a process known as hatching, much like a bird hatches from an egg. The trophectoderm can then stick to the uterus, allowing the embryo to nestle in the uterine wall; in humans this occurs 5 to 6 days after fertilization. This process of blastocyst formation and implantation is not necessarily identical in all placental mammals. In pigs, for example, the blastocyst lengthens about 300 times, changing from a hollow ball into a tubular structure that can end up being a meter or more long (Figure 2.15). In species such as chicken and frog, there is, of course, no implantation in the mother and the embryo gets all of the nutrition it needs to grow from the yolk which forms next to the embryo.

As just described, only two cell types can be distinguished in the early blastocyst: cells belonging to the inner cell mass and cells of the outer trophectodermal layer. After implantation in both mice and humans, three new cell types arise from the inner cell mass. The outer layer, called *ectoderm*, later forms skin, brain and nerve cells, parts of the eye including the lens, the epithelium of mouth and anus, the pituitary gland, part of the adrenal glands, and pigment cells (Figure 2.16). The embryonic "middle layer," or *mesoderm*, gives rise to skeletal muscle, heart, blood vessels, parts of the urogenital system (kidneys, urethra, gonads), bone marrow, blood, bone, cartilage, and fat. The inner layer, or *endoderm*, forms

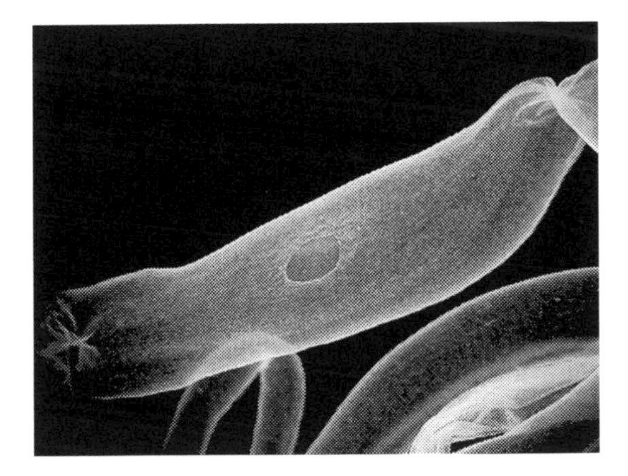

FIGURE 2.15 A pig embryo just before implantation in the uterus is shaped like a very thin tube, most of which form parts of the yolk sac and placenta. The actual embryo is visible as a small oval disc. *Source: Bernard Roelen/Stamcellen Veen Magazines.*

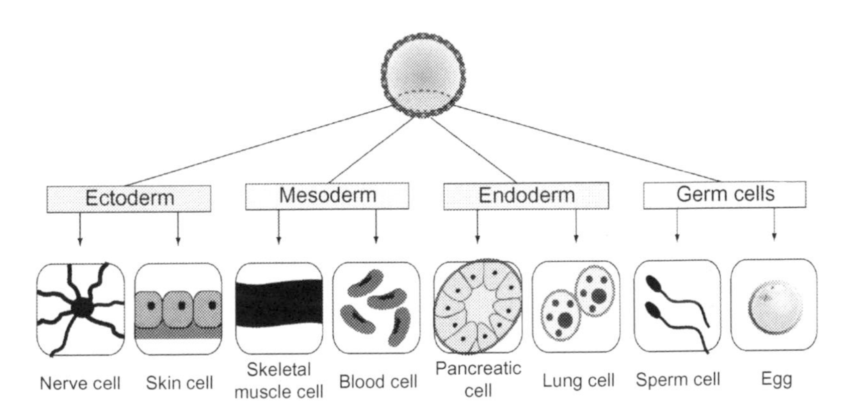

FIGURE 2.16 The inner cell mass of a blastocyst differentiates into the three germ layers (ectoderm, endoderm, and mesoderm) that contribute to all tissues and organs. In addition, the inner cell mass generates the precursors of the sperm and egg cells. *Source: Stamcellen Veen Magazines.*

the lining of the gastrointestinal and respiratory tracts, as well as liver, pancreas, thyroid, thymus, and bladder lining. It is important to realize that the extraembryonic (or nonembryonic) structures, such as parts of the placenta and umbilical cord, do not originate from the inner cell mass, but from the trophectoderm and yolk sac, and lie outside the "embryo proper." Thus, cells from the inner cell mass are not totipotent, like the zygote, but pluripotent. This means they can differentiate into all cells needed for an embryo, but not into all of the cells required to support its complete development. All mammalian embryos (with a few exceptions,

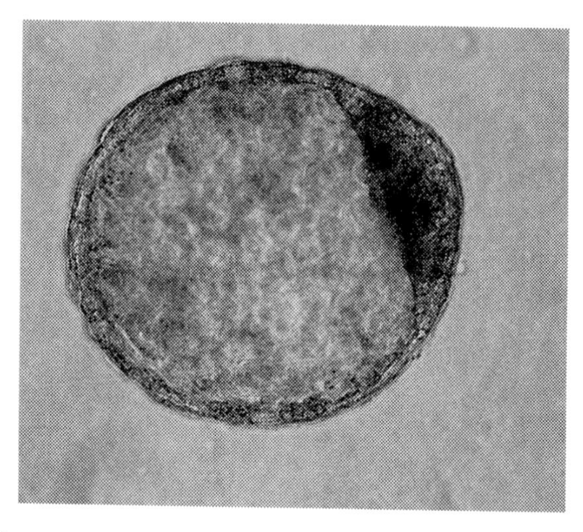

FIGURE 2.17 Micrograph of a pig blastocyst stage embryo with a clearly visible inner cell mass surrounded by a thin layer of trophectoderm. *Source: Ewart Kuijk/Leonie du Puy.*

such as the egg-laying platypus) need a mother in which to complete development; only the very early stages can take place in the laboratory, which is why *in vitro* fertilization (IVF) is possible (Figure 2.17).

2.2 SEX CELLS AND GERM CELL TUMORS

During embryonic development, the cells that make up the embryo differentiate into the different specialist tissues and, along the way, rapidly lose their pluripotent character. Only the sex cells (eggs and sperm) maintain the capacity to create new individuals of a next generation. Though the sex cells of an adult reside in the gonads (ovaries and testicles), that is not where they originated. They are actually formed as primordial germ cells long before the gonads develop. Over the course of a few days the primordial germ cells travel from the tail end of the developing embryo between the cells destined to form the intestines (the endoderm) toward the developing gonads. During this period of migration, they multiply to increase from an original population of a few dozen cells to a few thousand by the time they reach and colonize the gonads. At this stage they can still become either egg or sperm cells, but once they contact the developing gonads, which of the two cell types they will form becomes determined. This is dictated by whether the gonad is male or female, and, as a result, which specific hormones it produces (Figure 2.18).

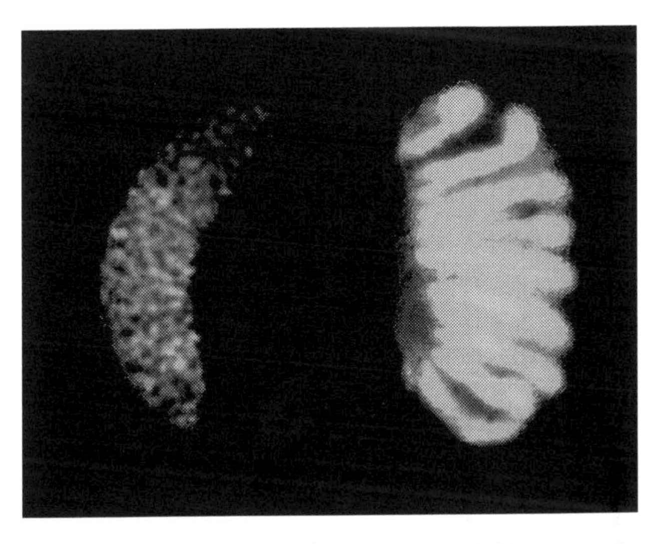

FIGURE 2.18 The gonads of a male (right) and female (left) mouse embryo. In these particular organs the germ cells (sperm on the left and egg in the right structure) express green fluorescent protein. The individual cells cannot be recognized as they are too small, but the presence of the cells at specific locations can be seen as a green glow. *Source: Susana Chuva de Sousa Lopes, Gurdon Institute, U.K./Stamcellen Veen Magazines.*

The developmental capacity of (primordial) germ cells is extremely high; after all, these are cells capable of forming a completely new individual after fertilization. This also means that development of these cells must be tightly choreographed. If this process is disrupted, for example if the cells do not reach the gonads during development but end up in another organ or tissue, or they start dividing in the gonads in an uncontrolled fashion. This can lead to the formation of a so-called germ cell tumor (a *teratoma* or *teratocarcinoma*) (Figure 2.19). In most cases, such a mistake happens in an ovary or testicle and then only rarely, but it can also occur in other parts of the body, such as the pineal gland, thymus, adrenal gland, pituitary, or peritoneum. Germ cell tumors are characterized by the presence of multiple tissue types, such as bone, tooth-like structures, and hair. This is not as odd as it may seem, since these tumors stem from totipotent cells that can specialize in all sorts of directions. This property of primordial germ cells can be exploited in a laboratory setting. Not only do they represent an alternative source of pluripotent stem cells, they can also teach us a great deal about the formation of sex cells and about the unique property of pluripotency.

The primordial germ cells in the embryo are derived from the inner cell mass, so it might seem logical that embryonic stem cells (or even

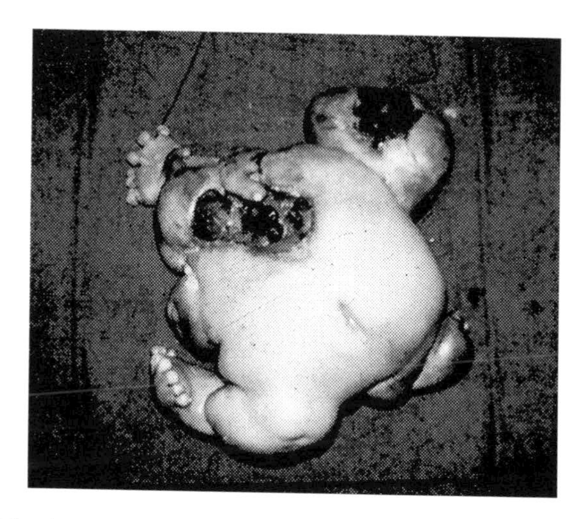

FIGURE 2.19 A germ cell tumor or teratoma can be composed of many different cell types and contain recognizable structures, such as hair, teeth, brain, and bone. If it contains undifferentiated stem cells, it is called a teratocarcinoma, and is malignant. *Source: Wolter Oosterhuis, Josephine Nefkens Institute Erasmus MC, Netherlands/Stamcellen Veen Magazines.*

induced pluripotent stem cells; see Chapter 4, "Of Mice and Men: The History of Embryonic Stem Cells" and Chapter 5, "Origins and Types of Stem Cells: What's in a Name?") could form primordial germ cells, as they are also pluripotent. This proved to be very difficult, however, and many early claims that germ cells had been generated from stem cells in the laboratory may not have been correct. Hayashi's group of researchers in Japan did, however, finally show that, at least from mouse embryonic stem cells, it was possible to produce primordial germ cells with the capacity to form eggs and sperm. They also showed that these cells can be induced to mature if transferred to the mouse gonads and even fertilized.[2] Though this has not yet been done in humans, there seems to be no principle reason, just technical ones, why the same procedure would not work. This is food for thought for medical specialists, infertile patients, homosexual couples, and ethicists; in fact for society as a whole.

[2]Hayashi K, Ohta H, Kurimoto K, Aramaki S, Saitou M. Reconstitution of the mouse germ cell specification pathway in culture by pluripotent stem cells. Cell 2011;146(4):519–32.
Hayashi K, Ogushi S, Kurimoto K, Shimamoto S, Ohta H, Saitou M. Offspring from oocytes derived from in vitro primordial germ cell-like cells in mice. Science 2012;338(6109):971–5.

CHAPTER

3

What Are Stem Cells?

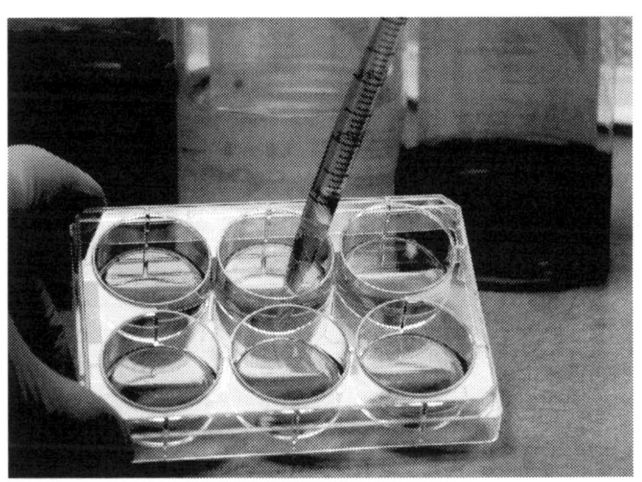

Jeffrey de Gier/ Stamcellen *Veen Magazines*

OUTLINE

The common definition of a stem cell is "a cell that can divide to give rise to both a new copy of itself and at least one specialized, differentiated, cell type." Although this broad definition provides a useful framework, much still remains to be discovered about the different types of stem cells, their common as well as unique properties, and how and for what applications we can best use them in research, medical practice, biotechnology, and pharmacology.

3.1 WHAT ARE THE PROPERTIES OF STEM CELLS THAT MAKE THEM DIFFERENT FROM OTHER CELLS?

Just like many normal cells in the body, stem cells are able to divide and produce new copies of themselves ("self renew"). In addition, stem cells can differentiate, or specialize, into other cell types of the body. If the stem cell is able to form all cell types of the embryo and adult, including germ cells and nonembryonic structures connected to the fetus such as the placenta, it is considered *totipotent*. A fertilized egg cell, for example, is totipotent, as are some of the early blastomeres, the first cells that result from cleavage or splitting of the fertilized egg; these cells are not stem cells, however, since they do not self-renew. If a stem cell can do all of these things but is unable to form the nonembryonic (also called *extra-embryonic*) structures (such as parts of the placenta and the yolk sac) then it is called *pluripotent*. Embryonic stem cells are pluripotent. All other stem cells found in specialized tissues of the fetus or adult are referred to as *multipotent*, meaning that they are able to form many but not all tissue cells of the body. Multipotent stem cells only form those cell types in the organs or tissue of the body in which they are normally located. Finally, there are *unipotent* stem cells; these are able to form just one other cell type. Spermatogonia, for instance, are unipotent and are only able to form sperm.

To be able to divide without losing the stem cell pool for later use, a stem cell is capable of multiplying through division, but after each cell division often one of the two daughter cells retains the original stem cell properties. The daughter cell that loses stem cell properties then becomes fully differentiated: a specialized cell that produces all of the proteins required for its proper function within the tissue or organ it belongs to. It can at best divide only a few times, but more often not at all; this is, for example, the case for (at least most) brain cells.

To simplify discussion, it is conceptually easiest to divide stem cells into two types: embryonic stem cells and adult stem cells. Embryonic stem cells are derived from a very early embryo just a few days after fertilization of the egg, and adult stem cells are found in postnatal tissues, not only of the body but also the umbilical cord at birth. Although some controversy still exists about the potency of adult stem cells and their ability to form different cell types, adult stem cells are generally considered to be either multipotent or unipotent. In the adult body *in vivo* there are no totipotent or pluripotent stem cells. This does not mean that the artificial conditions of culture cannot convert ("reprogram") some adult cells into stem cells with pluripotent properties. We will return to this issue of reprogramming and *induced pluripotent stem cells* in culture later. Under normal circumstances, however, a differentiated cell cannot change back into a stem cell since differentiation is, in principle, a unidirectional process. Fully differentiated cells can at best only divide a limited number of times to give rise to daughter cells with the same characteristics; differentiation into other cell types is not an option. Similarly, a multipotent or unipotent adult stem cell in an organ or tissue, such as the pancreas, cannot normally revert to the pluripotent state of an embryonic cell. The adult stem cell also "remembers" where it came from so that, for example, a pancreas stem cell can only become cells of the pancreas, *hematopoietic* (blood) stem cells in the bone marrow can only become blood cells, neural stem cells can only become cells of the nervous system, and so on. By contrast, embryonic stem cells are by definition pluripotent, and can in principle develop into any of the cell types in the human body.

During embryonic development, as all the different new cell types emerge and tissues are formed, most cells gradually lose their stem cell characteristics and with this their ability to differentiate in many different directions. This differentiation process starts when the totipotent fertilized egg moves on into the pluripotent state of the inner cell mass of a blastocyst-stage embryo, then subsequently into specialized cells and into the multipotent or unipotent stem cells of a specific organ or tissue. These multipotent or unipotent stem cells are also known as *progenitor cells*. Progenitor cells can divide, but this capacity is generally considered more limited than that of pluripotent stem cells. Progenitor cells are predestined to differentiate only into the cell types that are needed for the proper function in their own specific organ or tissue and they are thought to be important for turnover and repair of the organ. The identification of adult stem cells, which seem to have a greater capacity to divide than tissue progenitors but a similar differentiation capacity, has confused the issue somewhat and it may simply be that there are adult stem cells with different capacities to divide: some can undergo many divisions, others only a few.

The biology underlying the tendency for differentiation being in one direction only is based on the epigenetic modifications and mechanisms discussed in Chapter 1, "The Biology of the Cell." In normal development, an ever-increasing number of genes are more-or-less permanently altered by epigenetic modifications, which ensure the stability of the differentiated state and limit the ability of cells to divide indefinitely. These changes were thought to be irreversible, although reprogramming that was revealed by the birth of cloned animals (Chapter 6, "Cloning: History and Current Application"), the generation of induced pluripotent stem cells from adult cells (Chapter 4, "Of Mice and Men: The History of Embryonic Stem Cells" and Chapter 5, "Origins and Types of Stem Cells: What's in a Name?"), and identification of cancer stem cells (Chapter 12, "Cancer Stem Cells: Where Do They Come from and Where Are They Going?") has demonstrated this is not always the case. However, in general, the epigenetic modifications that take place in normal development reduce the differentiation options for the individual cells in later life and their ability to proliferate.

3.1

EPIGENETIC MODIFICATION OF THE SECOND X CHROMOSOME IN FEMALE CELLS

Women have two female sex chromosomes (XX), and men have one X and one Y chromosome (XY) in each cell. As with the autosomal chromosomes, females, therefore, have two copies of every gene present on the X chromosomes, and males have only one X chromosome and, therefore, only one copy of each gene on this chromosome. Two active copies of a gene could lead to overproduction of the corresponding protein and this, in turn, could cause major problems during embryo development and postnatal life. In mammals, including humans, nature has found a solution to this problem by permanently deactivating one of the two X chromosomes in the embryo during development. This occurs in every cell that a double X chromosome is found. As a result, the DNA remains in the cell but cannot be actively used any more. This process is called *dosage compensation*.

Other proteins ensure that when a cell divides, X-chromosome inactivity is reproduced in the two daughter cells. The X chromosome that is inactivated is randomly picked from the two during early embryonic development. As a consequence, in approximately half of the cells of a normal early embryo, the paternal X chromosome inherited from the father is switched off, and in the other half, the maternal X chromosome inherited from the mother is inactivated. This means that across the whole body, around half of the cells make use of the father's X chromosome, and in the other half, this X chromosome is permanently inactivated and the mother's is used. On close examination, for instance using a microscope, the inactivated X chromosome is visible in the cell nucleus as a dark spot, the *Barr body,* named after the Canadian physician Murray Barr who first described the structure. In mammals, in which a placenta develops to support the embryos, X-chromosome inactivation is random, while in marsupials, which do not develop a placenta, it is always the paternal X chromosome that is inactivated (Figure B3.1.1).

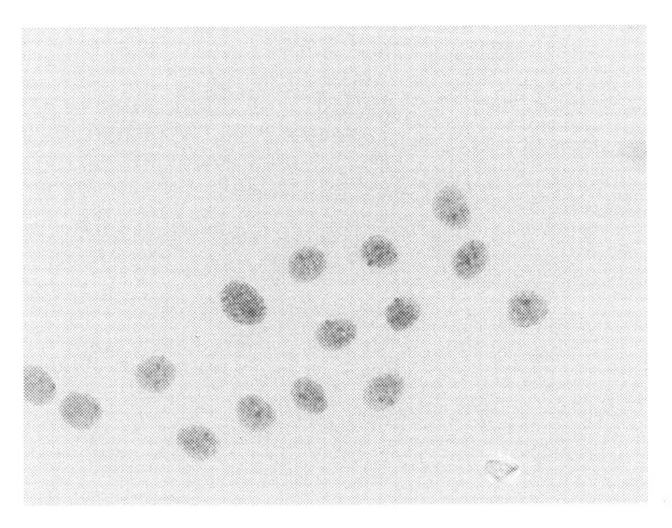

FIGURE B3.1.1 In female cells, one of the X chromosomes is inactivated. In the cell nucleus it can be seen as a dark spot called the Barr body. *Source: Susana Chuva de Sousa Lopes, Leiden University Medical Center, Netherlands.*

3.2 TOTIPOTENCY AND PLURIPOTENCY, AND EMBRYONIC STEM CELLS

Our body is made up of some 220 different types of cells grouped in different organs and tissues, all of which are descendants of a single fertilized egg. The DNA (or genome) of the egg contains the entire program for the development of the embryo. The fertilized egg first cleaves (or divides) into two, the two resulting cells (called blastomeres) being half the size of the egg. The same then happens again and at the second cleavage four blastomeres are formed, each being a quarter of the size of the egg. At each cleavage division, the cells get smaller and this

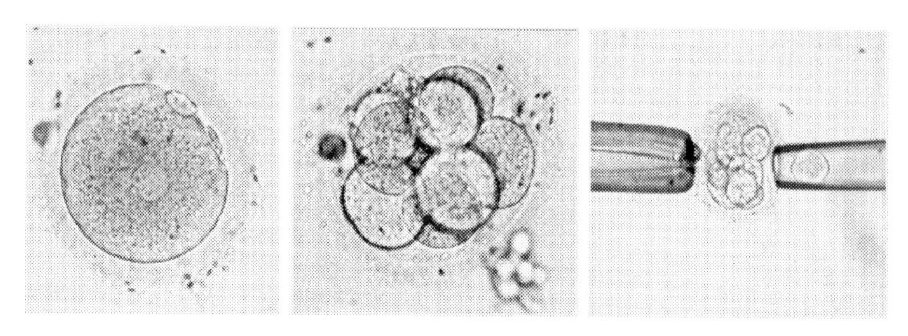

FIGURE 3.1 Preimplantation genetic diagnosis (PGD): detecting serious genetic disorders in early human embryos. A human egg (left) that has been fertilized in the laboratory of a fertility clinic develops into a multicellular embryo (middle). One cell is removed from the embryo using a glass pipette (right) and used to investigate whether a disease-causing mutation is present in the genomic DNA. In the right-hand image, the embryo is held by a "holding pipette" to keep it in place (to the left of the embryo) while a cell is carefully aspirated (or removed) with another pipette (on the right). When the genetic analysis is complete and the embryo found to be normal, it can be placed in the woman's womb to develop into a healthy baby. If it contains a genetic disease, it will be discarded or perhaps used for research. This may include generating human embryonic stem cells with the disorder for study. *Source: Joep Geraedts, Academic Hospital Maastricht, Netherlands/Stamcellen Veen Magazines.*

makes cleavage different from division. Within a few days of fertilization, three cleavage divisions have taken place and eight identical cells have formed (one in the center surrounded by seven others on the outside). At the eight-cell stage, the embryo is remarkably flexible: one single cell can actually be removed and used, for example, for prenatal genetic testing to diagnose specific disease-causing mutations in the embryo (Figure 3.1). The embryo itself, now consisting only of the remaining seven cells, can continue to develop entirely normally and be transferred to the uterus where, if all goes well, a healthy baby will be born. The missing cell is replaced without a problem and prenatal genetic testing thus provides a means of selecting genetically normal from abnormal embryos in families where specific, often untreatable, genetic diseases are prevalent. The procedure demonstrates that the cells of the embryo at this point are still entirely totipotent and the embryo is extremely "plastic." This is also the stage at which it can divide into two and form identical twin embryos.

The first division of function between groups of cells then takes place. The earliest totipotent cells have disappeared and the cells in the middle of the embryo divide to form a group of pluripotent cells that eventually form the fetus. The outer cells become multipotent cells and form the non-embryonic tissues, such as the placenta and umbilical cord, needed to support fetal development. The next embryonic stage occurs and the *blastocyst* is formed. Now the distinction between pluripotency and multipotency becomes clearer. The innermost population of cells is called the

FIGURE 3.2 A preserved and stuffed chimeric mouse generated at Vanderbilt University in Nashville, Tennessee. A chimeric (or genetically mixed) mouse can develop after introduction of embryonic stem cells from one strain into a blastocyst-stage embryo of another. If the embryo is then transferred to the uterus of a recipient female mouse the embryo can implant in the uterine wall and develop to term. The embryonic stem cells can participate in the formation of the embryo. If the blastocyst used came from albino (white coat color) parents and the stem cells originally came from a black mouse embryo, the resulting chimeric mouse has a black and white coat and it will be quite clear which parts originate from the embryonic stem cells and which from the host embryonic cells. *Source: Stamcellen Veen Magazines.*

inner cell mass, and the outermost cells form the *trophectoderm*. Cells from the inner cell mass have become pluripotent and can thus differentiate into all the different cell types that make up the body, but they can no longer contribute, for example, to placenta tissue. In contrast, the trophectoderm cells on the outside can no longer contribute to the embryo. In the pluripotent inner cell mass cells, almost all genes are still ready to be used for differentiation in whichever direction is dictated either by the position of the cells in the embryo or which cells are each other's neighbors. This pluripotent period only lasts for a short time, as cells quickly start to differentiate as development proceeds.

When embryonic cells are removed from the inner cell mass (say, from an albino mouse, which has a white coat color) and injected into a blastocyst from another mouse (say, with a black coat color), the resulting pups that are born will have coats with black and white patches, with cellular descendants of the injected inner cell mass cells present in all tissues and organs. Such experiments producing *chimeric* mice (i.e., from two different genetic backgrounds) have provided the evidence that cells of the inner cell mass indeed have the capacity to form all cell types in an adult mouse (Figure 3.2).

If instead of injecting the cells of the inner cell mass into a blastocyst, they are carefully removed from an embryo and cultured in a petri dish in a laboratory, embryonic stem cells can begin to grow out of the inner cell mass and form small colonies of cells, after a while visible to the naked eye. Under the right culture conditions, these cells will continue to divide outside the embryo for an unlimited period of time (in contrast to their temporary existence in the embryo) and remain pluripotent. Because they can grow indefinitely in culture, they are often referred to as an *embryonic stem cell line*. If cells from an embryonic stem cell line are derived from an albino (white) mouse and injected into a blastocyst from a black mouse, pups with black and white coats will also be born just as for freshly isolated inner cell mass cells, even if the embryonic stem cells have been grown in culture for more than 20 years. This ability to form chimeric mice is a remarkable demonstration that embryonic stem cells do not lose their pluripotent characteristics, even after very long periods of culture in the laboratory.

3.2

GENOMIC IMPRINTING

Normal cells are diploid, meaning that they have two pairs of chromosomes; one set from the mother and one set from the father. Most genes in cells are expressed from both chromosomes but there are exceptions to this rule. X-chromosome inactivation, discussed in Box 3.1, is one example. In addition, there are a number of genes that are only expressed from the chromosome inherited from the father, while other genes are always expressed from the chromosome inherited from the mother. This monoallelic expression in a "parent-of-origin-dependent" manner is referred to as genomic imprinting. This imprinting is regulated by silencing the genes. So, when an imprinted gene is only expressed from the father's chromosome, the expression from the mother's chromosome is actually inhibited; the imprinted genes are silenced. Inhibition of imprinted gene expression occurs via DNA methylation (see Chapter 1, "The Biology of the Cell") and modifications of the histone proteins. As a result, a normal individual only has one active copy of the imprinted gene. The exact number of genes that are imprinted is not known but it is known to be only a small fraction of all genes in the genome. Most imprinted genes have a role in the growth and development of either the embryo or the placenta.

Occasionally, the genes are not properly imprinted, and either the genes from both chromosomes are active or they are both inactive. These imprinting disorders frequently lead to prenatal overgrowth. This means the offspring are too big or have an excess of certain tissues. Examples include Angelman syndrome, Prader–Willi syndrome, and

Beckwith–Wiedemann syndrome. In animals the use of *in vitro* fertilization and subsequent embryo culture can lead to an imprinting disorder known as large offspring syndrome. Whether human *in vitro* fertilization also predisposes to imprinting disorders is unknown and still a matter of debate (Figure B3.2.1).

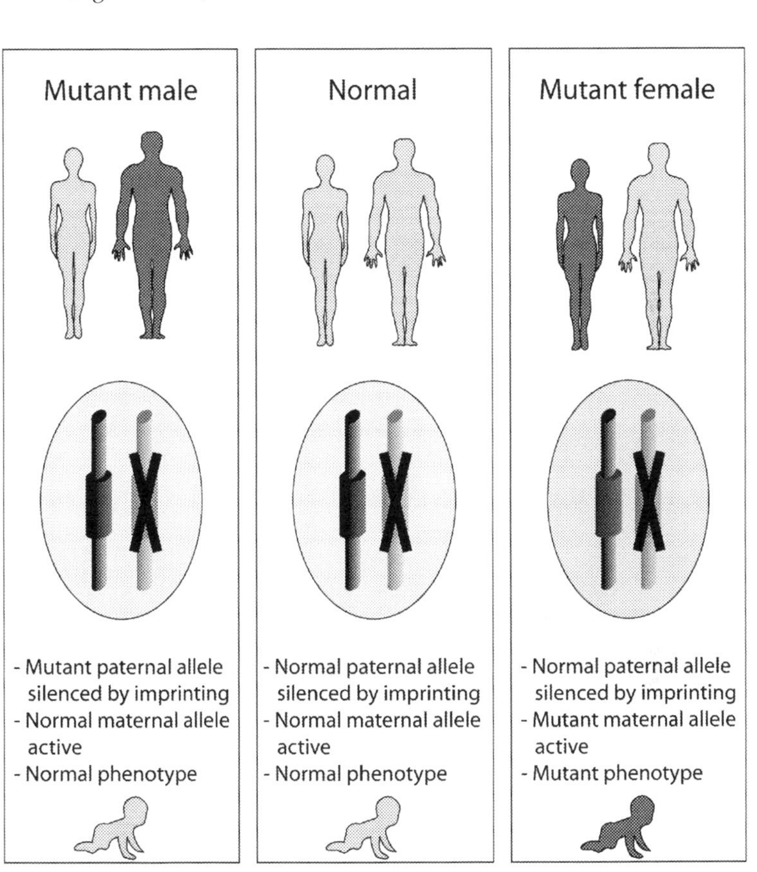

FIGURE B3.2.1 Imprinting is the mechanism by which expression of certain genes is activated or repressed depending on whether the chromosome originates from the sperm (father) or from the egg (mother). When a gene is imprinted so that it is only expressed from the paternal chromosome, a child from a father with a mutation in this gene that would lead to disease is not affected provided the mother does not carry a mutation: the gene from the father is imprinted and, although mutated, is not expressed. Since the gene from the maternal side is normal, the child will not be affected (Left). Similarly, if neither parent carries the mutation, the child will also not be affected (middle). However, when the father is normal but the mother carries the mutation (right), the child will be affected. In this case, although the father's gene contains a normal gene, its expression is repressed because of the imprinting; the gene is only expressed from the maternal side that carries the mutation.

3.3 MULTIPOTENCY, UNIPOTENCY, AND ADULT STEM CELLS

A curiosity of early development is that even though most cells differentiate and take up their specialized functions after the (pluripotent) blastocyst stage, a few cells somehow succeed in maintaining, or regaining, stem cell properties. In an adult, most or possibly all organs and tissues contain a small reserve of what we call adult stem cells. The most well-known of this type of stem cell are the blood stem cells in the bone marrow. These organ stem cells usually sit quietly doing little, but they have the capacity to divide and differentiate if necessary, for example, following damage to the organ or tissue or in its normal maintenance. These adult stem cells are multipotent and the spectrum of cell types they can form is generally limited to those normally present in the organ from which they derive, as mentioned earlier (Figure 3.3). Stem cells from the bone marrow can, for instance, form all cells that constitute the blood, but cannot form nerve cells, intestinal cells, or insulin-producing cells. The ability to differentiate to different cell types is even more limited in stem cells that can only give rise to one type of cell, such as spermatogonial stem cells that can only form sperm cells.

Adult stem cells are clearly different from embryonic stem cells as they maintain, in the organism, their ability to form either two stem cells or one stem cell and one differentiated daughter cell at each

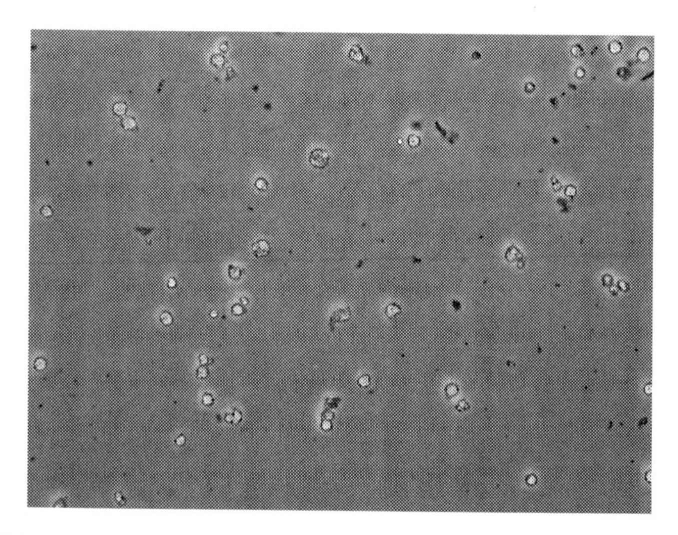

FIGURE 3.3 Bone marrow within the large bones of the body can be isolated and cells it contains cultured *in vitro*. This picture shows pig bone marrow cells that have been cultured. *Source: Ester Tjin, Utrecht University, Netherlands/Stamcellen Veen Magazines.*

division over the course of the entire lifetime of the species (80–90 years in humans). Pluripotent stem cells either form two stem cells (artificially in the laboratory) or two differentiated daughter cells, resulting in either a permanent cell population in culture or depletion of the stem cell pool in culture or in the embryo.

3.4 CELL DIVISION AND AGING: THE ROLE OF TELOMERASE

If normal cells are cultured outside the body in a petri dish, they can usually divide a limited number of times when placed under optimal growth conditions, but thereafter cell division ceases and the cells are said to become "quiescent." The term *Hayflick's limit* was coined to describe the empirical finding by Leonard Hayflick in 1961 that normal human fetal cells in culture will divide 40 to 60 times. Before his finding, it was believed that all cells were immortal and could in principle divide forever but we just did not know the right culture conditions to do it. Stem cells, some cancer cells, and some genetically or epigenetically modified cells are the only types of cells that can multiply indefinitely outside the body. The enzyme telomerase plays a key role in this process. Chromosomes in the nucleus of the cell have so-called telomeres at each end of the DNA chains, made up of well-defined, repeating nucleotide sequences. These telomeres function as protective caps for the chromosomes, but after every cell division they are shortened. Once they are "used up," errors can occur in the chromosomal DNA during the next cell division, most often leading to death of the cell. Thus, the shorter the telomere, the "older" the cell. Before the telomeric sequence becomes so short that errors in the rest of the chromosomal DNA occur, most normal cells respond by "committing suicide" through a special cellular mechanism known as *apoptosis*. Unlike normal body cells, embryonic stem cells make a relatively large amount of telomerase. This enzyme restores the original size of the telomeres after each cell division, so the chromosomal caps remain intact, uniquely providing stem cells with eternal life (Figure 3.4).

3.5 THE RELATIONSHIP BETWEEN CELL DIVISION AND DIFFERENTIATION: EPIGENETICS

Although the nucleotide sequence of the genome (DNA) of a fertilized egg is not different from that of differentiated cells, the fertilized egg does differ significantly from differentiated cells in one notable respect: epigenetic regulation (Chapter 1, "The Biology of the Cell").

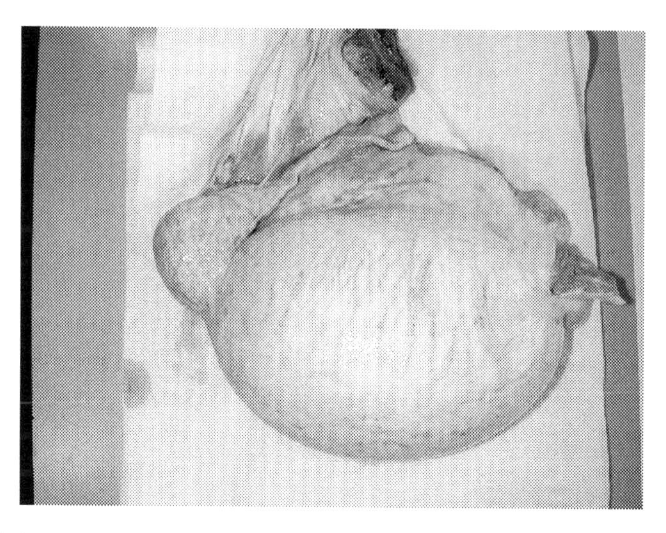

FIGURE 3.4 A pig testis. Sperm cells in the testis are continuously replenished by spermatogonial stem cells.

As mentioned earlier, totipotent and pluripotent cells in an early embryo are free to choose in which direction they will differentiate. These cells have yet to acquire cell type specific functions; for example, they do not produce insulin in response to glucose in the way a pancreas cell would. What they are good at is dividing: about once per day (or even more frequently in mice). Each cell divides into two daughter cells such that the embryo grows in size and cell number. As the cells start to differentiate and specialize, the ability of individual cells to proliferate by dividing decreases. A general rule is that the more differentiated a cell is, the less capable it is of dividing. This applies not only to cells in the living embryo or adult but also to cells in culture in the laboratory.

Each cell type has its own function and to fulfill this function it needs its own group of special genes to be active. As cells become more specialized, the number of genes it uses becomes limited to this essential set; other genes not required for the specialist function are blocked by dedicated chemical modification of the DNA and proteins, called histones, that together make up the chromatin. These are the epigenetic modifications; on the DNA this is seen as addition of methyl groups to certain C nucleotides in the promoter regions of the genes to be inactivated. In addition, the histone proteins that form molecular complexes with the DNA can be chemically altered, thereby changing the accessibility of the DNA to other proteins. With advancing differentiation, the genes required for the complex process of cell division are similarly blocked and become unavailable for use. Thus, fully differentiated and specialized cells, such as brain and heart muscle cells, are left with a limited ability to divide and can thus dedicate all of their time and energy to their assigned function in the body (Figure 3.5).

Another consequence of these epigenetic changes that come with differentiation is that a cell differentiated in one direction cannot simply change into another differentiated cell type; the repertoire of genes that other cell types have available has been made inaccessible. Less differentiated cells, such as multipotent stem cells, present in, for example, heart, pancreas, intestine, and bone marrow, can use their genome somewhat more extensively and choose to become one of several possible subtypes of cells in the organ they are part of. In addition, they can still multiply to some extent. Multipotent stem cells in an organ, such as the blood stem cells and mesenchymal stromal cells in the bone marrow, are cells with even fewer epigenetic limitations. Finally, stem cells present in the early embryo are characterized by minimal DNA blocking.

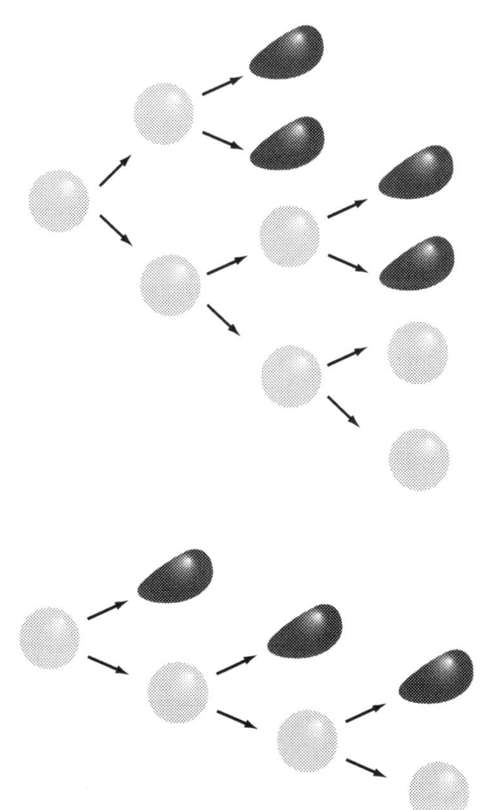

FIGURE 3.5 In general, there are three different ways by which stem cells can divide. At the top, stem cells (yellow) divide into two new stem cells that are identical. This is called symmetric cell division. The stem cells can also divide in such a way that one cell remains a stem cell and the other differentiates to give rise to a specialized cell (blue) that is not a stem cell. This is called asymmetric cell division. The lower part of the illustration shows stem cells dividing into one new stem cell and one specialized cell. In this way, the pool of stem cells remains unchanged. Finally, both daughter cells can differentiate. The stem cell pool is then lost. *Source: Stamcellen Veen Magazines.*

3.6 EPIGENETICS IN STEM CELLS

It has become clear that epigenetic changes, reflected in specific DNA methylation patterns, are often abnormal in embryonic stem cells that have been in culture in a laboratory for a long time. This implies that different embryonic stem cell lines are in this respect not necessarily identical, which probably explains why different cell lines often exhibit slightly different characteristics, for example, in growth rate or the ease with which they differentiate into different cell types. One reason for this is that during each cell division there is a chance of introducing an abnormality into the DNA being copied, which would therefore also be present in the two daughter cells. This way cells can accidentally acquire additional features, such as mutations or abnormally methylated genes, which provide them with a growth advantage. Since rapidly proliferating cells tend to dominate and overgrow more slowly proliferating cells, the slower cells are gradually selected out during culture at the expense of cells with the original characteristics. In this way, after multiple "passages" in culture, when cells from one culture dish are divided over multiple dishes to give them room to grow are only the very distant relatives of the original cells are present, slightly different cell lines may evolve.

3.3

PLANARIA

What's So Strange About Planaria?

Regeneration of most human tissues and organs is limited. A superficial wound will heal but leave a scar, and limbs or organs that are lost by accident, surgery, or disease do not regrow. This is not true for all animal species. Less complex organisms than humans often have remarkable regenerative capacities. Research into these species can teach us a great deal about how and why regeneration of tissues takes place and the plasticity of cells that take part. Planaria, a type of small flatworm, is an animal that has been investigated in depth in the past. Planaria is a collective name for a variety of flatworms, including *Schmidtea mediterranea* and *Dugesia japonica*. Planaria are simple organisms but they are, nevertheless, composed of cells from three germ cell layers, just like mammals: namely, the ectoderm, endoderm, and mesoderm. Therefore, they have a nervous system and eyes (formed from ectoderm), muscles (from mesoderm), and a digestive tract (from endoderm).

Planaria flatworms have been studied mostly because they are extraordinarily good at regenerating. If a Planaria flatworm is cut into two parts, each half will grow to form a complete flatworm. If the head is separated from the body, or the body is cut into a number of pieces, each piece will form a new worm with a head and tail, each animal being genetically identical to the original animal—they are clones of the original animal.

Oddly though, not all species of Planaria have such a remarkable regenerative capacity. In some species, such as *Schmidtea mediterranea* and *Dugesia japonica*, even isolated tail parts can grow a new head and become a complete worm again. Other species, such as the milkwhite flatworm (*Dendrocoelum lacteum*), have a more limited regenerative ability. In these worms, parts from the head half of the embryo can form both a new head and a new tail, but parts from the tail half can only form tail. How is this organized? In *Dendrocoelum*, the inhibitors of a signaling pathway called Wnt (pronounce "wint") are present in a concentration gradient. The concentration of Wnt inhibitors in the head is higher than in the tail, and it is the inhibition of the Wnt signaling pathway that increases the regeneration capacity. It is then completely logical that when Wnt signaling is artificially inhibited in the tail region, the capacity to grow a new head is restored. What is most interesting is that a piece of Planaria without a head will form a new head, but not an extra tail; Planaria without a tail will form a tail and no head. The cells that form new structures apparently also have positional information and somehow know exactly where they come from in the embryo and what they have to make (Figure B3.3.1).

Planaria's regenerative success is caused by a special group of cells with a large nucleus and little cytoplasm, called neoblasts. About 25 to 30% of Planaria cells are neoblasts. Little is known about these cells but it has become clear that neoblasts are self-renewing, pluripotent stem cells. Individual neoblasts can contribute to tissues of three germ layers (endoderm, ectoderm, and mesoderm). In intact Planaria, cell division takes place continuously, and new cells, formed from neoblasts, replace old cells that have served their function. If a Planaria is damaged, neoblasts migrate to the site of damage. Once there, they divide to increase their number very locally in the wound. The neoblasts then differentiate into the required cell types and replace the damaged tissue. Exactly which signals make the neoblasts migrate to the wound, while at the same time prevent them from differentiating into different cell types within the organism, and how they obtain their exact positional information, are questions that remain unanswered. In addition, it is not exactly clear how the cells know that they need to

3.3 (*cont'd*)

form the new tissues and organs of adult size, rather than like those in an embryo. Most likely, gradients of proteins allow cells to "read" their position in the body map within the animal (both in space and time) and to recreate the missing tissues. A better understanding of these mysteries may contribute to our knowledge of how human stem cells work and, indeed, whether controlled alteration of signaling pathways, much like Wnt pathway inhibition in Planaria, would help human tissues and organs to regenerate.

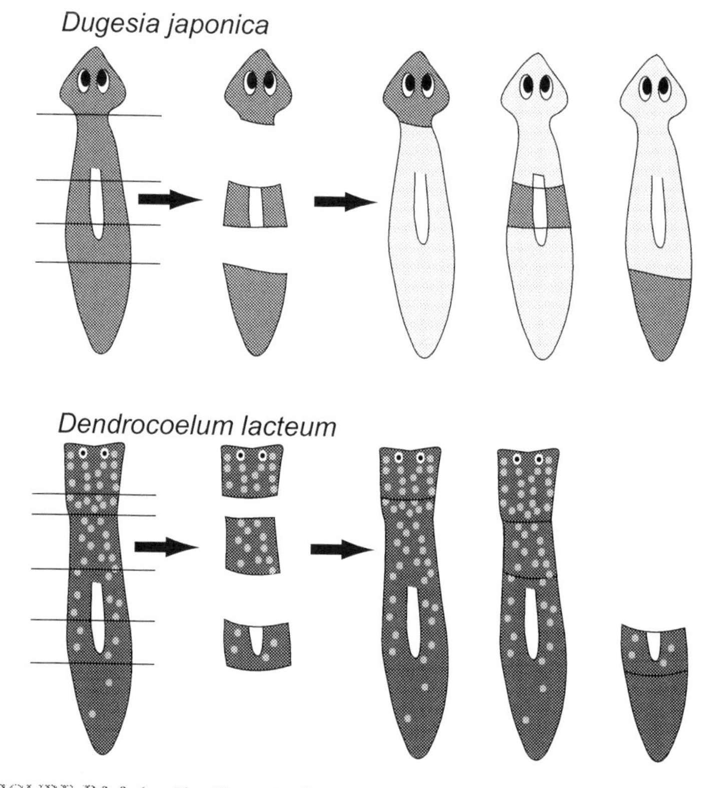

FIGURE B3.3.1 If a Planaria flatworm of the species *Dugesia japonica* is cut into several pieces, every part can reform (or regenerate) the missing part: within a few days, each individual piece has grown out to become a new flatworm. Other flatworm species, such as *Dendrocoelum lacteum*, have a more limited regenerative capacity: the head piece can form a new tail, but the tail alone is not able to form a new head.

CHAPTER

4

Of Mice and Men: The History of Embryonic Stem Cells

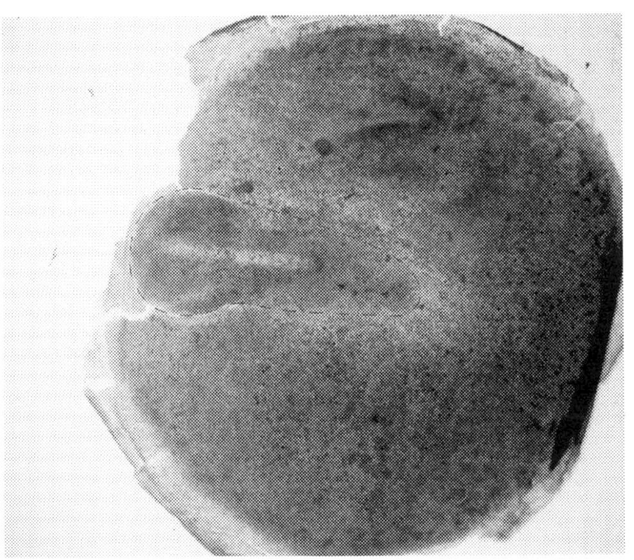

Bernard Roelen/Stamcellen *Veen Magazines*

OUTLINE

4.1 HOW IT ALL BEGAN: PLURIPOTENT CELLS IN EARLY EMBRYOS

It is sometimes difficult to say exactly when a discovery was made. Usually information accumulates over a long period of time and suddenly it becomes clear that something has been discovered. Such is the case with embryonic stem cells. These stem cells emerged as a way to explain findings in science and medicine that go as far back as ancient Greece. On ancient Greek tablets we find the first descriptions of teratomas, tumors that grew from the testes of baby boys or young men and looked like misformed embryos. Teratoma in Greek means "monster." We will describe these tumors and their importance in the history of embryonic stem cells later in this chapter, but first we describe the landmarks that led to our present understanding of these cells. In Chapter 5, "Origins and Types of Stem Cells: What's in a Name?" we describe other stem cells, collectively referred to as adult stem cells.

The recent origins of embryonic stem cell research can be traced back to the late nineteenth century, when the first attempts were made to keep rabbit embryos alive outside of the womb. The preference was for rabbits over mice, because rabbit embryos are larger and therefore more easily manipulated. In 1890, Walter Heape reported he that had successfully transferred a fertilized egg from the original rabbit mother, a fluffy white angora, into the womb of a surrogate mother, a brown rabbit that had been just been fertilized by a brown male. Of the six baby rabbits born, four were brown like the original parents but two were unmistakably fluffy and white. Twenty years later, Brussels embryologist Albert Brachet successfully kept a rabbit blastocyst alive for a day in blood plasma (blood with the cells removed) (Figure 4.1). These experiments were built upon by Lewis and Gregory, who filmed the development of a fertilized rabbit egg for a few days in a Petri dish until it reached the blastocyst stage. At this stage, the embryo normally consists of 32 to 64

4.1

MILESTONES IN STEM CELL BIOLOGY AND EMBRYONIC DEVELOPMENT: THE NOBEL PRIZE WINNERS

(Figure B4.1.1)

When?	Who?	Discovered what?
1935	Hans Spemann	The *organizer* in embryonic development
1962	James Watson, Francis Crick, Maurice Wilkins	The structure of DNA
1986	Stanley Cohen, Rita Levi-Montalcini	Growth factors
1995	Edward Lewis, Eric Wieschaus, Christiane Nüsslein-Volhard	Genetic control of early development
2007	Martin Evans, Mario Capecchi, Oliver Smithies	Genetic modification of mice using embryonic stem cells
2009	Elizabeth Blackburn, Carol Greider, Jack Szostak	*Telomere* ends of chromosomes and *telomerase*
2010	Robert Edwards	Development of *in vitro* fertilization
2012	John Gurdon, Shinya Yamanaka	Reprogramming of mature cells to *pluripotency*

FIGURE B4.1.1 List of Nobel Prize winners whose discoveries have been important for the stem cell field.

FIGURE 4.1 A three-day-old mouse embryo visualized using confocal laser scanning microscopy. The boundaries of individual trophectoderm cells (the outermost cell layer) can be recognized, as well as their nuclei. *Source: Stamcellen Veen Magazines.*

FIGURE 4.2 The first experiments that paved the road for the generation of embryonic stem cell lines were performed with rabbits. *Source: Leonie du Puy/Stamcellen Veen Magazines.*

cells and is a balloon-like structure with a fluid filled cavity (a cyst) on one side and cells that form the embryo itself on the other (see Chapter 2, "Embryonic Development"). In 1942, Nicholas and Hall discovered that both cells of a two-cell rabbit embryo are totipotent (can form a complete new embryo) by showing that when each of the two cells were placed singly into the womb, two rabbits (identical twins) and not just one, were born. In 1956, Wesley Whitten also grew an eight-cell embryo in a Petri dish to the blastocyst stage, the phase just before the embryo embeds in the uterus (or womb). Together, these studies showed that it was possible for the first stages of embryonic development in mammals to take place outside the body of the mother in the relatively simple culture conditions of the laboratory; and despite this period outside the mother, healthy offspring could be born.

This triggered an era of intensive research on early embryonic development. In the 1960s, Robert Edwards in Cambridge discovered that individual cells (called blastomeres), formed during the first cell divisions in a fertilized embryo, were all capable of dividing a few more times in a Petri dish. He would later go on to become the founding father of this type of research in humans which eventually led to *in vitro* fertilization (IVF) or test-tube babies (as we shall learn later). Together with colleague John Paul in Glasgow, Edwards then tried to culture rabbit blastocyst cells, stored them by freezing, and investigated which factors caused them to stop dividing and to differentiate (Figure 4.2). Paul and Edwards found that the smaller cells of the inner cell mass of the blastocyst stuck firmly to each other during the culture

FIGURE 4.3 Stem cell research in Cambridge, U.K. has always been at the forefront of stem cell science. *Source: Stamcellen Veen Magazines.*

process and formed structures we now call *embryoid bodies*. These embryoid bodies contained all kinds of different tissue types, including muscle and nerve cells, which suggested the inner cell mass consisted of pluripotent cells—cells with the capacity to differentiate into many cell types. To get rid of the rapidly dividing trophoblast cells from the outer layer of the cultured blastocyst so that they did not overgrow the inner cell mass cells, the areas of the culture with the smaller pluripotent cells were cut out using dissection needles and transferred to new culture dishes. As it was still unclear how further development of the embryonic cells could be inhibited and their growth sustained, these early cultures quickly differentiated, and were used to create various fibroblast and epithelial cell lines. Similar results were obtained with mice, and the two researchers gradually realized that if this could be done with human embryos, the embryonic stem cells derived from them might be useful for therapeutic purposes, such as tissue regeneration and repair. They also realized that what they had been learning about early development could have important consequences for understanding and treating infertility (Figure 4.3).

4.2 MOUSE EMBRYONAL CARCINOMA CELL LINES

Completely independently of these studies on early embryo development and reproduction, elsewhere in the world research was being carried out on one of the strangest tumors to form in the human body, a teratocarcinoma. This develops in the testes of young men and is fortunately very rare, although until cis-platinum was discovered as an effective chemotherapy, it was the second most common cause of death among young men, after motorcycle accidents. Lance Armstrong, the seven-time but now discredited winner of the Tour de France cycle race, was one of its best-known victims. In these tumors, a germ cell (or gamete, the precursor cell of eggs and sperm) present in the developing testis does not differentiate properly, and during puberty it can form a tumor in which many different kinds of differentiated cell types are present. This type of tumor is called a teratoma if it is benign, or nonmalignant, and a teratocarcinoma if it is malignant. The lack of organization within the tumor leads to the random development of all kinds of tissue types. For example, teeth, bones, cartilage, muscle, and hair form without a spatial relationship, making the tumor appear like a misformed embryo. In the 1950s, Leroy Stevens, working at the Jackson Memorial Laboratory in Bar Harbor, U.S.A., by chance came across a strain of mice during his research in which the males frequently developed teratomas (or teratocarcinomas) in their testes. From these tumors, he was able to harvest pluripotent stem cells—responsible for the development of the various differentiated cell types in the tumor—creating the first embryonal carcinoma cell line, so-called because of its similarity to cells of the early embryo. These cell lines were immortal and could be cultured indefinitely (Figure 4.4).

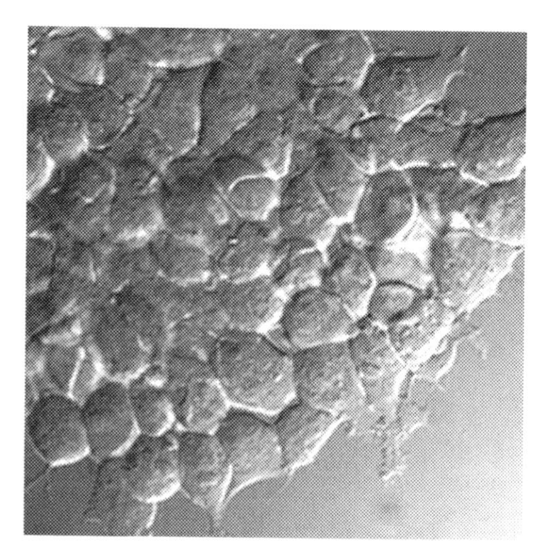

FIGURE 4.4 Embryonal carcinoma cells, such as these cells derived from a teratocarcinoma (a tumor induced in mice by placing a 4–7-day-old embryo under the kidney capsule) behave in some respects like genuine pluripotent stem cells, despite their tumor origin. *Source: Thorold Theunisen, Hubrecht Institute, Netherlands/Stamcellen Veen Magazines.*

4.2

ARMSTRONG

Lance Armstrong: Discredited but a Teratocarcinoma Survivor

Malignant germ cell tumors—teratocarcinomas—can occur spontaneously in human testes and ovaries. This form of cancer was the leading cause of death among young men until the 1980s, with the exception of motorcycle accidents. Major research efforts were made to identify ways to force the carcinoma cells to differentiate, with the hope that they would lose their malignant properties, but to no avail. Fortunately, the embryonal carcinoma cells in a teratocarcinoma proved extremely sensitive to cisplatin, a drug treatment that is usually sufficient to kill the cells.

American cyclist Lance Armstrong (b. 1971, Texas) is one of the most high-profile victims of this form of cancer. Armstrong first rose to fame as a cyclist at the Olympic Games in Barcelona in 1992, where he passed the finish line in 14th place. A year later, he became world cycling champion in Oslo's pouring rain (Figure B4.2.1).

FIGURE B4.2.1 A large part of the proceeds from the yellow Livestrong wristbands goes to cancer research. *Source: Stamcellen Veen Magazines.*

After the Atlanta Olympics in 1996, he suffered lasting fatigue, and a few months later a germ cell testicular tumor was diagnosed. The cancer was already in an advanced stage and metastases were found in

4.2 *(cont'd)*

his lungs and brain. In addition to chemotherapy, a testicle was removed and brain surgery was performed. During this period, he founded the Lance Armstrong Foundation, which provides support for cancer patients and donates funds to scientific research. Armstrong managed to beat the cancer and resumed his professional cycling career in 1998. The Texan went on to win the Tour de France for seven successive years (1999–2005), despite having faced a life-threatening disease, making him the most successful Tour de France cyclist of all times. Inspired by the yellow Tour de France registration bracelets for journalists and racing teams that became popular in 2004, his foundation introduced its yellow Livestrong bracelets in 2004. The proceeds from these bracelets mostly go to aid organizations for cancer patients, and to scientific research. Armstrong stopped racing in 2005, ran a few marathons but decided to pick up competitive cycling again in 2009. He was a source of inspiration for many, especially cancer patients, until he became associated with doping in the cycling sport. Accumulating allegations ultimately led to a confession of performance-enhancing drug use that related to most of his racing career. With the confession, Armstrong's reputation was ruined. He lost all his Tour de France victories and had to relinquish his bronze medal from the 2000 Sydney Olympics. Armstrong is no longer associated with Livestrong but the work of the organization raising funds for cancer research continues.

In 1975, Martin Evans and Gail Martin (then working at University College, London) showed that these embryonal carcinoma cells are not only immortal but also pluripotent. In embryoid bodies, or cell aggregates in suspension in the culture medium, they could differentiate into several cell types arising from all three of the primary germ layers: ectoderm (nerve cells), endoderm (yolk sac cells), and mesoderm (muscle cells) (Figure 4.5). Beatrice Mintz and Karl Illmensee of the Fox Chase Cancer Center in Philadelphia discovered that if teratocarcinoma stem cells were injected into a mouse blastocyst and the chimeric (or mixed) embryo was implanted into the womb of a surrogate mother, the stem cells could contribute to all tissues in the body of the resulting pup, with the exception of the gametes. This was most striking if the teratocarcinoma stem cells came from a brown mouse and the blastocyst from a white mouse: the coat of the pups was brown and white.

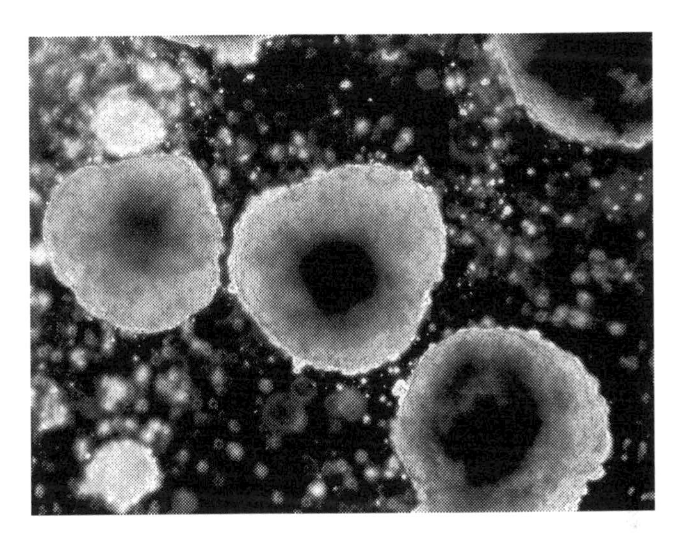

FIGURE 4.5 When embryonic stem cells are cultured in such a way that they cannot attach to a surface but can only cling to each other, they form a structure that in a way resembles an early embryo and in which various differentiated cell types can be formed. These structures are therefore called "embryoid bodies." Both human and mouse embryonic stem cells can form embryoid bodies, and this is the most commonly used way to induce differentiation. *Source: Stieneke van den Brink, Hubrecht Institute, Netherlands.*

These experiments provided the definite proof that teratocarcinoma cells can divide indefinitely and are pluripotent: they act like real stem cells. The major downside of these types of cells was that even though they could form many types of tissue and could also be derived from human teratocarcinomas, they were of tumor origin. This would make them unsuitable for any therapeutic applications no matter how well they differentiated, because their chromosomes were abnormal. Human teratocarcinoma stem cells have particularly severe chromosomal abnormalities, often with whole regions of DNA repeated or missing.

At the time of these discoveries, the teratocarcinomas just looked like an interesting curiosity for developmental biologists to study. For cancer researchers, mouse and human cells in culture were examined in the search for treatments.

4.3 PLURIPOTENT CELLS IN AN EARLY EMBRYO

In the 1960s, Robert Edwards in Cambridge, and Krzystof Tarkowsky and Richard Gardner in Oxford discovered that fusing two early embryos before the blastocyst stage, or transplanting a few cells from the inner cell mass of one blastocyst to another, could also result in a chimeric mouse, in which cells from both sources contributed to all tissue types. If cells or embryos come from different mouse strains, one brown and one white, the chimeras that are born have a mixture of brown and white patches on their coats and look very different from their biological parents. This led the researchers to conclude that undifferentiated cells were present in the embryo that could contribute to all tissues in the mouse, and were therefore pluripotent. This indicated the exciting potential of deriving cultured cell lines directly from embryos in the laboratory, similar to the embryonal carcinoma cell lines but perhaps without the abnormal chromosomes. The only obstacle was to work out the culture conditions that would actually support the growth of the embryonic cells in a laboratory Petri dish.

4.3

HOW TO OBTAIN EMBRYONIC STEM CELLS

Embryonic stem cells can be obtained from early embryos a few days after fertilization. In humans, this is after *in vitro* fertilization (IVF). By culturing cells from the inner cell mass of a blastocyst-stage embryo on a layer of fibroblasts, a permanent culture of embryonic cells can be derived, which we refer to as an *embryonic stem cell line*. The inner cell mass is a clump of cells inside the blastocyst attached to the inner surface of the *trophectoderm* cell layer. The trophectoderm is enclosed by a protein coat called the *zona pellucida*, which normally protects the embryo as it moves along the uterine tract. The zona pellucida is first removed by dipping the embryo in an acidic solution. The trophectoderm is then removed. This is done in one of two ways: (1) by incubating in a fluid containing a combination of antibodies that recognize human cells and "complement" from serum (these work together to destroy trophectodermal cells, a process called immunosurgery), or (2) the trophectoderm is removed mechanically using a small knife. The isolated inner cell mass is then transferred on to a fibroblast cell layer in a Petri dish. The fibroblast cells need to be irradiated so that they can no longer undergo cell division. The fibroblast cells are necessary to "feed" the newly formed embryonic stem cells as they grow out of the inner cell mass, enabling them to continue growing in an undifferentiated state.

4.4 MOUSE EMBRYONIC STEM CELL LINES

Building on the early work on teratocarcinoma stem cells, Martin Evans and Matt Kaufmann, both working at Cambridge University in the early 1980s, cultured the first pluripotent embryonic stem (ES) cell lines from mouse blastocysts, simply by trying a variety of culture conditions and stages of embryo development (Figure 4.6). Martin Evans' former collaborator Gail Martin, now working at the University of California in San Francisco, achieved the same result independently. Just like teratocarcinoma stem cells, these mouse embryonic stem cells could contribute to all somatic tissues in chimeric mice after injection into a blastocyst, but unlike teratocarcinoma stem cells, they could also form gametes (eggs and sperm) in the embryo. This means that their offspring could be entirely derived from the cultured cells. This "germ line transmission" was later adopted as important evidence that a cell was truly pluripotent, although, of course, it could never be used in humans (Figure 4.7).

Embryonic stem cells differentiated more easily in culture than teratocarcinoma stem cells and needed an unknown factor in their culture medium to remain pluripotent. This could be provided by fetal mouse fibroblasts, so that embryonic stem cells were usually cultured on a layer of other cells,

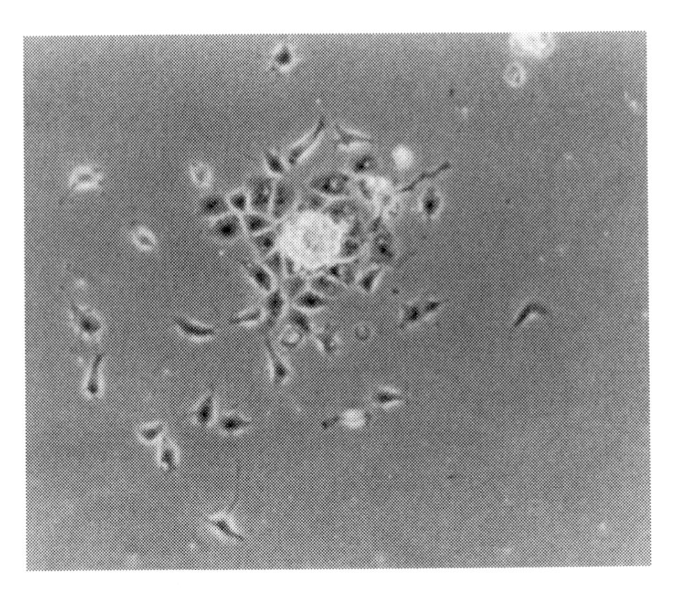

FIGURE 4.6 The first attempts to maintain cells of the inner cell mass in culture failed. As the proper culture conditions were not known, the cells started to differentiate rapidly. This limited their capacity to divide and thus prevented successful cell culture. This micrograph shows the inner cell mass of a mouse embryo from which various types of differentiated cells are formed and migrate to the periphery. *Source: Stamcellen Veen Magazines.*

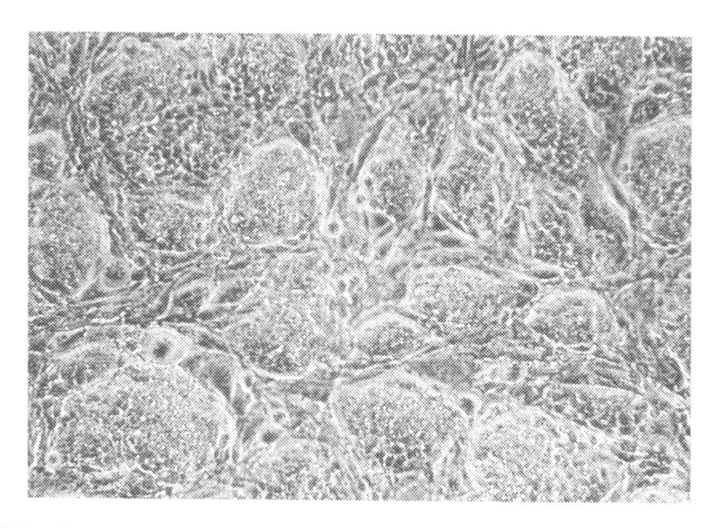

FIGURE 4.7 Mouse embryonic stem cells grow in small clumps. In these clumps no individual cells can be distinguished, which is a good indication that the cells are indeed undifferentiated. *Source: Susana Chuva de Sousa Lopes/Stamcellen Veen Magazines.*

known as *feeders*, which secreted the factor into the medium. Only in the late 1980s was this factor identified by three independent groups, in Australia and in the United Kingdom, as *leukemia inhibitory factor* (LIF). This was an important step forward, as the addition of LIF allowed the culture of embryonic stem cells without feeders. Austin Smith in Cambridge showed that if mouse embryonic stem cells were kept in a completely neutral environment with all of the signals that induce differentiation removed or blocked, they could grow essentially forever without losing their ability to form chimeras when injected into an embryo.

As a result of this work, it is now relatively easy to culture mouse embryonic stem cells, and they have proved to be excellent models for studying early embryonic development (Figure 4.8). More importantly, however, they became an essential tool in creating genetically manipulated mice, which could be used to study the function, in principle, of all genes because of their capacity to form gametes and transmit their genetic information to the next generation of mice. Many thousands of genes have been deleted or removed using a technique called homologous recombination. This technique allows genes to be removed entirely or just in specific tissues and at certain times in development or after birth. The function of that gene can then be investigated by studying the effect of its absence on the developing embryo or the adult mouse. It has proven one of the most powerful research tools of the last three decades in the life sciences. It earned Martin Evans, Mario Capecchi, and Oliver Smithies the Nobel Prize in Physiology or Medicine in 2007.

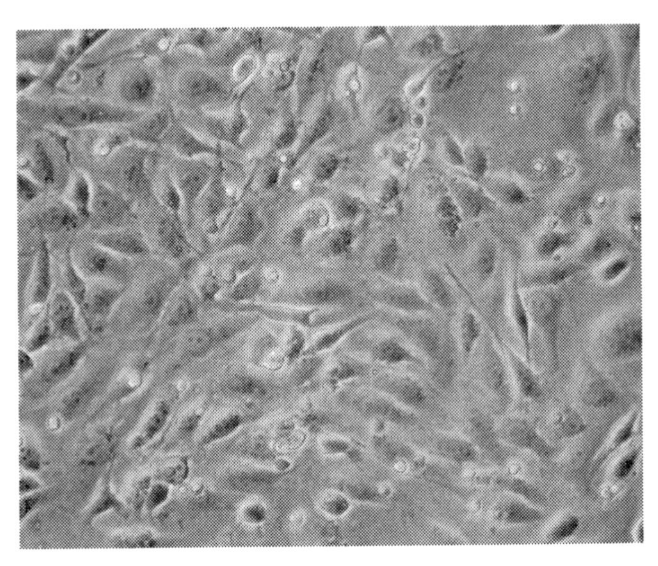

FIGURE 4.8 To maintain embryonic stem cells in an undifferentiated state, they are generally cultured on top of a layer of fibroblast cells. These cells secrete factors that are essential to keep the stem cells healthy and growing and prevent them from differentiating. *Source: Stamcellen Veen Magazines.*

4.4

GENETICALLY MODIFIED MICE

In the 1970s, there was a growing curiosity about the function of individual genes, not only what genes did in cells in culture, but also what they did in living organisms. To investigate this, in 1974, Rudolf Jaenisch and Beatrice Mintz created the first transgenic mouse. They injected DNA in which they were interested (DNA with the nucleotide sequence for a specific gene) into a blastocyst-stage embryo in culture, where it integrated into the DNA in the nucleus of a few cells. This allowed the genome of the mouse to be changed. A more reliable technique proved to be the direct injection of DNA into a newly fertilized egg. These genetically modified embryos were then transferred to the uterus of a surrogate mother mouse where they developed to term. The mice that were then born contained the gene (DNA sequence) of interest in some or all of their cells.

The amount of DNA injected that would actually be incorporated into the chromosomes using approaches such as this was, however, unpredictable. Techniques needed to be developed that would allow DNA to be inserted into the genome to replace the normal gene that was already there. This process, called homologous recombination, was known to occur in

4.4 (*cont'd*)

bacteria and it turned out to work very efficiently in mouse embryonic stem cells. It became known as gene targeting (Figure B4.4.1).

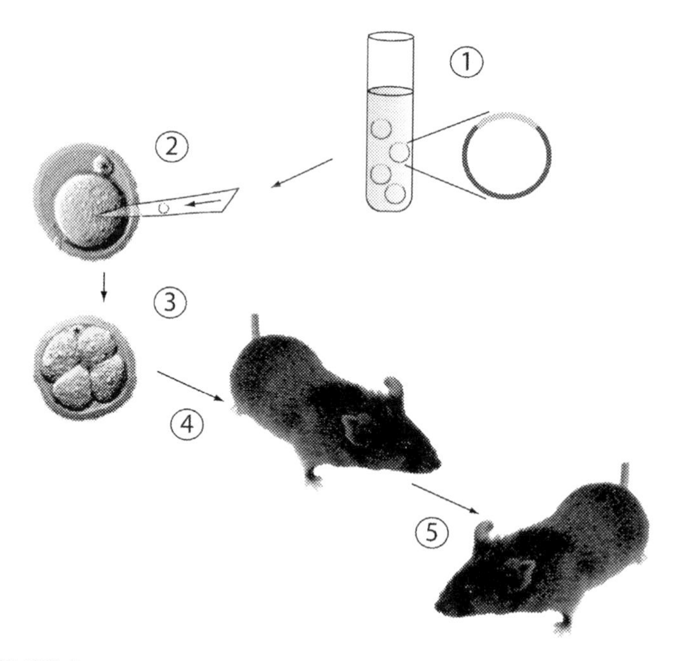

FIGURE B4.4.1 Generation of a transgenic mouse. (1) A DNA construct is made in the laboratory that ensures the expression of a gene (e.g., a human gene, here in green) at a specific location (e.g., a specific tissue or organ). (2) The DNA is injected into a fertilized egg. (3) The injected DNA integrates in the genome of the cells of the developing embryo. (4) The embryo is transferred to the uterus a pseudopregnant female mouse, where it can develop till term. (5) A mouse is born that is genetically altered: a transgenic mouse. *Source: Stamcellen Veen Magazines.*

Homologous recombination proved to be an easy way to replace an *endogenous* gene, a gene already present in a cell, with a gene variant that was defective. The stem cells would then have one normal copy of the gene and one mutant copy. If the modified stem cells were inserted into a blastocyst and the chimeric blastocyst was then transferred to a surrogate mother much like when making transgenic mice, with a little luck, the mouse then born would contain some cells with the abnormal gene on one chromosome. However, to make a mouse in which *all* cells contained one abnormal chromosome, these mutated stem cells also needed to contribute to the gametes. Although teratocarcinoma stem cells did not make gametes in chimeric mice, the remarkable thing was that embryonic stem cells did. If the chimeric mice reached adulthood and were mated, their offspring were heterozygous and contained one mutant gene copy in all cells. If these heterozygote mice were then mated again but with each other, both chromosomes on the offspring would carry the mutant gene. These mice became known as knockout mice (homozygotes): they lacked a particular gene entirely, so the effect of the absence of that gene could be studied both in development of the embryo and, if it was not lethal during development, in the adult mouse. Later refinements to the method have made it possible to modify genes "conditionally:" by placing small extra pieces of DNA on either side of the mutant DNA construct (called loxP sites), it has become possible to delete the endogenous gene and allow the mutation to become active at any time in development or adulthood.

The elegant genetics combined with embryonic stem cell technology resulted in the Nobel Prize in Physiology or Medicine in 2007 being awarded to Oliver Smithies and Mario Capecchi for homologous recombination techniques, and to Martin Evans for the discovery that embryonic stem cells could pass on their DNA to the next generation by forming gametes.

These two techniques—gain and loss of gene function—have together produced a huge amount of information about what individual genes in our genome do, both in embryos and adult individuals, because very often gene functions have been conserved between mice and humans

4.5 TOWARD HUMAN EMBRYONIC STEM CELLS

While all these developments in the mouse were taking place, progress on another front, assisted reproduction to treat fertility problems, was also underway. Research on *in vitro* fertilization (IVF), the fertilization of human ova/eggs in the laboratory to increase the chances of subfertile couples having a child, had reached the point that egg cells could be collected, fertilized with the partner's sperm in the laboratory, and allowed to develop to the blastocyst stage for transfer to the womb of the mother. In 1978, the first baby, Louise Brown, was born following IVF treatment.

Often, more egg cells were fertilized than there were embryos implanted in the uterus of the mother, and as there was no technology available for temporarily freezing embryos, the problem of what to do with poor quality embryos, or those simply left over, then arose. Attempts by Robert Edwards and Patrick Steptoe, the scientific fathers of Louise Brown, to interest politicians in the looming ethical dilemmas surrounding research involving human embryos proved fruitless. Politicians neither understood nor seemed to wish to understand the nature of the discussions that needed to take place. These leftover embryos thus became available for basic research, albeit on a limited scale. Several researchers, including Christopher Graham in Oxford, tried in the early 1980s to use these surplus embryos to derive human embryonic stem cell lines from blastocyst-stage embryos, much as Gail Martin, Martin Evans, and colleagues had done in mice. Shortly after, however, a procedure for freezing embryos and storing them in liquid nitrogen was developed. The flow of embryos for research stopped and frozen embryos began to accumulate in storage. The ethical discussions on the use of "spare" IVF embryos only then began in earnest. Further research deriving embryonic stem cells stopped while the ethical issues were debated. The United Kingdom led the discussion forum on the use of human embryos in research—including for stem cell derivation—long before the first human embryonic stem cells were actually derived (Figure 4.9). The legislation, position papers, and committee reports that arose formed the basis of later legislation in many other countries. In the meantime, excess embryos remained frozen for potential future use or were discarded, and research moved back toward mice.

In fact it would be 20 more years before attempts to create embryonic stem cell lines would resume in England under the jurisdiction of the Human Fertility and Embryo Authority. Other countries followed later. Stockpiles of excess embryos had grown exponentially, without any clear idea of what to do with them: few couples were interested in donating "their" unwanted embryos to other childless couples for adoption. Reasons given have varied from anxiety about meeting biological offspring unexpectedly to fears of liability claims should the child be handicapped.

Ariff Bongso, an expert IVF physician in Singapore, was the first to rekindle research into human stem cells. In 1994, he described the

FIGURE 4.9 A container filled with liquid nitrogen in which vials with stem cells are frozen and stored. A method has also been developed to store embryos for years by deep freezing, while maintaining viability after thawing. This is also how healthy surplus embryos are stored after *in vitro* fertilization for a future pregnancy or donation for stem cell derivation. This procedure of freezing was actually the start of discussion on what to do with those embryos that were frozen but would not be transferred later to a mother-to-be.

procedure for harvesting cells from the inner cell mass of a blastocyst and growing them on a feeder cell layer. Building on this publication,[1] which fell short of actually producing immortal cells, American scientist James Thompson created the first human embryonic stem cell line in 1998. This scientific breakthrough provided the field with a fresh impetus, and within a short time period various researchers, including Ariff Bongso himself, in collaboration with Alan Trounson, Martin Pera, and Ben Reubinoff in Australia, reported the creation of their own new human embryonic stem cell lines.

Much of the information and technology that had been obtained from the study of human embryonal and teratocarcinoma stem cell lines proved vital to the success of these first studies. However, the ethical discussions reared up again, this time because human embryonic stem cell lines and embryo use was a reality, not just a prediction. Many governments began to address legislation but in 2001, former President Bush took the unusual step of banning U.S. government-funded research on all embryonic stem cell lines created after August 9, 2001, by presidential decree in his State of the Union address that same evening. This limited embryonic stem cell research in the United States, with the exception of a few privately funded institutes, to the relatively few existing stem cell lines officially registered with the National Institutes of Health (NIH); the so-called "presidential list." Originally it was thought that there were some 70 cell lines on this list but many of these were difficult to culture; scaling up culture to create larger numbers of cells failed and many were abnormal, making them unsuitable for experimental research (Figure 4.10). In the end, it turned out that only around 12 lines were actually useful. These came under the ownership of an organization called WiCell, originally part of the University of Wisconsin where James Thomson derived the lines. WiCell was very restrictive about distributing the lines to other researchers in the field, had patent claims on anything discovered with them, and requested 5000 US$ for a tube of frozen cells.

The use of embryos to create stem cell lines was also banned in Europe for a long time. Many research groups moved research to less ethically sensitive research involving adult stem cells, particularly blood stem cells or germ line stem cells, causing explosive growth in this field. The antipathy to human embryonic stem cells in many countries was the motor behind the search for alternatives.

Great Britain, but also other countries including Australia, Singapore, Sweden, Finland, Israel, the Czech Republic, Korea, Spain, and China, quickly took less restrictive positions and introduced legislation that

[1]Bongso A, Fong CY, Ng SC, Ratnam S. Isolation and culture of inner cell mass cells from human blastocysts. Hum Reprod 1994;9(11):2110–7.

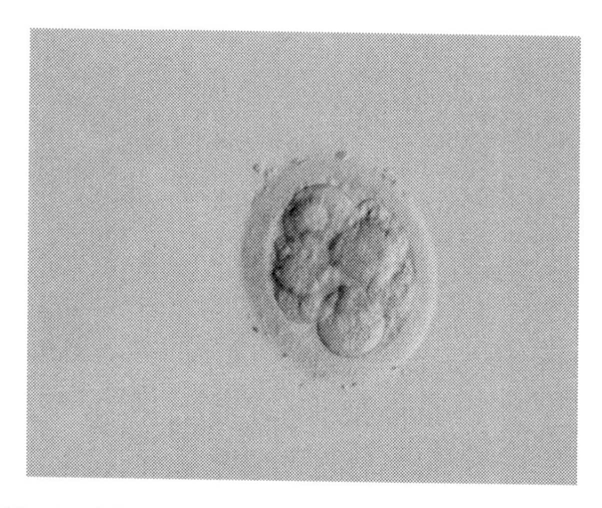

FIGURE 4.10 An eight-cell human embryo that had been frozen in liquid nitrogen, 1 hour after it had been thawed. Such embryos can give rise to perfectly healthy babies after transfer to their future mother's uterus. *Source: Dagmar Gutknecht, University Medical Center Utrecht, Netherlands.*

allowed the derivation of new lines from surplus embryos or even embryos created for research purposes. This shift to the "east" created the possibility that the "west"—Europe and the United States—would lose its leading role within stem cell research. Stem cell research did not disappear entirely in the United States, but it had to be funded by private institutions and grants; Douglas Melton, working at Harvard University, looked for human embryonic stem cell lines that would be easier to culture, a project financed by the Howard Hughes Medical Institute. In 2004, he reported that his lab had generated 17 new cell lines that could be grown almost as easily as mouse embryonic stem cells. Research groups in a growing number of countries within Europe, including Sweden, Spain, and The Netherlands, who, along with the United Kingdom had leading roles, also gathered experience creating and culturing human embryonic stem cell lines, now referred to as hESCs. Collectively, however, legal, religious, financial, funding, and patent issues, together with ethical questions by society, have restricted research on embryonic stem cells worldwide since their first discovery. Nevertheless, human embryonic stem cell research laid an important foundation for a new research field and an overall shift toward using human cells in culture for research whenever possible rather than rodent or other species.

Over the last few years, major developments in human embryonic stem cell research have included culture on feeder layers of human cells instead of on mouse cells, or no feeders at all but in defined commercial culture medium. Methods were developed to direct differentiation more precisely under defined conditions using specific growth and differentiation factors, or small molecules that mimic signals that take place in development. In addition, methods were developed to select differentiated cells from mixtures of different cell types formed in embryoid bodies or monolayer cultures. This has all been exceptionally costly in time and materials, and the human cells have proven much more difficult to deal with than mouse cells. Nevertheless, major progress has been made and, excitingly, many of the procedures developed have turned out to work just as well on human-induced pluripotent stem cells (see Chapter 5, "Origins and Types of Stem Cells: What's in a Name?"), allowing this area to advance much more quickly than it would otherwise have been expected to do.

One of the important puzzles remains: why are not all human embryonic stem cells the same and why do some apparently prefer to differentiate into one cell type and others into different cell types? The International Stem Cell Initiative (ISCI) has been, in part, founded to address this issue and, in 2005, it decided to compare all registered human embryonic stem cell lines to determine differences and similarities and to stimulate further research. Over 75 cell lines from 14 countries were involved in the study. Making comparable embryonic stem cell lines from animal species other than mice and humans had, until recently, only succeeded with rhesus monkeys, although a technique developed in Austin Smith's laboratory in Cambridge finally allowed derivation of cell lines from rat embryos. Why there is this species preference is not known.

4.5

SIR MARTIN EVANS AND THE NOBEL PRIZE IN 2007 FOR GENERATING MOUSE EMBRYONIC STEM CELLS, A CRUCIAL CONTRIBUTION TO GENETIC MODIFICATION OF MICE

Mario Capecchi and Oliver Smithies had the vision to see that mammalian cells could be genetically modified by a process called homologous recombination. Their experiments suggested that all genes could be modified in this way. However, the cells they studied could not be used to create lines of animals in which a specific gene had been modified. A new kind of cell was needed, which could give rise to germ cells and allow DNA modifications to be inherited.

4.5 *(cont'd)*

Sir Martin Evans' contribution was to discover these cells. Initial attempts with teratocarcinoma cells from mice were unsuccessful. However, he then turned to embryonic stem cells. He modified these cells genetically and showed that they could create mice that could pass on the new genes to the next generation (Figure B4.5.1).

FIGURE B4.5.1 Sir Martin Evans (right) shares his Nobel Prize with Professor Mario Capecchi (left) of the University of Utah and U.K.-born Professor Oliver Smithies (middle) of the University of North Carolina. *Source: © Martin Evans.*

The marriage of Professors Capecchi and Smithies' homologous recombination technique with Sir Martin's stem cell discoveries has created the highly versatile technology of gene targeting. It is now possible to introduce mice strains (known as knockout mice) in which specific genetic modifications can be activated at specific times, or in specific cells or organs. It is possible to study almost every aspect of mammalian physiology in this way. The technique has been used for research in fields as diverse as cancer, immunology, neurobiology, human genetic disorders, and endocrinology.

Sir Martin himself has used the technique in studying cystic fibrosis and breast cancer. Professor Capecchi has shed light on the causes of several human birth abnormalities, and Professor Smithies has worked on the blood disease thalassemia and hypertension.

Interview with Sir Martin Evans

When did you get the idea to try to isolate stem cells directly from embryos rather than teratocarcinomas?

Our studies using monoclonal antibodies showed that the cell surface of embryonal carcinoma cells was closely similar to that of early mouse embryos and, moreover, the dramatic results showing that these teratoma-derived cells could make extensive chimeras when added to a normal mouse embryo clearly led to the idea that the tumor-derived embryonal carcinoma cells were very similar to normal embryonic cells. It was also known that these tumors could be made by transplanting a normal embryo to an abnormal site.

I don't know exactly when the idea that it should be possible to grow them directly was conceived, but I do know that my first experiments trying to isolate equivalent cells from mouse morulae and blastocysts were in 1974 and 1975.

How did you feel when you saw the first colonies looking like embryonic stem cells appear?

I knew immediately that I had succeeded. I had spent so long observing colonies of embryonal carcinoma cells and I also knew all the properties I should expect. One very simple test, which was immediately applicable, was staining the colonies for alkaline phosphatase. I was then able to test the cells with all the monoclonal antibodies we had characterized.

Which experiment told you that you really had mouse embryonic stem cells?

There were a whole series of tests. Did the cells have a normal chromosome complement, i.e., karyotype? Could they differentiate *in vitro*? Did they form teratomas showing multiple tissue differentiation when injected into mice? To show that all these properties resided in a single cell type it was necessary to establish cultures from a single cell, i.e., clonally.

We then had to show that these cells would—like some of the embryonal carcinoma cells—contribute to healthy chimeric mice; and, finally, that these mice not only had contributions to all their somatic tissues from the tissue culture cells but also to their germ cells and that, when they bred, some of the offspring would inherit their genes from the tissue culture cells (Figure B4.5.2).

FIGURE B4.5.2 Sir Martin Evans. *Source:* © *Martin Evans.*

4.5 *(cont'd)*

How did you feel when the Nobel committee called you, and has it changed your life?

I was completely surprised by the call. I had a voicemail message from the secretary at Cardiff University asking me to call this number very urgently. When I called it was the secretary of the committee, who told me the news and then said "I'm sorry I can't tell you anything more at the moment because I have to announce it at the press conference in four minutes." I therefore heard just four minutes before the rest of the world!

It is a wonderful honor and has changed my life mainly in terms of recognition. I was about to retire and had just come to the end of my active scientific research. I have received many invitations to speak and numbers of honorary positions. My quiet retirement continues to be busier than expected!

4.6 ON THE ROAD TO STEM CELL THERAPY

The interest in human embryonic stem cells for cell therapy is driven largely by the shortage of donor organs, which could cure patients of chronic diseases of the heart, lungs, liver, kidney, and repair damage to eyes and ears that lead to blindness and deafness.

Now that a number of well-characterized hESC lines are available and several groups in countries where it is permitted have been deriving new cell lines under conditions compatible with clinical use, much research effort has been focused on finding the factors and conditions that can guide differentiation of human embryonic stem cells in a particular direction so that they could be used for cell replacement therapy. It has turned out that making cells of the ectoderm lineage, such as nerve, has been relatively straightforward; mesoderm derivatives, such as heart cells and vascular endothelial cells, are a little more difficult; but cells of the endoderm lineage, such as lung and pancreas, have been extraordinarily difficult to produce. Thus, the conundrum has arisen that the cells that are easiest to transplant, such as the insulin-producing pancreas β-islet cells (these cells can be collected from pancreas after death of the donor and transplanted to patients with diabetes, offering a cure that lasts for up to several years) are the most difficult to produce from human embryonic stem cells; and for the cells that are easy to produce, transplantation itself and the integration into the host tissue is fraught with problems and risks.

Just a few years after their first derivation, sufficient numbers of researchers were so convinced of the therapeutic potential of embryonic stem cell technology that the first commercial ventures for cell therapy based on hESC were launched. Various companies, including WiCell, Geron, and Advanced Cell Technology in the United States, ES International (ESI) in Singapore, and BresaGen in Singapore and Australia, tried to carve out a niche and focus on the development of embryonic stem cells for clinical purposes, such as insulin-producing cells, heart muscle cells, nerve cells, and liver cells. Trials in paralyzed rats injected with embryonic stem cells differentiated into oligodendrocytes, the cells that form the insulating cover of nerve cells, showed the rats could walk again after a few weeks. Based on these outcomes, Geron announced its application for U.S. Food and Drug Administration (FDA) approval to test these same cells in patients with acute crush lesions of the spinal cord. The crush damages the insulating sheath around the nerve cell and transplantation of stem cell derived oligodendrocytes could, in principle, restore lost function, but only before scar tissue has formed. Finally, after 22,000 pages of documentation on safety and feasibility, in 2009 Geron was granted approval to include 10 patients in a small trial. However, at its own request, because of some unexpected cyst-like lesions—but not teratomas—in the rats tested, Geron put its planned study on hold until the cause of the problem was found. Once found and solved, the first-in-man phase I (safety and feasibility) study was started. The FDA only allowed the same (small) number of cells to be injected in the patients as had been injected without teratoma formation in the rat, which created a safety and feasibility study with zero chance that there would be any clinical improvement in the patients' condition as there would be far too few cells to have an effect. Because of this, the trial was stopped after inclusion of five patients and Geron stopped all studies on human embryonic stem cell based therapies, including those on the heart.

Of the other potential clinical applications of human embryonic stem cells, however, macular degeneration (age-related blindness) is perhaps advancing the most rapidly. This still holds significant promise. Research groups in the United Kingdom, Israel, and the United States have all been able to generate retinal pigment epithelial (RPE) cells efficiently from most human embryonic stem cell lines. In Japan, this has been successful with human-induced pluripotent stem cells. The eye may be less sensitive to immune rejection than other tissues and should teratoma development start, the ophthalmologist will be able to spot it early. Furthermore, in the enclosed space of the eye any residual undifferentiated cells would be unable to escape and spread through the body. The retinal epithelium is destroyed by excess blood vessel formation in many millions of elderly people, causing blindness in the central portion of the field of view.

Although they rarely become completely blind, they see a black patch in their central field of view and can only see objects around the edge of the patch. Patients usually have to read with a magnifying glass. Replacing the RPE layer with new cells could restore sight. The operation itself, and the construct or scaffold on which to place the cells, is still a challenge, but several major pharmaceutical and small biotechnology companies are supporting the research, clear indications of their optimism that the technique will at some time in the future be clinically applicable for the treatment of this debilitating disease.

From the studies of Geron, a great deal was learnt about making cell lines that meet the stringent hygiene and safety criteria for use in man, so even though abandoned, these studies helped leapfrog some of the important issues for other clinical applications. First-in-man studies aimed at reversing macular degeneration have already taken place in London, and around 25 patients have been included in the first phase 1 studies (safety and feasibility) worldwide. It is still too soon to ask patients whether they can see better, because too few cell were used for injection (the potential risks of teratoma had to be minimized), but once these trials are complete, it may then be possible to test for effectiveness and then ask the patients "Can you see?"

Producing pure populations of the correct cell type, without contaminating undifferentiated cells, under conditions free of animal reagents remains a challenge and, in combination with tissue type matching/immune rejection and integration with the host tissue, still represents a major hurdle to the wider clinical application of human embryonic stem cells.

4.7 BIASED INTERPRETATION

In countries where embryonic stem cell research is banned, including Germany and Italy, research into other sources of stem cells has benefited, largely because research support for that purpose has often been expanded as compensation. Independently of the ethics, adult stem cells that could be autologous from the patient's own body have little risk of forming tumors and would usually be preferable for therapy over pluripotent stem cells in cases where they can form the correct cells or produce the right factors to induce the required effect. Results are now being obtained in these fields: the differentiation of mesenchymal stem cells from bone marrow into bone, skin cells into skin transplants, and mesenchymal cells for treating autoimmune diseases such as graft-versus-host disease or Crohn's disease. Most of the treatments, however, are still at the stage of clinical trials, although some commercial companies now provide cell products for these trials.

These commercial companies have been exceptionally welcomed in countries where there are strong objections to human embryonic stem cell research. However, this has often been at the expense of requiring experimental rigor in the treatments being offered. The Catholic Church has been strongly opposed to human embryonic stem cell research, for example, and is funding various trials on unproven therapies. One is called STAMINA and is based on mesenchymal cells; the other is being tested by a company called Neostem. Neostem has described some very small cells derived from bone marrow which they claim have the properties of embryonic stem cells. For this reason, they call them *very small embryonic like cells* or VSELs. Some researchers have said that the VSELs are not cells at all but pieces of cell debris. For the unwary patient looking for solutions to their complaint on the internet, this can be very confusing and something in which they can get lost for hours on end.

4.8 THE FUTURE: STEM CELL TRANSPLANTATION AS A CLINICAL TREATMENT

There are signs that the tides are changing in the United States. On November 2, 2004, California passed Proposition 71 (the California Stem Cell Research and Cures Act) via referendum, through which the state of California made $3 billion available for embryonic stem cell research over a period of 10 years. However, a number of long-running legal proceedings has meant these funds are slow in being released. With the election of President Obama in 2008, the U.S. government funding policy with respect to human embryonic stem cell lines was expected to change, and the Presidential NIH list has been reviewed with the idea that more cell lines could be included, although derivation of new lines is still not allowed. Proper and full informed consent of the gamete donors (biological parents of the donated embryos used for derivation of the human embryonic stem cells) was, however, required, and that created a problem. It turned out that for some of the original lines, the donors could not be found and the treating gynecologist of some 15 years earlier, was no longer alive to delve into his own records.

There were more delays in the United States after a private injunction prevented the use of government funds for research. Europe remains divided on funding because in the collective funds for research, some countries in which embryo and human embryonic stem cell research is not allowed do contribute. However, these countries do not wish to provide financial support, even indirectly. It is clear that research to develop clinical quality human embryonic stem cells and to address the safety issues is extremely expensive and that a single investigator driven program will not succeed. Large public funding bodies to support the research for clinical applications is needed for the research to reach its full potential. In addition, only the future will tell which stem cells are actually suitable for treating disease, which diseases might be treated, and whether embryonic stem cells will play a part at all. These are questions that have recently arisen again in the light of an exciting new development in the stem cell field: *induced pluripotent stem cells*.

4.9 BREAKTHROUGH OF THE DECADE IN THE TWENTY-FIRST CENTURY: INDUCED PLURIPOTENT STEM CELLS

However promising the potential of human embryonic stem cells for therapy might seem, ethical reservations associated with the use of embryos has spurred the search for alternative sources of pluripotent cells. As early as 1976, Carol Miller and Frank Ruddle from Yale University, U.S.A. demonstrated that normal differentiated cells can actually acquire pluripotency after artificial fusion with embryonal carcinoma cells derived from teratocarcinomas. They constructed a series of these hybrid cell lines that behaved like typical embryonal carcinoma cells even though each cell contained two nuclei with double the amount of DNA because of the fusion.[2] Later, Takashi Tada and colleagues in Kyoto, Japan discovered that something similar occurred when differentiated cells were fused with ES cells.[3] Their experiments demonstrated that both embryonal carcinoma and ES cells contain factors that can induce pluripotency in differentiated cells.

It was Shinya Yamanaka, however, also then from Kyoto, who identified the responsible proteins and their associated genes[4] that apparently can "reprogram" a differentiated cell to a pluripotent state. In an ingenious and incredibly labor intensive set of experiments, he developed a system that allowed him to identify the pluripotency of cells by their resistance to an antibiotic drug. Introduction of the candidate genes into fibroblast cells from a mouse, obtained from small biopsies from the skin, quite remarkably revealed that a set of only four genes was enough to reprogram a differentiated skin cell to a cell type closely

[2]Miller RA, Ruddle FH. Pluripotent teratocarcinoma-thymus somatic cell hybrids. Cell 1976;9(1):45–55.

[3]Tada M, Takahama Y, Abe K, Nakatsuji N, Tada T. Nuclear reprogramming of somatic cells by in vitro hybridization with ES cells. Curr Biol 2001;11(19):1553–8.

[4]Takahashi K, Yamanaka S. Induction of pluripotent stem cells from mouse embryonic and adult fibroblast cultures by defined factors. Cell 2006;126(4): 663–76.

resembling an ES cell, albeit at very low frequency. The four genes all encoded transcription factors that are normally expressed in ES cells. These modified cells, surprisingly, had all of the properties of ES cells, including the ability to grow indefinitely and form different cell types, such as nerve and heart cells. They are now known as induced pluripotent stem cells, or iPS cells, since pluripotency was artificially "induced" (Figure 4.11).

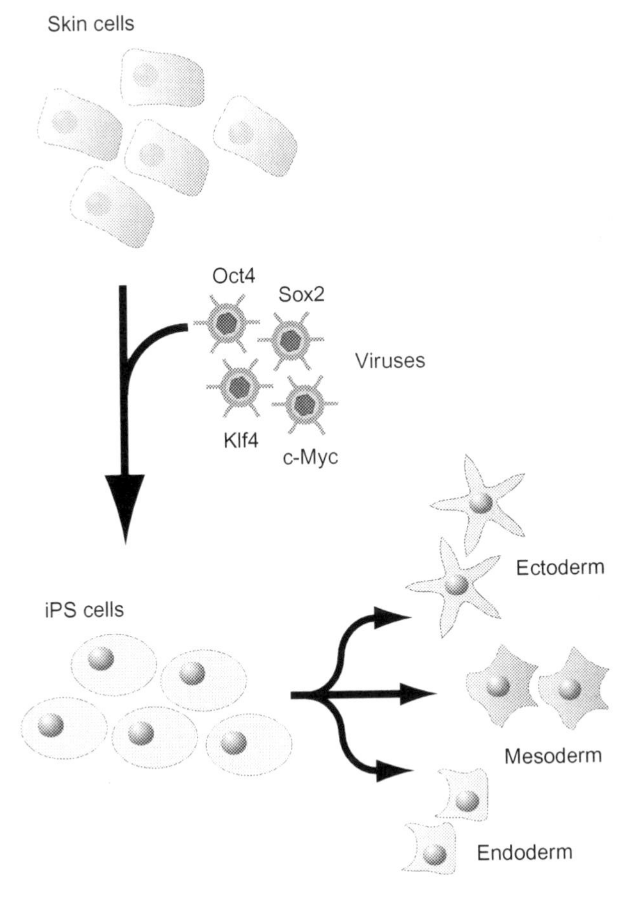

FIGURE 4.11 Schematic representation of the original method used to generate induced pluripotent (iPS) stem cells. Normal, differentiated skin cells are derived from a human individual and cultured as usual. Four new genes, coding for proteins that are required for reprogramming the cell to an embryonic state, are introduced (transfected) into the skin cells by infection with viruses that carry these genes. After a few weeks of culture a small fraction of transfected cells form colonies and become morphologically similar to embryonic stem cells. Indeed, it has been demonstrated in mice, at least, that these iPS cells are truly pluripotent and that they can differentiate to derivatives of the three germ layers ectoderm, mesoderm, and endoderm.

Subsequently, Yamanaka and other groups demonstrated that many types of differentiated cells, indeed also human cells, can be transformed to pluripotent stem cells using this technique.[5] This amazing feat caused a revolution in stem cell biology, as sacrificing embryos to obtain pluripotent human cells may no longer be needed in the future. Patient-specific pluripotent stem cells lines can quite easily be derived from a simple skin biopsy, enabling scientists in countries that were never allowed to carry out research on human embryonic stem cell lines to join in the search for therapeutic applications of stem cells. This work on reprogramming mature cells to pluripotency was awarded the Nobel Prize in Physiology or Medicine in 2012.

4.6

DR. JAMES THOMSON

Interview with Dr. Thomson

How did you start working on human embryonic stem cells?

In the early 1990s, my laboratory derived embryonic stem cells from an Old World monkey (the rhesus macaque) and a New World monkey (the common marmoset), work that led to the derivation of human embryonic stem cells. Much of the initial work in my laboratory after that derivation focused on establishing human embryonic stem cells as an accepted, practical model system. We developed, for example, improved culture conditions, methods for genetic manipulation, and approaches for the *in vitro* differentiation to key lineages of clinical importance. (Figure B4.6.1)

How did you feel when you realized that you had derived the first stable human embryonic stem cells? Which experiment convinced you that you had succeeded?

I don't recollect having the kind of eureka moment that people sometimes expect of scientific discovery. I felt more relief than enthusiasm. I had to keep the cells growing for six months and doing so involved a continuous series of minute steps and a good deal of uncertainty—the cells could have been lost at anytime. Once finished, I also had to redirect my efforts to publishing what I'd done. A more final satisfaction came when the results appeared in *Science*.

What are your present research interests in human embryonic stem cells?

[5]Takahashi K, Tanabe K, Ohnuki M, Narita M, Ichisaka T, Tomoda K, et al. Induction of pluripotent stem cells from adult human fibroblasts by defined factors. Cell 2007;131(5):861–72.

FIGURE B4.6.1 Professor James Thomson. *Source: Photograph courtesy of James Thomson.*

We are now focused on using these tools to understand the basic biology of pluripotency. For example, we use several conditions that induce uniform differentiation to specific lineages to study in detail how embryonic stem cells decide to exit the pluripotent state and become restricted in their potential, and we use a hematopoietic model system to study how that process of restriction can be reversed.

Do you feel enormously pressured by the expectations of human embryonic stem cells and iPS cells?

I feel less pressure than I once did. Early on, I had to be the representative of the field and, as such, I testified before the U.S. Senate and gave many interviews. The discussion around stem cell research at that time was very political, so I felt pressure to be a clear and careful spokesperson, exact with my words, and capable of keeping the field alive and well.

In terms of stem cell research's potential for curing diseases, I've always known that those kinds of achievements are going to take decades of research. Since many scientists have now joined me in advancing the field, I'm also no longer the lone standard bearer. Where stem cells are concerned, many minds are at work.

4.7

DR. SHINYA YAMANAKA AND THE CIRA INSTITUTE

Interview Dr. Yamanaka

When did you get the idea to try to reprogram cells directly?

By the time I had my laboratory for the first time in my academic career as an associate professor at the Nara Institute of Science and Technology in 2000, I became interested in how embryonic stem cells maintain their differentiation ability while rapidly proliferating. At the time, many laboratories were trying to differentiate embryonic stem cells into various functional cells. In contrast, I tried to differentiate somatic cells back to the embryonic state and made the idea the theme of my laboratory.

Were the ethical issues associated with human embryonic stem cells important in your motivation?

Yes, I was well aware of the ethical issues over the human embryonic stem cells and wanted to find a method to circumvent the problem. However, it is true that studying human embryonic stem cells is essential to advancing research on human-induced pluripotent stem cells. (Figure B4.7.1)

How did you feel when the first colonies looking like embryonic stem cells appeared?

FIGURE B4.7.1 Professor Shinya Yamanaka. *Source: Center for iPS Research and Application (CIRA), Kyoto, Japan.*

At first, I couldn't believe that mouse fibroblasts turned into the embryonic state so easily. So we carefully analyzed that the colonies had similar characteristics to those of embryonic stem cells and repeated the same experiments several times. It took us nearly half a year to confirm that embryonic-like stem cells were generated from mouse fibroblasts by introducing the four transcription factors: Oct-3/4, Klf4, Sox-2, and c-Myc.

When did you realize that iPS cells could be a real alternative to human embryonic stem cells?

I am hoping that iPS cells can eventually replace embryonic stem cells in the future, but it is too early to tell whether we can really do it. Many iPS cells are the same as human embryonic stem cells, but not all. You need to avoid the bad ones. If you use the good ones, they are the same.

Do you feel enormously pressured by the expectations of iPS cells?

I feel pressured as I know that many patients with intractable diseases expect that iPS cells could help create new cures for their diseases. I am optimistic, as many researchers around the world are working hard to advance iPS cell technology to bring it to the [patient] as early as possible. I believe that iPS cells will become an effective research tool for drug screening in a few years, but at least several years of basic research will be needed before we can overcome various obstacles, such as tumor formation, and realize regenerative medicine such as cell transplantation. (Figure B4.7.2)

How has it been since you won the Nobel Prize?

FIGURE B4.7.2 The Center for iPS Research and Application (CIRA) in Kyoto, Japan. *Source: Center for iPS Research and Application (CIRA), Kyoto, Japan.*

4.7 *(cont'd)*

In Japan it has been hectic. There has been a lot of attention from the Japanese press. But I still wanted to finish some papers so it was very busy.

My goals over the [next] decade include to develop new drugs to treat intractable diseases by using iPS cell technology and to conduct clinical trials using it on a few patients with Parkinson's diseases, diabetes, or blood diseases. ***Shinya Yamanaka***

4.8

STEM CELL RESEARCH HISTORY

(Figure B4.8.1)

1890	Transfer of rabbit embryo to 'surrogate mother'
1893	'Germ plasm theory' which shows all heritable trais are via the gametes
1933	Development of a fertilised rebbit embryo to blastocyst stage in the laboratory
1940	Totipotency of a cell from a two-cell embryo demenstrated
1951	First human cell line cultured: HeLa cells
1954	Discovery of a strain of mice with a very high incidence of teratomas (tumors) in the testis
1958	First nuclear transfer experiments in forgs
1959	Fusion of two embryos results in a chimeric mouse
1964	First pluripotent (embryonal carcinoma) cells isolated from teratocarcinoma in mice
1965	Cells from the inner cell mass form embryoid bodies and are discovered to be pluripotent
1968	Injection of a blastomere in a blastocyst leads to e chimeric mouose
196?	First stem cell lines isolated from human teratocarcinomas
1971	Human blastocysts cultured
1974	Injection of a teratocarcinoma cell into a blastocyst leads to a chimeric mouse
1975	Embryoid bodies formed from taratocarcinoma cells
1978	Birth of the first IVF baby, Louise Brown
1980	Fusion between mouse embryonal carcinoma cells and differentiated (T-) cells gives a pluripotent hybrid cell
1981	First mouse embryonic stem cell lines
1985	Genetic modification of a mouse and homologous recombination
1985	Differentiation into heart muscle cells shown in an embryoid body from mouse embryonic stem cells
1988	Discovery that leukemia inhibitory factor (LIF) can replace the feeder layer in mouse embryonic stem cell culture
1994	Isolation and culture of inner cell mass cells from a human embryo in the blastocyst stage
1995	First rhesus monkey embryonic stem cell line
1997	Sheep Dolly cloned using a differentiated cell from the udder
1998	First human embryonc stem cell line
1998	Culture of human primordial germ cells in the laboratory
2000	Human genome sequenced: the nucleotide order of human DNA is known
2000	Differentiation of human embryonic stem cells into neural cells in culture
2001	Differentiation of human embryonic stem cells into heart muscle cells in culture
2004	Facorts in the nucleus of mouse pluripotent stem cells shown to make fusion hybrids with T-cells pluripotent
2006	First mouse induced pluripotent stem cells
2007	First human induced pluripotent stem cells
2008	Generation of many pluripotent stem cell lines from patients with genetic diseases
2013	Nuclear transfer in humans and derivation of stem cells from cloned embryos

FIGURE B4.8.1 Stem cell research timeline.

CHAPTER

5

Origins and Types of Stem Cells: What's in a Name?

Stamcellen Veen Magazines

OUTLINE

Many organs and tissue types in embryos and adult individuals, both animals and human, contain stem cells. In Chapter 3, "What Are Stem Cells?" we explained that embryonic stem cells, derived from embryos three to five days after fertilization of an egg in a laboratory test tube, are considered to be *pluripotent*; i.e., able to make all cells of the fetus, whereas stem cells derived from tissues after birth are generally considered to be *multipotent* or *unipotent*. Literally, multipotency and pluripotency actually mean the same: the ability to become many cell types by differentiation. However, conventionally we use the term *multipotency* for adult (or "postnatal") stem cells, to refer to the more limited range of cell types that these stem cells can form. Of note however, is that while "adult" stem cells might make only a limited range of cell types when still in the tissue itself, they may be able to make more cell types in the laboratory. How many and what, exactly, is still being explored.

Unipotency usually refers to spermatogonia, cells that can divide but only for sperm cells. *Totipotency*, by contrast, is the exact opposite: not only can totipotent cells form all cells of the body (just like pluripotent cells), they can also form all of the cells types needed to support fetal development in the mother: the yolk sac, placenta, and umbilical cord. We will give examples of these different types of "developmental potency" and exactly what they mean, in the following sections.

5.1 PLURIPOTENT STEM CELLS

Stem cells that have the capacity to form all cells of the fetus and adult are called pluripotent. There are various types of pluripotent stem cells.

5.1.1 Embryonal Carcinoma Cells

Our understanding of stem cells made a great leap forward in the 1960s after the discovery that teratocarcinomas, tumors that form spontaneously in the testes of some strains of mice, can be grown indefinitely by repeated transplantation to syngeneic (or genetically identical) animals. These tumors look remarkably like disorganized embryos, and contain many cell and tissue types. This led to experiments in the 1970s that showed that teratocarcinomas could be induced in mice simply by transplanting normal embryos still early in development to tissues and organs other than the uterus: an embryo placed in the space underneath the kidney capsule, for instance, will form a teratocarcinoma that looks just like the one that forms spontaneously in the testis. It also turned out that it was possible to isolate stem cells from both the spontaneous- and embryo-derived teratocarcinomas, and that these could also be maintained in an undifferentiated state in culture dishes (Figure 5.1). This discovery made the use of animals to propagate the stem cells superfluous: they could be grown in the laboratory without continuous transfer to new mice. These stem cells are the cells that we now know as embryonal carcinoma (EC) cells.

Embryonal carcinoma cells were quickly seen to resemble undifferentiated embryonic cells of the mouse blastocyst stage embryo in many respects, expressing the same proteins on their cell surface and the same enzymes. They were also characteristically small, with a very large nucleus and little cytoplasm (Figure 5.2). Human embryonal carcinoma cell lines have also been derived from spontaneous teratocarcinomas that occur in young men; they are thought to result from germ cell development going wrong before birth. Research into these cell lines has provided important knowledge about the properties of teratocarcinomas and the differentiation of the stem cells they contain. When all cells in a teratocarcinoma have differentiated, they are referred to as *teratomas*. They are no longer malignant because all of the tumor (embryonal carcinoma) stem cells have differentiated.

Differentiation was long thought to be a way to possibly cure the tumor, because if you could induce all of the cells to differentiate, they

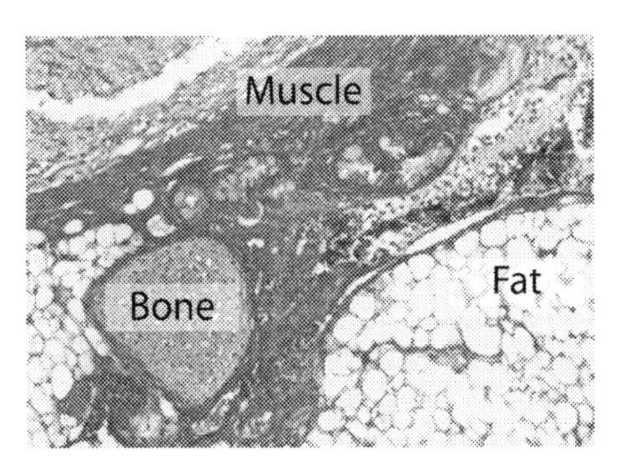

FIGURE 5.1 A section cut through a mouse teratoma that was induced by injecting embryonic stem cells into the testis of a mouse. Various tissue types can be recognized that originate from the stem cells, not from the testis. *Source: Rui Monteiro, Hubrecht Institute, Netherlands.*

FIGURE 5.2 Stem cells are cultured in dishes of various shapes and sizes. The cells are "fed" with a liquid culture medium that contains all of the components necessary to keep the cells proliferating and to maintain them in an undifferentiated state. *Source: Stamcellen Veen Magazines.*

would become benign. This did not work, however, and chemotherapy that preferentially killed the EC cells worked much better. The tumor is now curable. Nevertheless, this research provided the intellectual framework for the later derivation and culture of embryonic stem cells directly from embryos without an intervening teratocarcinoma stage, both in mice and in humans.

5.1.2 Embryonic Stem Cells

The observation that early embryos can form teratocarcinomas when transplanted into tissues outside the uterus in animals led to the hypothesis that intact embryos may contain cells that are, or can become, pluripotent stem cells. The search was then on for conditions that would allow culture of these cells *in vitro* directly from the embryo, without the need to form teratocarcinomas in live animals first. Within a few years, scientists were able to culture stem cells from blastocyst-stage mouse embryos, using conditions similar to those developed earlier for the culture of EC cells. These cells were called embryonic stem (ES) cells and, just like embryonal carcinoma cells, they were immortal, meaning that they could be cultured indefinitely. In this case, however, they had never been part of a tumor and their genes were perfectly normal (Figure 5.3). Their immortality had been uniquely achieved without malignant transformation; which means they had not changed from a normal cell into a cancer cell in culture, even though when injected into mice they could form teratomas (the nonmalignant form of teratocarcinomas).

To generate the first embryonic stem cell lines, the inner cell mass was separated from the trophectoderm of a blastocyst-stage embryo using a technique called *immunosurgery*. As the name suggests, factors from the immune system are used rather ingeniously for the selective

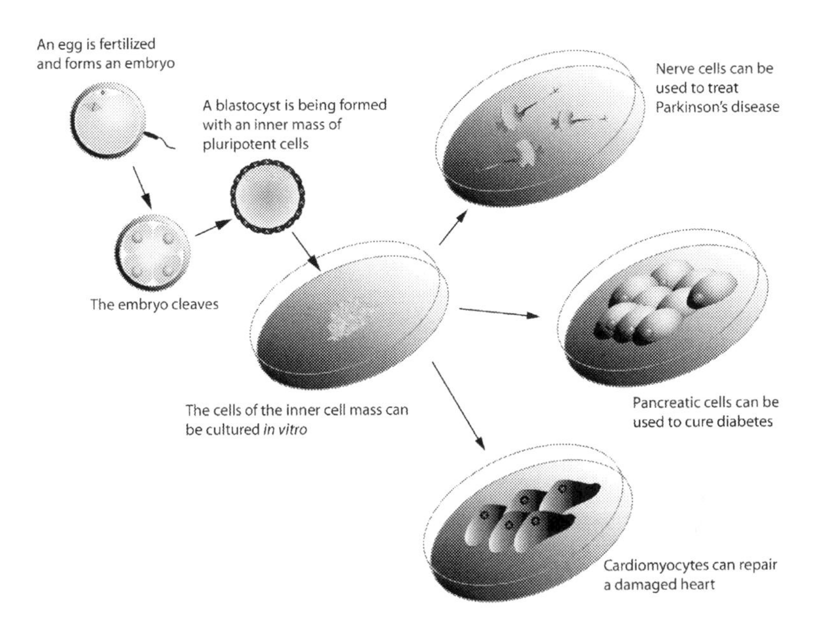

FIGURE 5.3 Generation of embryonic stem cells from blastocyst-stage embryos. *Source: Stamcellen Veen Magazines.*

isolation of cells from the inner cell mass. First, the blastocyst is incubated as a whole in a solution containing antibodies against the animal from which the embryo is derived. For mouse embryos this means antibodies against mouse cells; for human embryos, antibodies against human cells. The antibodies bind to the trophectoderm cells of the embryos but not to the cells of the inner cell mass, because the antibodies cannot penetrate the epithelial trophectoderm cell layer that surrounds the inner cell mass. The cells of the inner cell mass are thus protected. If the embryo is subsequently exposed to factors derived from blood serum called *complement*, these recognize the bound antibodies and activate an immune reaction in which the complement starts to act as a *protease* (Figure 5.4). Proteases are enzymes that break down proteins. The trophectoderm is, thus, specifically destroyed by the protease action, but the cells of the inner cell mass remain intact as they have not bound to antibodies and, therefore, do not activate proteases (Figure 5.5).

The inner cell mass cells can be collected in a pipette and transferred to a plastic dish for culture *in vitro*. However, these cells have an innate tendency to differentiate into the specialized cell types that comprise the fetus. To do anything useful with the cells and to obtain enough for experimentation, it is essential to inhibit this process. It turns out that

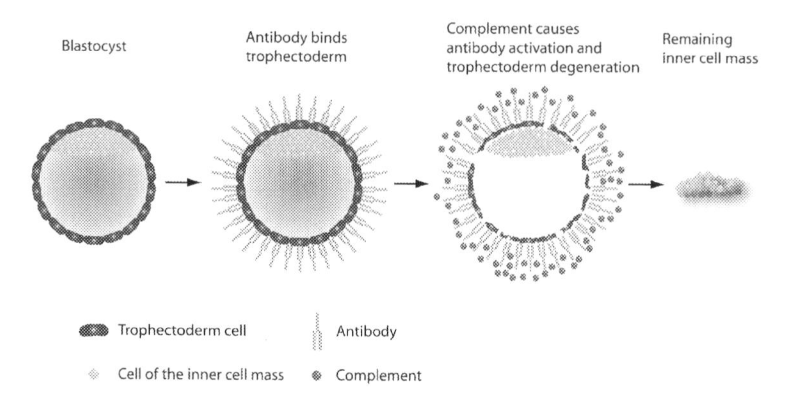

FIGURE 5.4 Isolation of the inner cell mass from blastocyst-stage embryos can be done using immunosurgery. This technique makes use of antibodies that bind to the outer trophectoderm cells but not to the cells of the inner cells mass. After exposure of the embryo to complement factors in blood (a mixture of proteins that are part of the immune system) the cells that have bound antibodies are destroyed as the complement attaches to the antibodies, leaving the inner cell mass intact. This can subsequently be separated into single cells and cultured. *Source: Stamcellen Veen Magazines.*

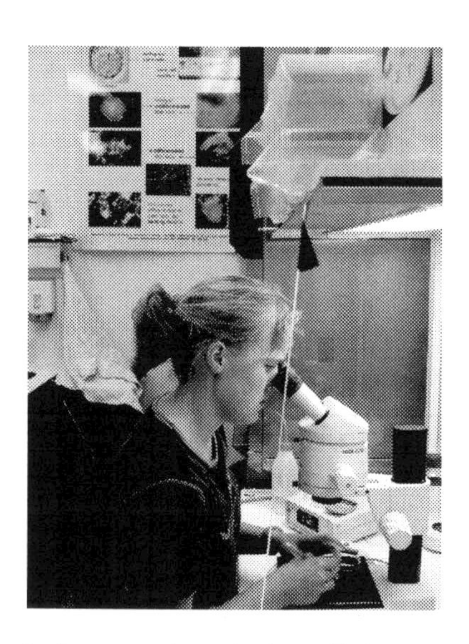

FIGURE 5.5 To transfer human embryonic stem cells to new culture dishes, groups of cells (called colonies) are divided carefully into smaller groups of cells with small glass knives. This is somewhat comparable to slicing a pizza, only at a microscale. The procedure can only be done with the aid of a microscope (and a steady hand!). *Source: Stamcellen Veen Magazines.*

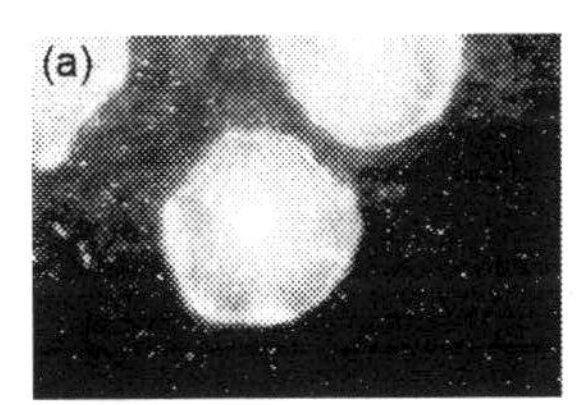

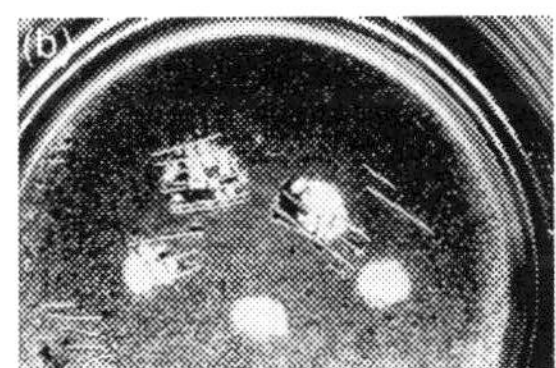

FIGURE 5.6 Human embryonic stem cells grow in colonies, as can be seen in (a). Each colony consists of thousands of cells. As the cells start to divide, the colony surface area increases. If the colony becomes too big, the cells in the middle die and those at the edge differentiate. To make sure that this does not take place, the colonies have to be "split" or divided regularly over a number of new dishes. Using small glass knives, the colonies are cut into several pieces (b) after which the fragments are transferred to new dishes; the cells start to divide again and form new colonies (c). After several days the colonies have increased in size and the procedure needs to be repeated. This procedure is called mechanical "passaging" of the stem cells. *Source: Stieneke van den Brink, Hubrecht Institute, Netherlands/Stamcellen Veen Magazines.*

differentiation is inhibited if the inner cell mass cells are cultured on top of a layer of so-called "feeder" cells, usually rather ill-defined fibroblast cells that can, for instance, be derived from mouse embryos around mid-gestation (halfway through pregnancy).

Experiments with embryonal carcinoma cells had revealed that these embryonic fibroblasts, or mouse embryonic fibroblasts (MEFs), secrete one or more factors that maintain stem cells in a state of pluripotency, which means that the cells remain undifferentiated and can self-renew indefinitely (Figure 5.6). To prevent the feeder cells from overgrowing the early stem cell colonies, their growth is usually inhibited, either by irradiation or specific chemicals, before the stem cells are "seeded" on top of them. The inner cell mass cells will start to divide while maintaining their stem cell state and, as time progresses, small colonies will form which steadily grow larger, until the cells in the middle of the colonies start to die. The colonies then need to be split up into pieces, and these pieces are transferred to multiple new culture dishes; a process called *passaging*. The cells are passaged every few days or so to new dishes containing fresh feeder cells. Initially, all mouse embryonic stem cells were cultured on feeder cells, but in the late 1980s it was discovered that the spontaneous differentiation of embryonic stem cells could be prevented, just by using culture fluid exposed for a few days to the feeder cells.

The factor secreted by the feeder cells was later identified as *leukemia inhibitory factor*, or LIF. This protein had originally been identified as a hematopoietic regulator that induces the differentiation of certain leukemic cells, hence its name. The knowledge that LIF can maintain mouse

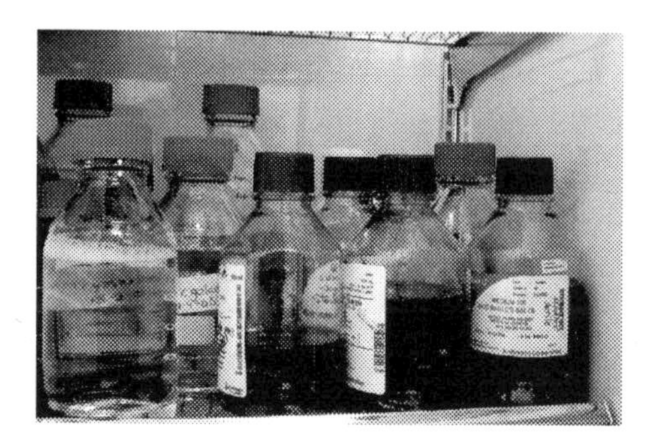

FIGURE 5.7 To culture cells *in vitro*, large amounts of liquid culture fluid (or "media") are needed. The culture media usually have a red color because of a coloring agent added that indicates the pH, or acidity, of the culture medium. When cells such as pluripotent stem cells are metabolically active, they secrete waste products such as lactic acid that lower the pH of the medium. As a result, the liquid turns yellow, an indication that the culture medium needs to be refreshed. *Source: Stamcellen Veen Magazines.*

embryonic stem cells in an undifferentiated state, provided that fetal calf serum was also present, was a big leap forward (Figure 5.7). The availability of recombinant LIF protein greatly facilitated research on mouse embryonic stem cells and allowed the development of fully defined culture conditions using an additional factor called bone morphogenetic protein (BMP). Largely through the work of a research group in Cambridge, United Kingdom, it is now possible to grow mouse embryonic stem cells extremely successfully in a completely chemically defined medium, which has made the whole process much easier and less costly.

Many years later, in 1998, the first human embryonic stem cell (hESC) lines were derived, and it turned out that these cells were also best kept in an undifferentiated state when cultured on feeder cells in the presence of fetal calf serum, just like mouse embryonic stem cells. However, feeder cells could not be replaced by LIF and BMP for human embryonic stem cells. Instead, two other proteins were required (in addition to an appropriate extracellular matrix protein to "coat" the plastic dish): (1) fibroblast growth factor (FGF) and (2) nodal or activin. If one wanted to use human embryonic stem cells for cell and tissue transplantation, it would be important to avoid the use of animal (or xenogeneic) components in the culture medium. Human embryonic stem cells have, for this reason, been cultured on human feeder fibroblast cells instead of mouse fibroblasts. More recently, culture conditions have been fully defined and this now enables culturing the cells without feeder cells or animal supplements to the culture medium. This has opened the way to clinical applications that have now been started.

5.1

EUROPEAN HUMAN EMBRYONIC STEM CELL REGISTRY

There are several databases worldwide that have been set up to offer the research community, legislators, regulators, and the general public at large an in-depth overview on the current status of human embryonic stem cell research, most particularly which human embryonic stem cell lines have been derived worldwide, what their characteristics are, and whether they are available to the research community. The European Human Embryonic Stem Cell Registry (hESCreg) is one well-established and much-used database in Europe. Its present content is the result of a survey by the project team and its partners, and the input from numerous hESC line providers. The registry was originally funded by the European Union (EU) primarily to provide documentation on the embryonic stem cell lines that could be used in EU-funded research projects. When the funding to keep the database updated ran out, there was some concern on whether access by researchers could be maintained. However, with the increasing availability of human-induced pluripotent stem cell lines, new funding was provided to cover documentation on these cells. The future of the database was thus guaranteed, at least for the foreseeable future (Figure B5.1.1).

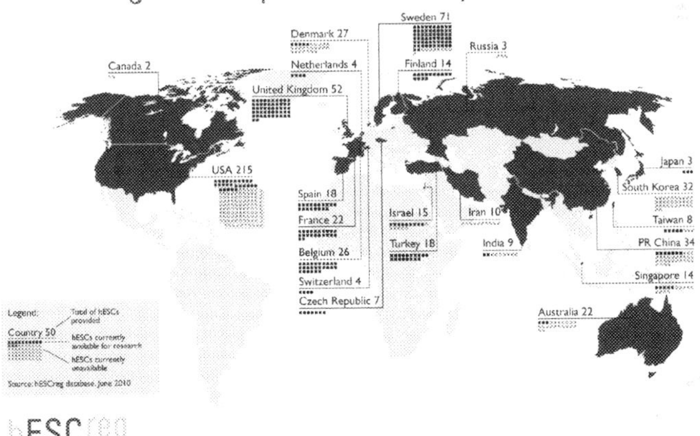

FIGURE B5.1.1 World map showing the countries in which human embryonic stem cell lines have been generated and how many have been registered. *Source: Alexander Damaschun, hESCreg, June 2010.*

5.1 (*cont'd*)

The database contains information on more than 660 human embryonic stem cell lines that have been derived in Europe and beyond, and is freely accessible. Providers of human embryonic stem cell lines and researchers who work with them are invited to register free of charge and provide detailed characteristics of their cell lines or information on their research. Literature references are also registered when available.

The registry acts as a platform for coordination and cooperation between 74 hESC providers from 24 countries worldwide. Nearly one-third of the human embryonic stem cell lines registered (i.e., some 250 lines) are available for research. More than 100 human embryonic stem cell lines in the registry carry a genetic modification. The vast majority of these possess inherent genetic defects that cause common diseases such as cystic fibrosis or hemophilia. Others carry myotonic dystrophy type 1, Huntington's disease, and fragile X syndrome. These cell lines represent prime candidates for future disease-specific lines of research and are complementary to human-induced pluripotent stem cell lines, for example, for diseases that are fatal during embryonic life or are usually aborted after prenatal genetic diagnosis because the genetic abnormality has too great an impact on quality of life.

Other lines in the registry carry experimentally induced modifications such as the expression of *reporter genes*. Cells that are marked by reporter genes, such as a gene coding for green fluorescent protein, will produce green fluorescence when differentiated to a particular cell type. Nerve cells or heart cells might be marked, for example. It makes it easy to pick out these cell types during differentiation in mixed populations of cells and see exactly how many have become heart or nerve cells without having to disturb the living cell culture as it grows.

Derivation of Human Embryonic Stem Cell Lines Worldwide

A question often asked by legislators and researchers is how many human embryonic stem cell lines are there in the world? Each cell line is derived from one donated embryo. This was in part answered in a study in 2006, based on a systematic analysis of the entire peer-reviewed literature on human embryonic stem cells published at that time. It showed that there are more than 1000 publicly disclosed human embryonic stem cell lines that had been established in 87 institutions from 24 countries. More than 40% of the nearly 1000 research papers published at that time came from U.S. groups. With respect to the use of specific human embryonic stem cell lines in research, there was a predominance of studies using the original cell lines of James Thomson derived in 1998. At least one of these hESC lines was used in 75% of studies from the United States but these same cell lines were only used in less than 5% of studies performed by Australian, Swedish, or Korean groups. In the last several years however, this has changed, and many human embryonic stem cell lines are now established in India and the Asian continent, more European countries (Belgium, in particular), and some in Africa.

Interestingly, comparative analysis of the cell lines from different ethnic backgrounds seems to show they are not all the same. It will be interesting to know whether these differences impact studies to understand the ethnic diversity of disease. The collection and dissemination of data for as many human embryonic stem cell lines as possible by international registries may be an important step toward diversification of human embryonic stem cell use, and may help in deciding whether human embryonic stem cell derivation and research can be replaced by human-induced pluripotent stem cells. For the present, there seem to be enough human embryonic stem cell lines available from different ethnic backgrounds for research purposes, and relatively few research groups now use embryos to derive new stem cell lines.

5.1.3 Embryonic Germ Cells

The germ cells (eggs and sperm cells) are extraordinary as they need to specialize into functional germ cells or gametes while at the same time maintain the capacity to form an entirely new individual after fusing with a germ cell of the opposite sex. Indeed, in the early embryo the precursors of these cells are already formed even before the gonads (testes for sperm cells and ovaries for the eggs) are present. This requires that the precursors of the germ cells move within the developing embryo to the gonads once these have been formed. As the precursors make their epic journey toward the newly formed gonads, they can be isolated from the embryos and cultured *in vitro* in the laboratory, using techniques similar to those used to culture embryonic stem cells. Cells in this state are known as *embryonic germ* (EG) cells. Human embryonic germ cells also exist, derived for the first time in 1998, the same year as the first human embryonic stem cells (Figure 5.8). These human embryonic germ cell lines had been derived from five- to nine-week-old fetuses obtained after therapeutic abortion. For some, research on these cells is ethically more acceptable than on human embryonic stem cell lines since they are derived from a fetus that no longer has the potential for "life." However, human embryonic germ cells differ in one important biological respect from their

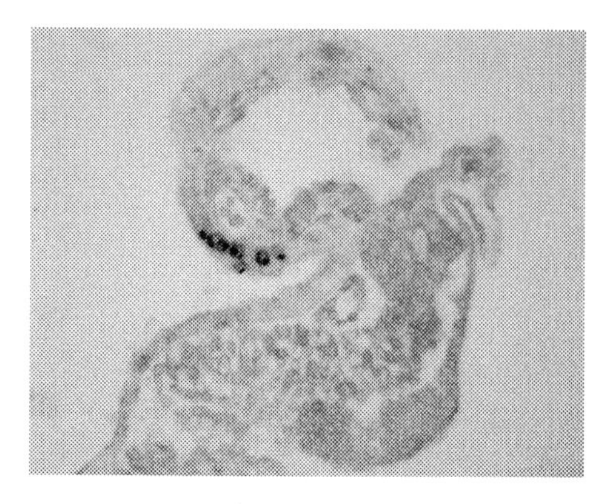

FIGURE 5.8 Photomicrograph of a section cut through a mouse embryo. In this particular section, individual primordial germ cells, the precursors of the germ cells, have been stained red and can be recognized by a distinctive dot. At this stage in embryonic development, the primordial germ cells are still on their journey to the gonads. It is at this stage that they can be isolated and cultured to form embryonic germ cells. *Source: Susana Chuva de Sousa Lopes, Gurdon Institute, U.K./Stamcellen Veen Magazines.*

mouse counterparts (apart from being derived from germ cells after they had colonized the newly formed gonads): they have a very limited capacity for self-renewal and do not appear to be immortal. For this reason, they have fallen into disuse and are rarely used for research.

5.2

GERM CELLS FROM STEM CELLS

Both embryonic stem cells and induced pluripotent stem cells are pluripotent. This means that they have the potential to form any type of cell in the body. Germ cells are one cell type in the body; they can form eggs in females and sperm in males. It should then be possible to make germ cells from pluripotent stem cells, but is this indeed so? Could we really make eggs and sperm from stem cells cultured in a dish? A group of Japanese researchers from the Department of Anatomy and Cell Biology of Kyoto University has demonstrated quite amazingly that, at least in the mouse, they can actually do this. They even showed that healthy mouse pups could be born from these stem cell derived germ cells if the embryos were transferred to a mother mouse after they had been made by *in vitro* fertilization (IVF).

How did they do this? Basically they "simply" copied processes that occur in early embryos. This sounds easier than it actually is: it requires, for example, very detailed knowledge about how germ cells are formed in the early embryo. This means knowing which molecules are involved and when they are needed in the developmental process. Only after the Japanese researchers had obtained a clear picture of how germ cells are formed in mouse embryos did they try to recapitulate this starting with stem cells. The first important step was to produce the precursors of germ cells, so-called *primordial germ cells*. These immature cells are not capable of being fertilized when first formed but need to "ripen." To become functional germ cells they have to undergo a series of reduction divisions (meiosis) so that they end up having only one set of chromosomes (they become *haploid*) instead of a pair of chromosomes (*diploid*) like normal cells of the body. To make proper sperm, the Japanese researchers transplanted the primitive cells into the testes of adult male mice that could not produce sperm, but were otherwise healthy. For *oocytes* (or eggs) a similar approach was taken, but the primitive cells were transferred to the ovaries of a mouse with a different coat color. In both cases healthy mouse pups were born from the stem cell derived eggs and sperm. They did not have the coat color of their biological mother but the coat color of the mouse from which the stem cells were derived.

Could this also work for humans? And what would be the implications? Could healthy eggs be made from, say, the skin of an 80-year-old woman? Or could eggs be created from induced pluripotent stem cells made from males, or sperm from female cells? Could this be an option in the future for gay couples to have their own genetically related children or a solution for infertility in patients who cannot produce healthy eggs or sperm themselves, for instance, because of a childhood cancer? Answers to these questions cannot be easily given. The mechanisms responsible for germ cell formation in mice may be quite different from those in the human embryo. In fact, little is known about germ cell formation in human embryos. Maybe more importantly, testing germ cell function in humans is a lot more complicated than in mice. Many countries have laws prohibiting human embryos being made for research purposes. Is it ethical and would it be technically safe to transfer these embryos to a surrogate mother? These issues still need to be discussed and debated in society even though it will most likely be decades before eggs or sperm can be created from human pluripotent stem cells.

The two key scientists that created eggs and sperm in the Japanese laboratory are Katsuhiko Hayashi and his senior professor Mitinori Saitou. Both received a large part of their education and training in the laboratory of the germ cell pioneer Professor Azim Surani in Cambridge, United Kingdom.

5.2 *(cont'd)*

An Interview with Katsuhiko Hayashi

How did you become interested in germ cells?

When I was a student at high school in Japan, I had a chance to become introduced to various types of biotechnology, such as transgenic animals and also cloned animals. From that time, I became really interested in biotechnology, particularly the applications for livestock animals. I therefore entered the agriculture department of the university and learned a lot about eggs, sperm, early embryos, and cloning. During my study at the university, my interest gradually shifted to the germ cells themselves and to the mechanisms that regulate development and pluripotency (Figure B5.2.1).

FIGURE B5.2.1 Dr. Katsuhiko Hayashi. *Source: Katsuhiko Hayashi.*

At what stage did you realize that you had indeed made functional sperm and eggs from stem cells?

When I had analyzed the primordial germ cell-like cells that I had differentiated from stem cells I saw that they had a very similar gene expression pattern to the primordial germ cells that are present in early embryos. I was at that stage already quite confident that these cells could to give rise to functional sperm or eggs. But I was obviously really pleased when I saw that in the mice real sperm and eggs were formed from the transplanted cells

Do you think that in the future functional eggs and sperm can be made from human skin cells?

First of all, since I cannot cover all ethical issues, I can only speculate on this from a purely scientific point of view. In theory, it would be possible to make eggs and sperm from human stem cells. However, it seems that it may take many years, maybe 20 to 40 years. This is because we do not have any concrete strategy for testing whether the germ cells are really functional in humans: it would be impossible to transfer "anonymous cells" to a human. To test potential as germ cells from stem cells, we need to use some animals very similar to humans, such as monkeys.

Do you think that in the future babies will be born completely from in vitro procedures (fertilization and pregnancy all in a test tube or bioreactor)?

It may take 100 years. But I cannot say it is impossible. Indeed history shows that many scientific technologies have come true from people's imagination and creativity. I think, to complete a baby in a tube, a large number of breakthroughs have to be made, not only in biology but also in, for instance, engineering to enable to reconstitute circulation and long-term cell culture devices.

Can you describe what impact the birth of mouse pups from stem cells have had on your life and career? For instance, are you often approached by (popular) media for radio, TV, magazines?

The impact on my life was not so big, to be honest. Indeed, there have been approaches from various media, but it has not changed what I am doing in my work. In the lab I continue to culture cells and analyze gene expression, and think about novel experiments to do. All from a rather fundamental point of view.

5.1.4 Spermatogonial Stem Cells

The *germline* is a term used to describe the sequential series of cell types that eventually results in the formation of either egg cells or sperm. Eggs and sperm are known as germ cells: if germ cells of the opposite sex fuse, fertilization is said to take place and a new individual arises from the totipotent cell that results. With respect to developmental potency, germ cells represent a conundrum by defying the standard pluri-/totipotency description: a germ cell on its own has limited cellular potency, meaning that it is neither pluripotent or totipotent but unipotent, yet after fusion with another germ cell that is also unipotent, a totipotent cell is created. Male germ cells, the sperm, are continuously formed throughout adult life by stem cells in the testes. By sharp contrast, cells of the female germline, the egg cells, are thought to be fixed in number during development of the embryo with no new egg cells being formed after birth.

When the stem cells responsible for the formation of sperm cells in the adult testes, so-called *spermatogonial stem cells*, are isolated from mouse testes and cultured *in vitro* in the laboratory, they can be cultured for a long period (over six months) without losing their stem cell

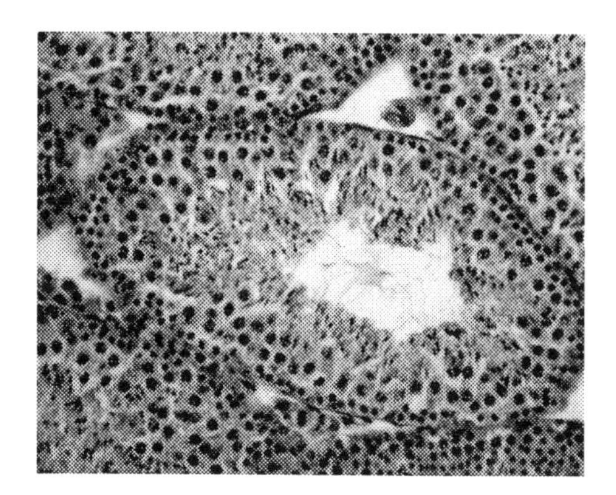

FIGURE 5.9 Section of a mouse testis. Sperm cells are continuously formed in the testis by spermatogonial stem cells. *In vivo* in the testis, these progenitor stem cells can only form sperm cells. They are referred to a "unipotent." *Source: Susana Chuva de Sousa Lopes, Gurdon Institute, U.K./Stamcellen Veen Magazines.*

characteristics (Figure 5.9). These cultured stem cells have been termed *germline stem cells* (or GS cells) and found to have a variable cellular potency: a subset of these germline stem cells remains unipotent when cultured and can differentiate into just one specific cell type: a sperm cell. This has been elegantly demonstrated in a complex set of experiments. The cultured cells were transplanted into the testes of genetically infertile mice that have no germ cells at all, so-called W mice. After transplantation with the cultured unipotent germline stem cells, the mice exhibited normal spermatogenesis and could produce offspring derived from the transplanted cells if mated with fertile females. Stem cell transplantation had restored this type of male infertility caused by lack of sperm, a breakthrough with important medical implications in fertility treatment. Quite remarkably and as yet unexplained, a very small percentage of the germline stem cells is capable of spontaneously changing morphology and, instead of remaining unipotent, these cells acquire embryonic stem cell-like properties, and become pluripotent. Only about one in 15,000,000 cells on average will make this transformation: a very rare event! During the transformation, the cells also become dependent on factors that are normally needed to maintain mouse embryonic stem cells in an undifferentiated state: LIF and fetal bovine serum.

Remarkably, this subset of germline stem cells was no longer able to produce normal sperm after transplantation into testes but instead formed tumors, teratomas, composed of many different cell types just as if they were embryonic stem cells, another indication that the cells had become pluripotent. As final proof of their pluripotency, after injection into a mouse blastocyst stage embryo later transferred to the uterus of a mother mouse, the cells contributed to the formation of all tissues in the newborn mouse: this animal was thus *chimeric*. Such pluripotent germline stem cells, referred to as multipotent germline stem (mGS) cells, have not only been derived from the testes of neonatal mice, but also from the testes of adult mice. Various reports have claimed the derivation of similar

FIGURE 5.10 When mouse embryonic stem cells are injected into a mouse blastocyst, a "chimeric" embryo is generated. If these chimeric embryos are transferred into the uterus of a recipient mouse they develop further to pups born just over two weeks later. Without transfer, development stops at the blastocyst stage. The transfer is usually done with a narrow glass pipette. In the pipette, 12 embryos are visible as white dots. *Source: Stamcellen Veen Magazines.*

pluripotent cells from adult human testes, but these results remain controversial to date and it is unclear whether these cells are indeed truly pluripotent. Obviously, autologous stem cell lines such as these can only be generated from males (Figure 5.10).

5.3

THE SALAMANDER: NATURE'S TISSUE ENGINEER

If, in the future, we were to try and grow new organs in the lab, we could learn a lot from the salamander. Salamanders can regenerate many parts of their bodies, retaining their original complex structure, if they are accidentally lost. Tail, foot, jaw, eye, intestines, even parts of the heart can all be regenerated almost perfectly. Loss of a limb, for example, leads to the creation of a supply of special (mesenchymal) stem cells on the limb stump, in a structure called a blastema. The entire structure of the limb is rebuilt from there, including all of the bones, ligaments, muscles, skin, blood vessels, and nerves. For other injuries, for example to the eye, another blastema forms but different to that of the limb: each blastema can only form the organ that was originally at the wound site at which it develops.

The stem cells in the blastema need to be able to create a large number of different cell types. The pattern for growing a new limb is

5.3 (*cont'd*)

probably largely comparable to that present in a developing salamander embryo. However, there are also key differences, including a major question on how size is regulated. Why is the new body part exactly the same size as it was originally and why, for example, is a new limb exactly the same length as the one on the opposite side of the body that is undamaged? (Figure B5.3.1)

FIGURE B5.3.1 Salamanders such as the axolotl (*Ambystoma mexicanum*) have unique regenerative capacities.

The mesenchymal stem cells in the blastema do not require any additional information from the salamander's body to make a new limb. If this lump of tissue is moved to another part of the salamander's body, for example the tail, it will still grow into a limb, not a tail. Apparently, all of the information to create a leg is encoded in this small group of stem cells. How does this regenerative process work? A great deal remains to be discovered, but now that the genomes of various salamanders have been mapped, research has been initiated to identify genes and corresponding proteins involved in the regeneration process. Comparison with genes known to play a role in development of a human fetal arm or leg may reveal factors that are unique to the formation of a blastema and the customized regenerative process.

If we are able, in the far future, to create a human variant of the salamander blastema, could we use it to grow an entire arm or leg, or any desired organ? A first indication that regeneration in mammals may be better than we think is evidence that the heart of a newborn mouse with an experimental myocardial infarction can grow back without a scar. And the little finger of a newborn human baby can also grow back without a scar. But what is the "evolutionary cost" of regeneration? Ask the salamander!

5.1.5 Induced Pluripotent Stem Cells

Induced pluripotent stem cells were first generated from normal adult mouse cells (actually skin fibroblasts) by introducing just four genes. These genes had already been associated with pluripotency in mouse ES cells, and were identified after a very labor-intensive screen of many other pluripotency candidates. The four genes are called Oct4, Sox2, Klf4, and c-Myc, and they are all genes that encode proteins that can bind DNA and affect the transcription of the genome. These types of genes are known collectively as transcription factors. The original method described putting the genes into fibroblasts using retroviruses. Firstly, recombinant DNA of the four genes was introduced into cultured cells from a special cell line that had already been infected with retroviruses. Within these cells the DNA was transcribed to RNA, which was packaged into viral particles. This could then be isolated to infect normal adult fibroblast cells. Crucially, retroviruses have genomes made up of RNA instead of DNA, which in infected cells are reverse transcribed to produce DNA. This DNA is subsequently integrated into the infected cell's DNA genome in the nucleus by an integrase enzyme. The viruses can be extremely harmful, as they change the genome of the cells they infect; researchers using retroviruses, therefore, have to be very careful in carrying out the experiments so they do not infect themselves. Several serious diseases are caused by retroviruses, one example being the retrovirus HIV that can cause AIDS.

After integration of the four genes in the cell's genome, they are transcribed and activate the same endogenous genes (i.e., the Oct4, c-Myc, Klf4, and Sox2 genes of the cells themselves). They also activate many other endogenous genes associated with pluripotency, in fact, the whole repertoire of pluripotency genes. As a result, the fibroblast cells change their morphology, gene expression pattern, and growth properties to be like that of embryonic stem cells. Even more excitingly, these cells also change from being nullipotent (no ability to differentiate) to being pluripotent. The cells became known as *induced pluripotent stem cells* (or iPS cells) because pluripotency is induced by forced ectopic expression of the four transcription factors.

There are still many things that are not known about iPS cells. For instance, among the cells that express the four proteins, less than 1% actually form true iPS cell lines. Their differentiation efficiencies are sometimes not quite as high as embryonic stem cells but the reasons for this are unclear. Sometimes they acquire chromosomal abnormalities during reprogramming and it is necessary to select those iPS cells that have as few of these as possible, or preferably none, for further research.

The use of retroviruses was initially indispensable for the generation of iPS cells, making them unsafe for any future clinical application in cell therapy: any uncontrolled genome changes caused by the retroviruses might make them tumorigenic, meaning that their clinical use would be associated with an increased risk of developing cancer. Scientists have, therefore, been keen to find alternative methods to reprogram adult cells to pluripotent stem cells, eliminating the need for retroviruses or other methods that lead to the integration of foreign DNA into the genome of the cell. For example, one could directly deliver proteins into somatic cells, but this is not easy since the four factors need to function in the nucleus of the cell and protein transfer across cell membranes is rather difficult. It has, however, been possible to reprogram adult mouse cells to pluripotent stem cells in this way, albeit at very low efficiency: only about 0.006% of the fibroblasts reprogrammed (less than one in every 16,600 cells) compared with ~1% using viral methods. Because this presented an important hurdle to their clinical and experimental use, a huge amount of effort has gone into finding efficient, nonintegrating methods to induce pluripotency, largely in Japan, where they were first discovered, and in Cambridge, U.S.A., where a large number of researchers work at the Harvard Stem Cell Institute. Nonintegrating methods now include Sendai virus, small molecules, messenger ribonucleic acid (mRNA) delivered directly to cells, and small interfering RNA (siRNA). In addition, there are alternatives to the four initial transcription factors to induce pluripotency. Commercial kits are available that now make it possible for almost any cell culture laboratory, with a little practice, to make pluripotent stem cells from mouse and human tissues as diverse as skin, blood, tooth pulp, keratinocytes from plucked hair, and even small numbers of kidney cells present in urine.

Cells that are generated in any of these ways behave very similarly to normal embryonic stem cells, in the sense that they show indefinite self-renewal and, if from mice, are pluripotent when injected into blastocyst-stage embryos: a chimeric animal results, in which each tissue type contains some descendants of the pluripotent stem cells that were injected in the blastocyst. Mouse iPS cells can even form germ cells in a chimeric mouse. This shows that the injected cells indeed have the capacity to become all kinds of different cell types.

5.2 MULTIPOTENT STEM CELLS

Most tissues and organs contain specific cells that can help repair tissue that is lost or damaged during daily "use" or after disease or injury. Tissues with a high turnover rate, such as the intestine, bone marrow, and skin, are thought to have more stem cells than quiescent tissues such as the brain and heart. True stem cells within a tissue manage its "turnover" (replacement of cells lost at the end of their natural lifespan); they have a capacity for self-renewal and can differentiate into at least one other cell type. They are known under the rather broad collective term *adult stem cells*, although they actually comprise many different cell types. Blood-forming cells that reside in the bone marrow are perhaps the best known and these cells have been used in transplantation therapies for many years to reconstitute bone marrow that has, for instance, been destroyed by anticancer therapy. Despite evidence, based on tissue repair and wound healing, that adult stem cells exist in many tissues, their isolation and subsequent culture *in vitro* has been difficult, in part because there are so few and in part because of their enormous diversity and lack of clear identity. Some very original experiments in mice, however, have shown that it is possible to identify stem cells in many more organs than previously thought, particularly organs of endodermal origin in development. These include the large and small intestine, the stomach, the lungs, the prostate, and the pancreas. Thanks to this pioneering work in which the cells in mice were "marked" by green fluorescent protein that allowed their isolation for growth in culture, our whole concept of the limited utility of adult stem cells other than from blood and bone marrow has changed.

For some tissues, the location of the stem cells is known precisely. Stem cells in skeletal muscle, for instance, are located underneath the outside membrane covering the muscle and are called satellite cells (Figure 5.11). The cells are normally dormant in the muscle fiber but become activated after injury, when they can generate large numbers

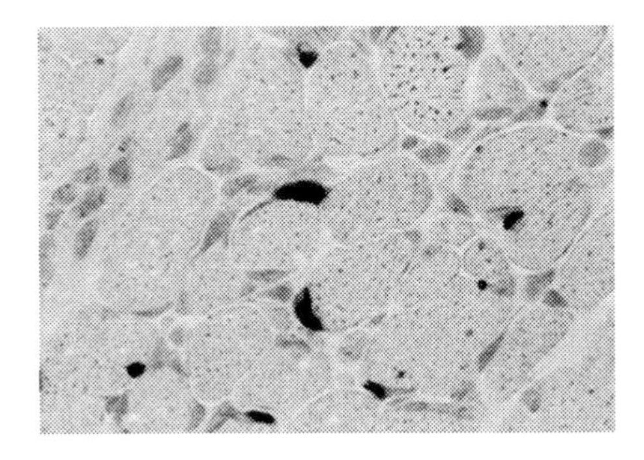

FIGURE 5.11 Stem cells in muscle tissue can be easily identified by their peripheral position within the muscle. In this cross section through muscle tissue from a pig, the stem cells are visible as large red cells. The muscle is composed of many cells that together form fibers. It is thought that stem cells in organs help to repair tissue after damage caused, for instance, by disease or trauma. *Source: Karlijn Wilschut, Utrecht University, Netherlands.*

of new muscle fibers in a relatively short period of time. Other stem cells are much more difficult to identify, and are usually found by virtue of a characteristic gene or protein expression profile. Depending on the tissue type, there are various methods to isolate and culture them *in vitro* in a way that maintains their stem cell properties. The most common procedure of isolating adult stem cells is to take a biopsy, dissociate the small piece of tissue into individual cells, and subsequently identify and isolate the stem cells by virtue of the proteins they make and expose on their cell surface. This does not work for all adult stem cells, however, and we often only know there has been a stem cell present after we see the evidence in the form of the differentiated cells it gave rise to.

Adult stem cells have also sometimes been described as "plastic" and are able to differentiate to unexpected cell types by "crossing lineage boundaries." This means that a mesodermal stem cell, such as in bone marrow, would be able to become an ectodermal cell type, such as a nerve cell. Early reports of adult stem cell plasticity *in vivo* in animals are now thought, in fact, to be an experimental *artifact*: fusion of adult stem cells with other body cells after transplantation so that the stem cells acquire the characteristics of the adult cell and appear to have "transdifferentiated" to the adult cell type, although they do not actually do so. It is now thought that adult stem cells can only differentiate into specialized cells of their own lineage. Stem cells from endodermal origin (such as liver) cannot, for instance, form nerve cells (of ectodermal origin) or blood cells (of mesodermal origin).

Though we now think that this degree of plasticity in adult stem cells does not take place in tissues of the body, it may take place artificially in culture in the laboratory when certain transcription factors are expressed in the cells. For example, it was shown many years ago that expression of a master transcription factor called MyoD in fibroblasts turned them into the skeletal muscle cells that normally express MyoD naturally. More recently, the expression of a combination of transcription factors normally expressed in nerve cells has also been shown to induce transdifferentiation of fibroblasts into nerve cells (Figure 5.12). The same trick may also work for the heart, where the injection of a combination of cardiac transcription factors into the hearts of mice has been shown to turn the fibroblast cells in the heart into cardiomyocytes. This direct reprogramming does not require the fibroblast cells to go through an in-between step as a pluripotent stem cell.

Despite much research, the origin of the adult stem cells in the body is still unclear. Whether they arise in the embryo and are retained after birth or are formed *de novo* during the lifetime of the postnatal individual is not known in most cases. To solve this mystery, stem cells in the embryo will need to be marked in some way and followed during development to see where in the organism they end up, and whether they then still

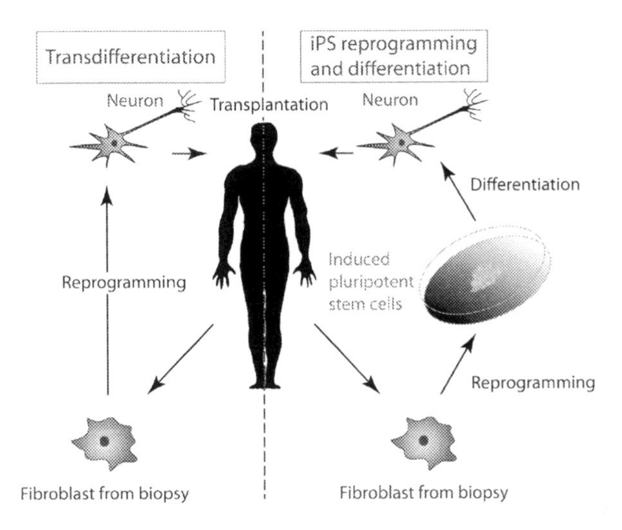

FIGURE 5.12 Two scenarios for the formation of patient-specific differentiated cells. Cells can be isolated from a patient and reprogrammed to become induced pluripotent stem cells (iPS cells), which can be cultured in the lab as a cell line. These cells can subsequently be differentiated to specialized cells such as neurons (right part of the figure). Alternatively (left part of the figure) isolated cells from a patient can be directly reprogrammed to specialized cells such as neurons, without the pluripotent stem cell intermediate stage. The advantage of the first method is that the iPS cells created allow repeated differentiation experiments to specialized cells, because the stem cell line is immortal and remains intact.

behave like stem cells or not. This type of study is called *lineage tracing*, and can be done in genetically modified mice (Figure 5.13).

5.2.1 Bone Marrow

Bone marrow contains two populations of stem cells, each with a distinct lineage. They are called hematopoietic (or blood) stem cells and mesenchymal stem cells (also called stromal cells because there is some controversy on whether they are truly stem cells). The hematopoietic stem cells take care of the constant repopulation of all types of blood cells, most of which are short lived. The mesenchymal stem cells can form fat, bone, and cartilage in culture when exposed to certain growth factors. Bone marrow can best be collected in humans from the iliac crest (the curved ridge on the top of the pelvic bone) by aspiration using a thick needle (Figure 5.14). This procedure is usually performed under anesthetic in an operating theater. The mesenchymal and hematopoietic cells from the bone marrow can be dissociated by gentle mechanical disruption. After plating the cells on tissue culture plastic, the mesenchymal cells rapidly adhere while the hematopoietic cells remain unattached, which allows the

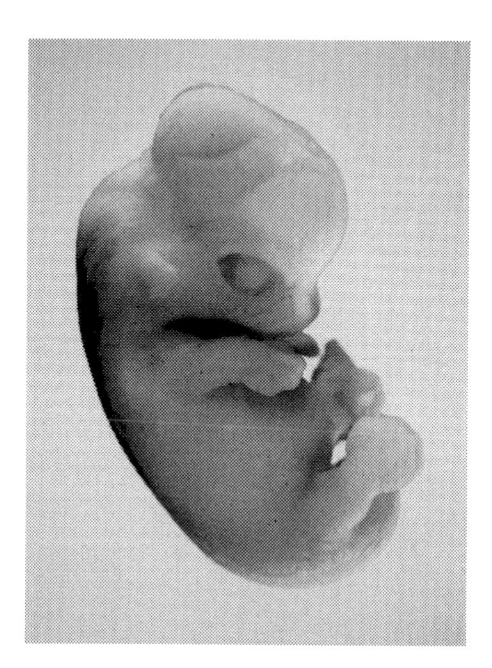

FIGURE 5.13 The organs of fetuses such as this rabbit fetus contain a relatively large number of stem cells. *Source: Stamcellen Veen Magazines.*

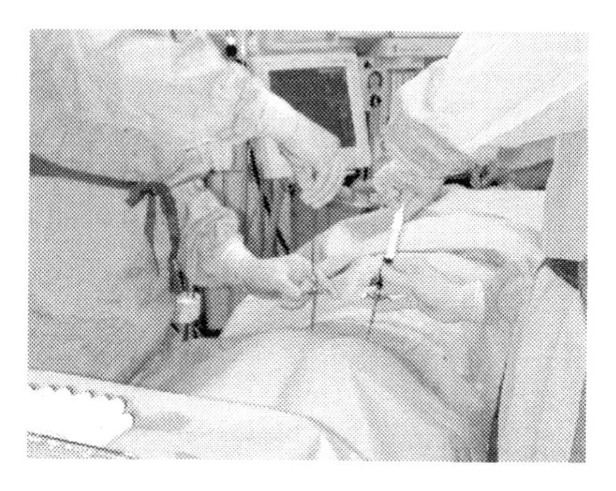

FIGURE 5.14 Bone marrow can be aspirated (or collected) with thick hollow needles that reach the center of the bone, usually from the back part of the hip (posterior superior iliac crest). The cells isolated from the bone marrow can be used to engraft recipients whose own bone marrow has been destroyed, for instance, by anti-cancer therapy. *Source: Reinier Raymakers, University Medical Center Nijmegen, Netherlands/Stamcellen Veen Magazines.*

two cell types to be distinguished and separated easily. Not all of the attached mesenchymal cells are multipotent stem cells, but the identification of the stem cells within the whole population is extremely difficult; the stem cells are often only identified in hindsight after analysis of the cell colonies that grow out from individual cells (Figure 5.15).

There is still much to be learnt about the potency and function of stem cells resident in the bone marrow *in vivo*. Most of the mesenchymal cells, for instance, will not divide any more once growth of the skeleton has ceased, yet when these cells are cultured *in vitro* they acquire "stem cell behavior" and not only divide but also differentiate (e.g., to bone, fat, and

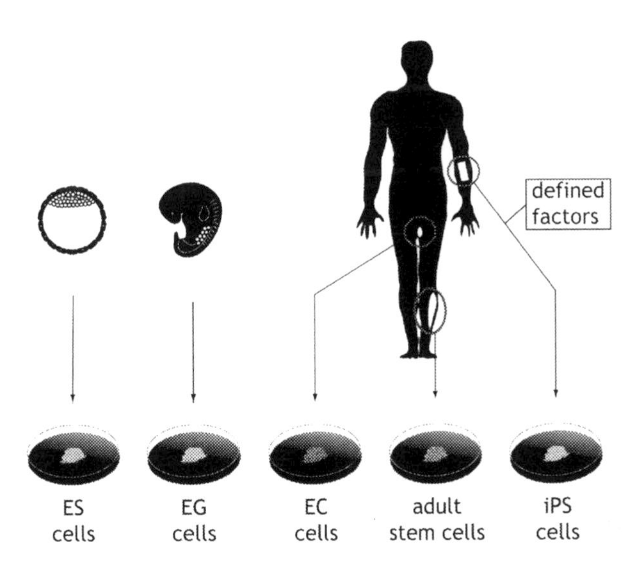

FIGURE 5.15 Stem cells of different origins can be cultured. Pluripotent stem cells are depicted in orange, multipotent cells in green. Embryonal carcinoma cells (in red) can also be pluripotent but their behavior can change rapidly due to their malignant (cancer) characteristics, which are associated with genomic instability and abnormal chromosomes.

cartilage) if given the right growth factors. The mechanisms behind this phenomenon remain to be elucidated. The extent of the potency of bone marrow stem cells to differentiate to cell types from other lineages is a matter of much debate: some researchers argue that stem cells in the bone marrow have a very limited plasticity, while others claim that at least some are even pluripotent and can form cells from several different lineages. The idea of pluripotency of bone marrow cells is not widely supported by experimental evidence, however, and has largely been abandoned. It is worth noting that although pluripotent embryonic stem and induced stem cells can make all cells of the body in a chimeric mouse, it has been very difficult to differentiate these cells into hematopoietic stem cells that repopulate the bone marrow in culture. This means that one cell could form all bone marrow cells after transplantation back into the mouse. If research on this is successful in the future and it also works in human cells, it would mean that some hematological (blood) disorders could be treated using pluripotent stem cells.

5.2.2 Umbilical Cord Blood

Umbilical cord blood has been demonstrated as a very useful and rich source of stem and progenitor cells, capable of restoring blood

FIGURE 5.16 The blood in the placenta and umbilical cord is a rich source of stem and progenitor cells. *Source: Susana Chuva de Sousa Lopes.*

formation and immunological functions after transplantation. Cord blood is the blood from the baby that remains in the placenta (afterbirth) and umbilical cord after delivery (Figure 5.16). Following delivery of a baby, this blood can be harvested from the umbilical cord after the attached placenta has also been delivered and thus is outside the body of the mother, which means that there is no risk to either mother or baby from the collection. For cord blood collection, the placenta is suspended and the blood is collected by puncture of the umbilical vein with a needle. The blood is collected in a sterile bag over a period of approximately nine minutes. Experienced collectors can harvest on average around 110 ml of blood from a single placenta and umbilical cord in this manner. After harvesting, the blood volume is usually reduced (simply because of space needed for storage) and the resulting sample frozen and stored at a temperature below −135°C.

The fact that cord blood can be frozen and stored for later use led, in 1991, to the establishment of the first cord blood bank from voluntary donors in New York. To date, there are over 54 public cord blood banks in different parts of the world with more than 350,000 units frozen and ready to be used. Indeed cord blood transplantation is being used as an alternative to bone marrow transplantation, and more than 14,000 transplants have been documented.

Private cord blood banks have also emerged, and these store cord blood for families in case the baby needs stem cell therapy during its life. Parents (and sometimes grandparents) who pay for these commercial services may see this as a kind of health insurance for later life. Generally families pay an initial storage and collection fee of between US$1000 and $1500, followed by a yearly storage fee of around US$100.

Although the reasons that families consider storing their baby's cord blood are understandable, it is important to realize that the numbers of diseases for which cord blood can currently be of help are limited and the chances of the babies or infants acquiring such disease are extremely small indeed. Actually, most families store their baby's blood with the idea of treating severe degenerative diseases sometime in the future, but there is no evidence at all that cord blood could be useful for these applications, even assuming it can be stored for so long: the onset of most degenerative diseases is middle to old age. Private cord blood banks may make claims to the contrary, but these, as yet, have no experimental basis.

Currently, donor cord blood is used as an alternative source of hematopoietic stem cells (e.g., bone marrow) for patients that suffer from genetic blood diseases or have undergone therapy to treat various forms of blood cancer. There is some evidence that cord blood may be used for the treatment of diseases other than those that are blood-related, particularly those that may have an (auto)immune basis.

5.4

THREE-PARENT EMBRYOS AND THE LAW

Cells contain genomic DNA, the blueprint of an individual, in chromosomes that are found inside the nucleus. Cells also contain DNA in many thousands of mitochondria, organelles (or small "organs") in the cytoplasm that are considered the powerhouses of cells that provide them with energy. The DNA in mitochondria (mitochondrial DNA or mtDNA) has very special functions and is only passed down with the mitochondria from mother to child, via the egg. Few mitochondria enter the egg with the sperm after fertilization, and those that do are rapidly eliminated by the egg. Mitochondrial DNA is only used for functions carried out by the mitochondria; it does not affect the appearance or any other characteristics of the individual. Like genomic DNA, however, mutations in the mitochondrial DNA can cause a variety of ailments, some life-threatening. Indeed over 150 mutations in mtDNA have been identified that are associated with human diseases and disorders. Currently, no treatments for mitochondrial diseases exist but this might change sometime in the future, using technology similar to cloning.

In the summer of 2013, the U.K. government approved a law that would allow an IVF-based technique aimed at preventing inherited mitochondrial disease to be carried out. By replacing the mitochondria

5.4 (*cont'd*)

from a diseased egg or embryo with those from a healthy egg, normal mitochondrial function could be restored. It would work as follows: the chromosomes from an egg with diseased mitochondria would be removed and placed into an egg with healthy mitochondria from which the nucleus had been removed. The second egg could then be fertilized normally. Alternatively, an egg with defective mitochondria would be fertilized and the resulting nucleus (containing DNA from the father and mother) would be transferred to another fertilized egg from which the chromosomes had been removed. Either way, the resulting embryo would have the genetic material from the father (sperm donor) and the mother with defective mitochondria, and

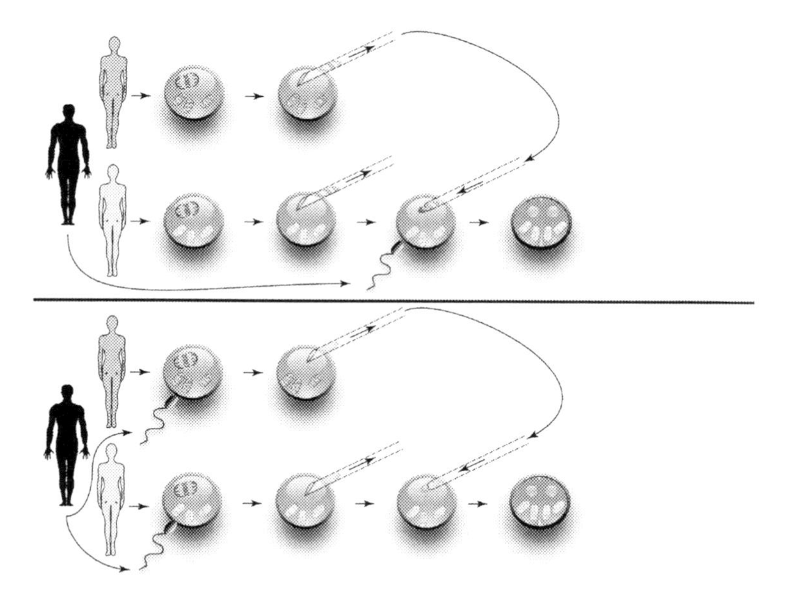

FIGURE B5.4.1 (top) A three-parent embryo can originate from an oocyte of a female patient carrying a disease on the mitochondrial DNA. The genomic DNA is carefully removed from the egg and injected into an oocyte from a donor with healthy mitochondria from which the genomic DNA has been removed. The egg is subsequently fertilized using *in vitro* fertilization. (bottom) Alternatively, an egg from a woman with diseased mitochondria is fertilized via *in vitro* fertilization. The cell nucleus from the single-cell embryo is removed and injected into the cytoplasm of an embryo that has been formed by *in vitro* fertilization of an egg with healthy mitochondria from which the cell nucleus has been removed. In both cases a three-parent embryo has been formed. Diseased mitochondria are depicted in red, healthy mitochondria are in green.

mitochondrial DNA from a second woman, the egg donor, with properly functioning mitochondria. The embryo and resulting baby would then have only healthy mitochondria but have genetic material from three individuals. Technically the baby would therefore have three parents. This has been welcomed as a possibility to cure, or rather prevent, previously incurable (mitochondrial) disease. The technique has also attracted opposition, however, and people have expressed concern that techniques like these will eventually lead to eugenic designer babies.

The U.K. government is drafting new regulations for the fertility law that includes the three-parent IVF technique in 2013. If Members of Parliament vote these in, Britain will be the first country that legalizes the use of DNA from three parents to create a baby. This would for the first time that women with affected mitochondria could have their own children free of mitochondrial disease (Figure B5.4.1).

CHAPTER

6

Cloning: History and Current Applications

Bernard Roelen

OUTLINE

6.1 BEFORE DOLLY

If an organ such as a kidney becomes diseased, one option is to transplant a new kidney, hoping to cure the patient. In a similar way, but on a smaller scale, stem cells could be used to replace not the whole organ but sick or dying cells in diseased tissues. However, in both cases the immune cells of the patient will recognize that the transplanted cells or tissues do not belong to the recipient patient: they are not "self," and the immune system will initiate a reaction to reject the foreign cells from the body unless this is suppressed with specific medication. Recognition of the immune "self" is mediated by a large number of genes that code for human leukocyte antigen (HLA) proteins, present on all cells of the body. If stem cells could be generated that had the same deoxyribonucleic acid (DNA) as that of a patient, this would enable treatment with cell or tissue transplantation without the necessity for immune suppression with all its accompanying side effects, since the transplanted cells would be genetically identical to the patient. Stem cells that express HLA proteins on their surface identical to those expressed by the patient would, in theory, be recognized by the patient's immune system as "self" and therefore not rejected.

One way to obtain immunologically matched stem cells is by using a method called *therapeutic cloning*. This is based on the technique of nuclear transfer, where the DNA of an egg cell is replaced by that of a normally differentiated *somatic* cell, for example, from the skin. In principle, any cell of the body, except a sperm or egg cell, can act as a donor of the DNA. This technique has more recently become known as *somatic cell nuclear transfer*, abbreviated as SCNT (Figure 6.1).

A fertilized egg can give rise to a complete new individual and is, therefore, totipotent (i.e., gives rise to all cells of the body), so it had been hypothesized that the cytoplasm of the egg cell contains factors that can reprogram a differentiated cell to the same totipotent state. In principle this would mean that when the nucleus of a differentiated cell is introduced into egg cytoplasm, the genomic DNA would be reprogrammed and a new organism could be formed from this cell (Figure 6.2). This organism would then be genetically identical to the organism from which the differentiated donor cell was derived, and, thus, represent a true clone of this cell donor. In the late 1950s, John Gurdon from Oxford University in the United Kingdom first performed exactly these experiments with frogs. He was able to generate cloned frogs from cells he collected from a tadpole. The importance of these findings was clear to researchers from the moment they were published, but when it was crowned by the award of the Nobel Prize in

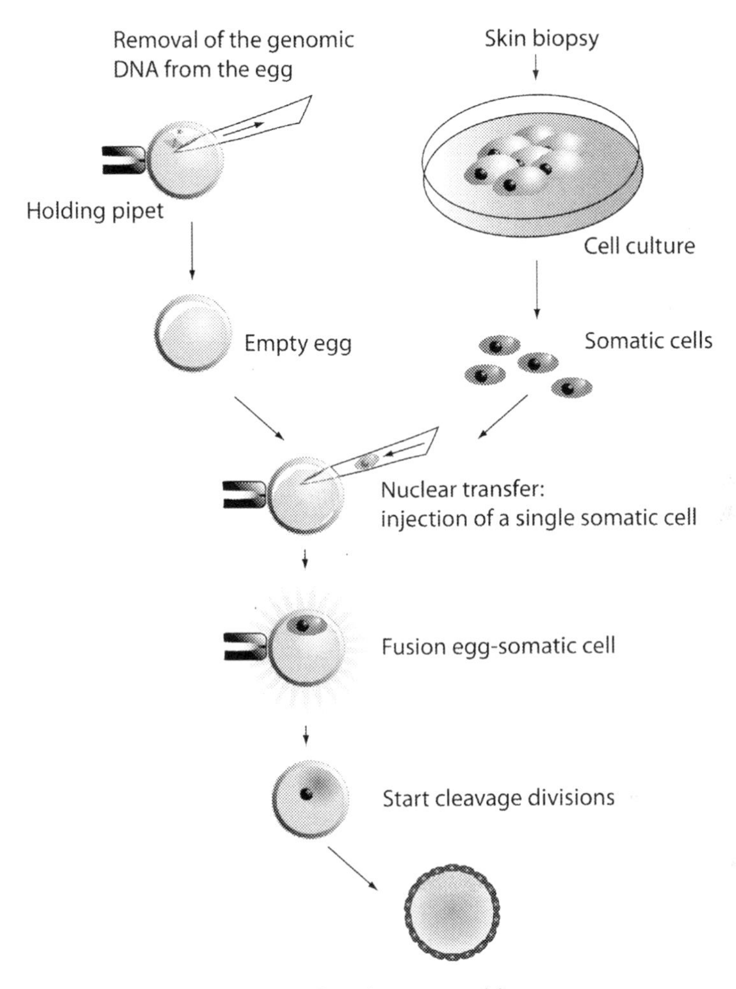

FIGURE 6.1 Cloning by somatic cell nuclear transfer. Cells from a skin biopsy are cultured in a culture dish for a short period. A mature egg is immobilized with a pipette and the genomic DNA (the chromosomes) is removed using a sharp hollow needle. A skin cell is picked up from the cells in culture using a thin glass pipette and carefully injected under the zona pellucida to lie adjacent to the egg. The egg cell and skin cell are then induced to fuse using an electrical pulse or virus and the newly formed hybrid cell starts cleavage divisions to form a blastocyst, rather like a fertilized egg, even though no fertilization has taken place. The inner cell mass of the resultant blastocyst embryo can, in principle, be used for the generation of embryonic stem cells or (in the case of animals) to create a cloned individual if placed in the uterus of a female of the same species. Various sorts of cloned animals have been created in this way (dogs, sheep, cows, even camels) but never humans. *Source: Stamcellen Veen Magazines.*

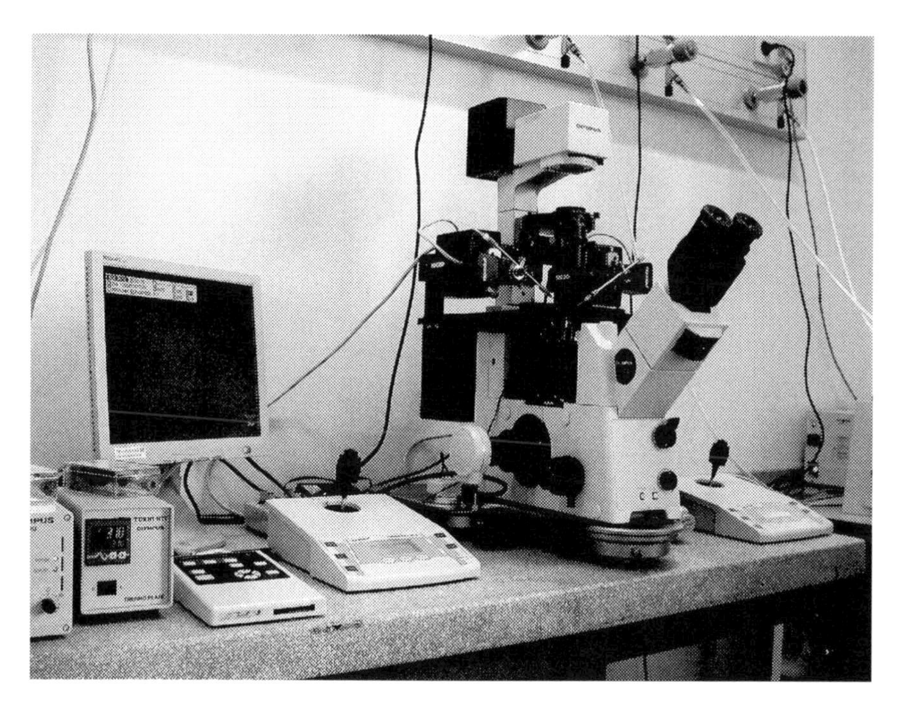

FIGURE 6.2 A special microscope is used to transfer a differentiated cell into an egg. With hydraulic manipulators, a hollow needle can be carefully inserted into the egg.

Physiology or Medicine to Gurdon in 2012, it was evident to the whole world what a breakthrough it had been.

There are still, however, some puzzles not solved: when Gurdon used cells from adult frogs, the tadpoles that developed were genetically identical to the frogs from which the donor cells were taken, but these tadpoles did not undergo metamorphosis to turn into frogs. This, and other similar experiments with mice, led to the conclusion that differentiated cells from adults cannot be reprogrammed (Figure 6.3).

This dogma was maintained until 1996, when the now-world-famous sheep Dolly (born July 5, 1996, died February 14, 2003) was born from an embryo "produced" using SCNT. Dolly was "identical" to the animal from which a cell was taken to create a cloned embryo, much like monozygotic twins are "identical." Only Dolly was six years younger than the animal from which the cell was taken, and born from a different mother. The birth of Dolly had a huge impact on science and society because she was the first mammal to be cloned using a cell derived from an adult animal. This application of the cloning technique

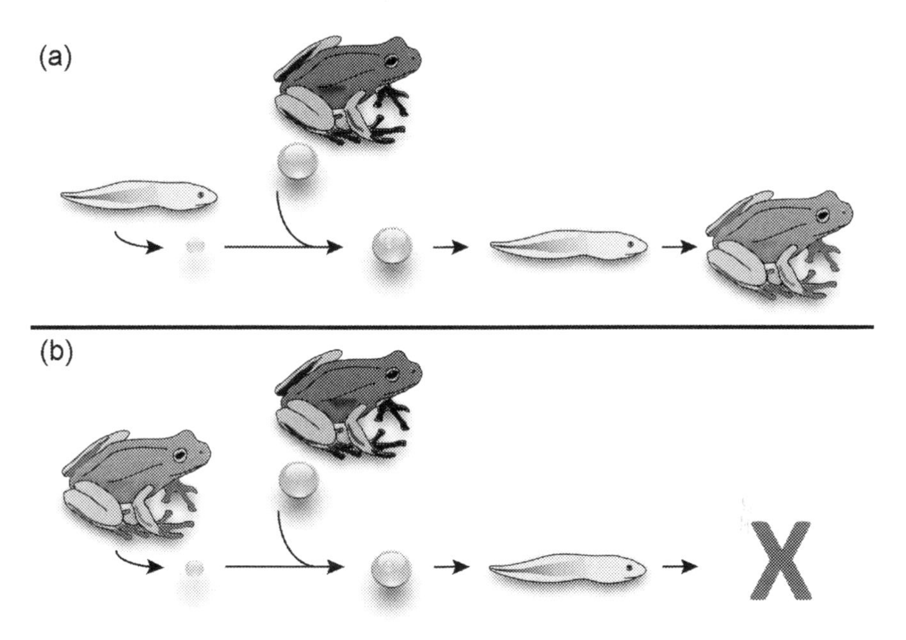

FIGURE 6.3 The first somatic cell nuclear transfer experiments were performed with frogs. When a cell of a tadpole was injected into a frog's egg from which the nuclear material had been removed, the cloned cell could grow out to form a tadpole that was genetically identical to the tadpole from which the cell was isolated. The tadpole was a clone of the original tadpole. Importantly, this tadpole could become a frog after metamorphosis (a). In contrast, when a cell of an adult frog was isolated and injected into the egg of a different frog from which the genomic DNA had been removed, the cell could grow out to become a tadpole, but the tadpole was incapable of undergoing metamorphosis to become a frog (b). This work in frogs was the first to show that an adult nucleus could be reprogrammed to an earlier developmental stage. It was the work that resulted in Sir John Gurdon becoming the joint winner of the Nobel Prize in 2012.

is known as *reproductive* cloning: a new animal is "created" using the genetic material from any cell, in the case of Dolly from the mammary gland, of an animal for which one or more cloned duplicates are required. Reproductive cloning of humans is, for many reasons, ethically unacceptable and attempts to carry out reproductive cloning of humans are illegal in many countries, even though, in principle, clones would simply be equivalent to identical twins but just with an age difference (Figure 6.4). Since Dolly, various domestic and nondomestic animals have been duplicated by reproductive cloning; prized or much-loved dogs, cats, pigs, cows, horses, and camels being just a few of the widely publicized examples (Figure 6.5).

FIGURE 6.4 The sheep Dolly was the first mammal cloned from an adult cell, an experiment that significantly changed our view of cell behaviour, differentiation, and how life begins. After her death in 2003, Dolly was stuffed and exhibited in the National Museum of Scotland in Edinburgh.

FIGURE 6.5 Dolly the sheep was cloned by researchers at the Roslin Institute, near Edinburgh.

6.1

IS THERE A FUTURE FOR THERAPEUTIC CLONING?

Aside from the problem of inefficiency when reprogramming an adult nucleus by SCNT, research on human therapeutic cloning has been seriously hampered by the scarcity of donated human eggs. Women are understandably reluctant to volunteer for invasive medical treatments like hormone treatment and egg collection for research only. Obtaining human eggs for research also has its own ethical issues. First, researchers are required by law in most countries to be sure that the donation of the eggs for any type of medical procedure is voluntary, without undue pressure, the donors are properly informed on what the eggs will be used for, and the health risks associated with egg retrieval are explained. Even though most countries (the United States being a notable exception) do not allow payment for donation of any tissue, blood, or cells, including eggs, to prevent possible exploitation of donors, compensation for time and travel expenses for instance is usually acceptable. Combined with the necessity of a whole medical team, the costs of a single egg has been estimated to be around $3600. To overcome the egg shortage, researchers have looked for alternatives to human eggs and have considered animal eggs as possible alternatives for reprogramming human donor cells, as they are easy to obtain in large numbers, for example, from animal ovaries discarded in abattoirs. This form of cloning has been called *interspecies somatic cell nuclear transfer* (iSCNT). The resulting embryo, a human–animal cytoplasmic hybrid embryo or cybrid, could, in principle, be used for the generation of an embryonic stem (ES) cell line, for instance, from a patient suffering from a specific disease. They would have a human nucleus but some cytoplasm and mitochondria from the animal egg donor. These cells could be valuable for the development of new treatments for diseases in much the same way as iPS cells.

In the United Kingdom, the Human Fertilization and Embryology Authority (HFEA) oversees IVF treatment and research using human embryos. In 2008, King's College in London and Newcastle University received approval from the HFEA to carry out research using human animal cybrids. This did not have any experimental success. However, as iPS cell technology has advanced and the use of human oocytes for reprogramming to derive pluripotent stem cell lines became obsolete, the interest in human–animal cybrid ES cells waned and research has been largely abandoned.

6.1 *(cont'd)*

A remaining question however, is how similar embryonic stem cells from cloned embryos are to those derived from "normal" embryos. Cells from cloned embryos are generated through reprogramming of adult cells. Human embryo clone-derived stem cells have not yet been rigorously tested and a good comparison of their (epi)genetic state with that of conventional, embryo-derived stem cells has not been made. It is possible that differences may exist even between iPSC cells and cloned embryo-derived embryonic stem cells obtained from the same patient, because from animal experiments it is clear that there are differences in the viability of cloned embryos.

The principal goal of somatic cell nuclear transfer in humans, however, would not be for reproductive cloning at all but primarily to generate new embryonic stem cell lines from patients. These stem cell lines would be genetically matched to the patient and could therefore be used to regenerate diseased or otherwise malfunctioning tissues without immune suppressive drugs, hence the name "therapeutic cloning." At present, first-in-man clinical research using embryonic stem cells are only just beginning, so the "therapeutic" aspect still remains to be proven. Nevertheless, the first step has now been taken: human embryonic stem cells have really been generated using embryos derived from cloning by somatic cell nuclear transfer. Before being successful, however, this research had a tumultuous history.

In 2006, a South Korean research group led by professor Woo-Suk Hwang, then at the prestigious Seoul National University, claimed to have cloned human embryos and to have used these at the blastocyst stage to generate 11 human embryonic stem cell lines that were matched to the patients. However, soon after publication, the key papers in top scientific journals that had described the research by Hwang and colleagues were found to contain faked data. Hwang was fired from Seoul National University, had to resign as head of the newly formed Center for Embryonic Stem Cell Research, and was later prosecuted for fraud. The editorial team of *Science* decided to retract (withdraw) the two papers concerned unconditionally. Hwang first accused his colleagues of improper actions and maintained his innocence, but later admitted his guilt. He was also charged with fraud, embezzlement, and bioethical violations and in court he admitted having falsified results and some of the pictures in his paper "to save time" (Figure 6.6).

FIGURE 6.6 Curiously, the nine-banded armadillo (*Dasypus novemcinctus*) always gives birth to monozygotic quadruplets. Each of the newborn animals thus has three clones (genetically identical). This particular specimen is on display at the Natural History Museum in London. *Source: Stamcellen Veen Magazines.*

Independent investigation later however showed that Hwang and colleagues had actually made scientific breakthroughs (they had generated a human embryo containing only maternal DNA) and were the first to clone a dog, an Afghan they named Snuppy. On October 26, 2009, Seoul Central District Court sentenced Hwang to two years in prison after being found guilty of the embezzlement accusation and violation of bioethics laws. The embezzlement included approximately US $700,000 of government money for which he had used over 60 different bank accounts, and by buying human eggs, he had violated bioethics laws. He was cleared of fraud, however, because it was not proven that he had used fraudulent data to gain grants.

The Hwang feuilleton has been one of the biggest cases of scientific fraud in the history of modern life science. It was also a significant blow for those conducting research on embryonic stem cells and applications, and set back the field considerably at the time, although with the advent of induced pluripotent stem cells in which adult nuclei are directly reprogrammed to pluripotency without cloning, the long-term impact is minimal. However, public trust in stem cell research was damaged, and this extended, at least temporarily, to science as a whole.

Hwang's scientific fraud unfortunately demonstrated how the "peer review" system used by scientific journals, where anonymous colleagues in the field evaluate the data, can sometimes fail. It also highlighted the importance in these particular studies of making sure that the DNA of a cloned cell line is really identical to that of the donor.

One of the techniques for doing this is known as *DNA fingerprinting*, which and is based on comparing the pattern of nucleotide variations in the DNA. These DNA sequence variations are specific and characteristic for every individual person, except for twins that have an identical sequence profile and are, indeed, true clones of each other.

After Hwang's unfortunate human cloning experiments, it turned out to be difficult to take the research further. There were several reasons for this, not least the ethical sensitivity of creating human embryos for research and the use of donated eggs. In 2006 and 2007, researchers from Harvard University in Boston advertised in local newspapers, on public transportation, and on the internet for women willing to donate their eggs for research. This would require hormone treatment followed by surgical egg retrieval and should be "altruistic," that is, the women would not be paid for their eggs or undergoing the procedure and would simply receive remuneration of direct expenses such as travel to the hospital. Over 200 potential donors initially responded, but fewer than 100 met criteria such as proper age, normal menstruation, and good overall health. In the end, only one woman actually donated eggs (six in total). In contrast, egg donors for *in vitro* fertilization clinics to assist women without healthy oocytes to have a child may typically receive between \$3000 and \$6000.[1] It seems clear that egg donation for research purposes, such as the generation of embryonic stem cell lines, without compensation similar to that received by egg donors for assisted reproduction, is impractical. On the other hand, ethics assessment by regulatory authorities needs to ensure that payment received is not "undue inducement" so that potential donors do not feel coerced or pressured by financial gain (Figure 6.7).

For cloning by somatic cell nuclear transfer, the genomic material (DNA) has to be removed from the oocyte and replaced by a donor cell, which is introduced by injection. If transfer of the nucleus is successful, it is remodeled in the cytoplasm of the oocyte and will be reprogrammed. The cytoplasm of the oocyte needs to be in a specific cellular state for it to be capable of performing the difficult task of reprogramming. Research with eggs of nonhuman primates (monkeys) revealed that removal of the *meiotic spindle*, including the DNA, causes premature meiosis exit in these cells. As a result, the egg's cytoplasm loses the ability to reprogram the DNA from a transferred nucleus and the egg cannot develop any further. By contrast, in experiments where the nucleus was transferred into an egg from which the DNA had not

[1]U.S. survey, American Society for Reproductive Medicine. Egli D, Chen AE, Saphier G, Powers D, Alper M, Katz K, Berger B, Goland R, Leibel RL, Melton DA, Eggan K. Impracticality of egg donor recruitment in the absence of compensation. Cell Stem Cell 2011;9(4):293–4.

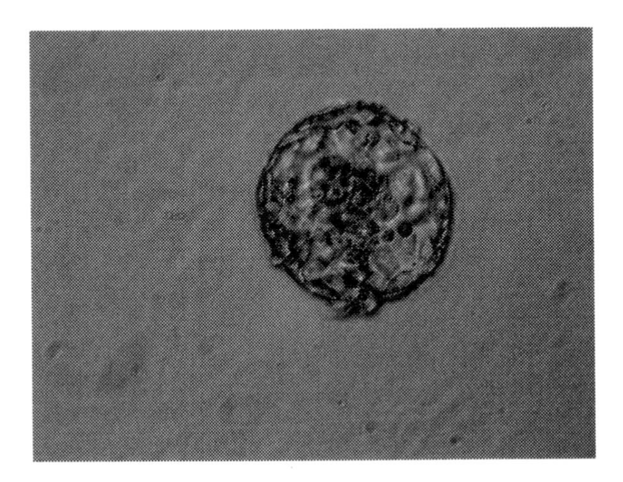

FIGURE 6.7 A human embryo at the blastocyst stage. Embryonic stem cells can be derived at this developmental stage when the embryo is only a tiny clump of cells hardly visible with the naked eye. *Source: Marga van Rooijen, Hubrecht Institute, Netherlands/ Stamcellen Veen Magazines.*

been removed at all, the resulting egg developed to a blastocyst stage embryo after activation. This embryo does not, however, carry the correct number of chromosomes. Instead of having one set of chromosomes from the father and another set of chromosomes from the mother (two sets of chromosomes, diploid, or 2n), the embryo now has two sets of chromosomes from the transferred nucleus and one set of chromosomes from the egg; the cells of the embryo are 3n (triploid) and the embryo would therefore not be able to form a normal fetus. Importantly, cell lines derived from these embryos are pluripotent even though they are triploid, suggesting that, indeed, the nucleus of the differentiated cell had been reprogrammed. The triploid nature of the cells makes them unsuitable for tissue regeneration purposes, but the experiments demonstrated that human egg cytoplasm can reprogram the nucleus of a differentiated cell.

In 2013, the long-awaited experiment attempted by Hwang was successful: a research group from the Oregon Health & Science University finally succeeded in generating embryonic stem cell lines from cloned human embryos. The scientists' technique worked because they slowed down the egg's exit from meiosis using caffeine. The reproducibility of these experiments, as well as the characteristics and stability of the stem cells obtained need further investigation, but in principle, however, it is now possible to create disease models for further study with this technique. Before 2006 this would have been a major advance in the field, but today we are rather inclined to see it as a clever solution to a technical problem. Why is that? The simple reason is that the

FIGURE 6.8 Winners of the 2012 Nobel Prize in Physiology or Medicine: Sir John B. Gurdon (left) and Shinya Yamanaka (right). The photo was taken at the ISSCR-Roddenberry International Symposium on Cellular Reprogramming only 10 days after the announcement of the laureates for 2012. *Source: Chris Goodfellow/ Gladstone Institutes.*

Japanese scientist Shinya Yamanaka discovered in 2006 how to reprogram an adult nucleus to an embryonic state in the absence of an egg (see Chapter 3 "What Are Stem Cells" and Chapter 4 "Of Mice and Men: The History of Embryonic Stem Cells"). He showed that introducing just four transcription factors into somatic cells could induce pluripotency: induced pluripotent stem cells (iPSC) in humans were the breakthrough of 2007 and earned Yamanaka the shared Nobel Prize in 2012, together with John Gurdon. It is now possible to make pluripotent stem cell lines from any individual of choice without needing the ethically sensitive step of creating embryos from oocytes as intermediates. The generation of iPSCs from patients with different diseases is now routine in many labs throughout the world without the need for the practical and regulatory difficulties of egg donation (Figure 6.8).

Although it is now possible to generate patient-specific stem cell lines from cloned embryos, this does not mean these cells could also be used for cell transplantations to cure the patient from whom the clone was derived. The first developmental processes that occur in most early cloned embryos, from the fertilized egg to the blastocyst stage, are rather abnormal. Thus, anomalies could also occur in the cells derived from them that may only become apparent much later, for example, long after transplantation. This could be the result of abnormalities in epigenetic programming of the DNA. The DNA in the nuclei of differentiated cells used for cloning normally carries a specific epigenetic signature, which needs to be "reset." Unfortunately there is only a short time window after nuclear transfer to the egg in which this essential process can take place. If the signature is not correctly reset to the totipotent state by the egg cytoplasm, some genes might remain abnormally active while other genes might be incorrectly blocked. This could have many different and detrimental consequences, such as failure of the cells to differentiate properly or maintain control over cell division, possibly leading to tumor formation.

6.2

THE HWANG FRAUD

In March 2004, the South Korean researcher Woo-Suk Hwang and his colleagues at Seoul National University published an article in *Science* magazine in which they claimed to have cloned 30 human blastocysts and used those for the derivation of human embryonic stem cell lines. This was regarded as an enormous scientific breakthrough because in this article a proof of principle was demonstrated of the generation of patient-own embryonic stem cells, which until then had only been theoretically possible. In June 2005 Hwang's team published another story in the same journal in which they demonstrated the derivation of 11 stem cell lines from skin cells of various patients with a disease that in the future might be curable with stem cell therapy. In Korea, Hwang obtained a movie star-like status because of these results. He was proclaimed "scientist of the year 2005" by *Scientific American*, and a new stem cell center of which Hwang was appointed president was opened by the South-Korean president Moo-Hyung Roh. The South-Korean government donated approximately US$20 million for stem cell research; indeed many envisioned Hwang to be a future Nobel Prize winner. From November 2005, however, cracks started to appear in Hwang's aura of ingenuity. Cloning requires eggs, and human eggs are not readily available but are collected from healthy volunteers. The procedure of egg collection is not entirely without health risks for the donor; between 0.3 and 5% of the women experience ovarian hyperstimulation syndrome after procurement of eggs. This syndrome can cause pain and occasionally leads to serious complications that require hospitalization and can even lead to death.

One of Hwang's own colleagues admitted that the eggs used for the research were improperly obtained. According to his statement, some donors allegedly received money for their donated eggs (approximately US$1400 per egg) and one of Hwangs's students, a co-author of the 2005 article, had donated her own eggs. According to South Korean legislation, women may not receive financial indemnification, nor are they allowed provision of personal benefits for the donation of eggs. In the 2004 article the authors do, however, and thus incorrectly, clearly state that there was no financial payment for the eggs.

After the first allegation, Hwang denied coercing his colleagues into donating eggs and claimed that he was unaware that the eggs

6.2 (*cont'd*)

were improperly obtained, but later he admitted his wrongdoing. This turned out to be only the tip of the iceberg. In December 2005, it became clear that Hwang had also engaged in research misconduct and that both papers from 2004 and 2005 contain fabricated data. At the time the 2005 paper was submitted for publication, on March 15, 2005, not 11 but only two cell lines were still viable. Four cell lines had become useless after a fungal infection, three cell lines appeared not to proliferate and were therefore not true cell lines, and of two cell lines any evidence that they had ever existed was unavailable. Research concluded that the two remaining cell lines had not originated from cloned blastocysts, despite the fact that Hwang and colleagues had shown (fabricated) data in their article that they had genuine cell lines from cloned blastocysts in culture. Seoul National University concluded that Hwang and his team had engaged in research misconduct, that there is no scientific basis that any of the 11 proclaimed "patient-specific" cell lines had ever existed, and that some of the data in the 2004 article had also been fabricated. In March 2006 Hwang was expelled from Seoul National University.

The editorial team of *Science* decided to unconditionally retract the two papers. Hwang first accused his colleagues of improper actions and maintained his innocence but later had to admit his culpability. He was also charged with fraud, embezzlement, and bioethical violations and in court he admitted having faked results such as certain images, to save time. On October 26, 2009, Seoul Central District Court sentenced Hwang to two years in prison after being found guilty of embezzlement and violation of bioethics laws. The embezzlement included approximately US$700,000 of government money for which he had used over 60 different bank accounts, and by buying human eggs he had violated bioethics laws. He was cleared of fraud, however, because it was not proven that he had used fraudulent data to gain grants.

Although the Hwang case was one of the biggest cases of scientific fraud in history at the time and a blow not only for those conducting research on embryonic stem cells but also for public opinion on stem cell research and patients hoping for cures, Hwang was later partly "rehabilitated" as a scientist. He was and is extremely good at cloning animals, and that is what he continues to do to this day.

Despite these issues and perhaps in contrast to expectations of scientists in the field, normal embryonic stem cells were generated from mouse embryos that had been cloned using the SCNT technique, and no major differences were observed when these cells were compared with the embryonic stem cells derived from normal mouse embryos. The SCNT embryonic stem cells could form normal chimeric mice when introduced into an early mouse embryo, and the mice that were born had germ cells (sperm and eggs) that were derived from the nucleus donor. As a possible explanation for the abnormalities in the human cloned embryos, it is thought that cloned embryos may have cells with more than one epigenetic signature and this could cause the death of some developing embryos. By contrast, during culture in the laboratory only the cells with a normal epigenetic signature stay alive.

It is, of course, important to realize that extrapolating results obtained with cells from mouse embryos to humans is not straightforward. Differences have been found between mice and cloned animals of other species so there is no reason to believe that humans and mice would be identical. For example, there are differences in the efficiency of the procedure and characteristics of the cells that make up the clone, such as the lengths of telomeres at the ends of chromosomes. Nevertheless, most knowledge on reproductive cloning comes from experiments with laboratory mice.

Although cloning was initially heralded as the future of medicine by some, others expressed serious doubts about the ethics of cloning in humans. This is because embryos have to be created and later destroyed to derive embryonic stem cell lines. It has been called by ethicists *instrumental use* of human embryos. In addition, concerns remain on the low success rate of cloning. In the United States, embryonic stem cell research was restricted anyway under President George W. Bush, so by definition so was cloning to derive stem cells, at least using state funding. In many countries, research on both human cloning and human embryonic stem cells are either restricted or illegal. This worldwide antipathy to human cloning for deriving specific donor embryonic stem cells, however, spurred research to discover alternative sources of human pluripotent cells. This search first focused on trying to find factors in the egg cytoplasm that could directly "reprogram" an adult cell so that it returned to an embryonic or pluripotent state. This approach has failed to date, even though we know that these factors must exist. However, some clever thinking by Shinya Yamanaka brought a revolutionary breakthrough: he and a student discovered a method for direct reprogramming, simply by introducing genes that code for the most important proteins in embryonic stem cells (see Chapter 3 "What Are Stem Cells" and Chapter 4 "Of Mice and Men: The History of Embryonic Stem Cells" for details).

Although the properties of these induced pluripotent stem (iPS) cells are still being explored, the generation of patient-specific pluripotent cells using this technique has been heralded as the solution to the ethical controversy surrounding therapeutic cloning: neither egg donation nor embryo creation are required. The way that the egg cytoplasm reprograms the adult genome remains unknown, and there is much curiosity about whether reprogramming by SCNT is similar to that taking place during the generation of iPS cells from somatic cells. Now that SCNT in humans has been successful and embryonic stem cell lines derived, this will undoubtedly be one of the questions that scientists will try to answer.

Irrespective of the source of pluripotent stem cells, tissue regeneration using a patient's own cells would not be possible for diseases that require acute treatment. Producing the pluripotent cells and differentiating them to obtain enough new cells for transplantation is a very slow process. Furthermore, if the disease is caused by a genetic defect, cell transplantation using a patient's own cells would introduce cells with the same defect, implying that some kind of gene repair would also be necessary for the transplanted cells to function normally.

6.3

DARWIN, MAX, AND SPARROW

In 1996 the world was taken by surprise at the birth of the sheep Dolly, the first cloned mammal, which was the result of the work of Ian Wilmut and Keith Campbell, then at the Roslin Institute in Edinburgh, Scotland. Dolly demonstrated for the first time that cells from adult mammals can be used to create a genetically identical animal with modern reproductive techniques. But Wilmut and Campbell were not the first to study cloning. Even in the 1960s and 1970s, scientists had tried to clone various animal species. The most successful in this regard was John Gurdon, then at Oxford University, who succeeded in obtaining adult frogs after nuclear transplantation of intestinal cells from tadpoles in 1962. Later he also succeeded in generating tadpoles using the same technique but cells from adult frogs, however, these tadpoles were not able to transform to adult frogs.

In this period the first experiments on *in vitro* fertilization in humans were conducted. On March 30, 1978, even before the first human "test-tube" baby was born, a remarkable book was published that came as a bombshell to the scientific community. The book entitled *In His Image: The Cloning of a Man* was written by David M. Rorvik and published by J.B. Lippincott Company, itself a well-regarded publisher of medical books. In the book, David Rorvik claims that he had helped in a project that was aimed at the cloning of an eccentric millionaire. David Rorvik was a well-respected freelance science journalist who had worked for *Time, Esquire,* and *The New York Times Magazine.* He had also written several books about modern reproductive techniques, among them the book *Brave New Baby,* published in 1972, about genetic modification, asexual reproduction, and test-tube babies.

In the book *In His Image,* Rorvik claims that in 1973 he was approached by a 67-year-old West Coast millionaire, known only by the code name "Max," who had read several of Rorvik's articles and asked Rorvik for help in producing a perfect genetic copy of himself. As motivation he said that he was aging but without a wife and desperately wanted a clone of himself as his heir. The man was prepared to pay millions to fulfill his dream, and Rorvik was asked to help because of his knowledge of the field and contacts in gynecology. He describes questioning whether it is scientifically possible to clone a human being and, if so, if this is ethically justifiable and whether he should then be involved. Eventually he decided to participate and help Max to locate a gynecologist, pseudonym "Darwin," who would conduct the experiments. They set off to a hospital on an undisclosed tropical island where Max owned rubber plantations and start experimenting in April 1974. After several attempts with different surrogate mothers, a 16-year-old island resident called "Sparrow" became pregnant and in December 1976 gave birth to a baby boy in an American hospital. According to the book, the baby is alive and healthy and tests confirmed that he is a genetic copy of Max. At that time, the first test-tube baby had yet to be born.

Already, before the book was published, it caused a sensation and was reported in nearly every newspaper in the United States. In an editorial of the March 24, 1978 issue of *Science,* it is reported that more than a dozen knowledgable researchers said "No" when asked whether the story could possibly be true, although the majority agreed that it would be possible in theory. As the book was published not as fiction but as the truth, a number of questions arose; for instance, why a scientist that had achieved this would like to remain anonymous rather than publish the results in scientific journals. Other scientists found the idea so preposterous that they refused to talk about it, afraid of generating free publicity. Rorvik kept to his story that the events described in the book were real, and the book became a bestseller.

6.3 *(cont'd)*

Several months after publication, Rorvik and his publisher were sued for defamation of character by J. Derek Bromhall, an embryologist from Oxford University. Bromhall charged that the book was a hoax, that the cloning technique described in the book had in fact been developed by Bromhall, and that Rorvik had used his name and technique in the book without permission. The court ruled that the book was "a fraud and a hoax" since Rorvik failed to demonstrate the veracity of his book. In April 1982, an out-of-court settlement between Lippincott and Bromhall was made for a payment of around U$100,000 to Bromhall. J.B. Lippincott Company conceded that they believed the story of the alleged boy clone was untrue, but Rorvik stood by his account of events. It is unclear why Rorvik would perpetrate such a hoax and go to such lengths to maintain that it is true. Maybe he wanted to stimulate the ethical debate around progress in biotechnological techniques. Fact or fiction, and partly due to the book, the social debate around cloning but also modern molecular and cell biological research deepened at the end of the 1970s.

6.2 CLONING PETS: SNUPPY, MISSY, AND COPYCAT

One application of reproductive cloning considered by some to have commercial potential is the cloning of pets. Many pet owners, it seems, would like to have an exact duplicate of their beloved cat or dog, particularly when the animal dies. But how realistic is this? What would be the cost of cloning a pet? And what would be the consequences? After the birth of the cloned sheep Dolly it was likely that mammals such as cats and dogs could be cloned. The American billionaire John Sperling thought along the same lines and in 1997, together with Bay Area entrepreneur Lou Hawthorne, decided to clone a dog. They coined the name "Missyplicity" for their enterprise, after the dog they wanted to clone, Missy, a mixed-breed Border Collie and Siberian Husky. Sperling and Hawthorne sought scientific expertise in Texas A&M University and invested approximately US$3.7 million in the Missyplicity project. However, dog cloning turned out to be much more difficult than imagined, not least because the ovulated eggs of dogs are not yet ready to be fertilized, in contrast to ovulated eggs of many other mammals. Instead, these eggs undergo a large part of meiosis in the oviduct, an environment

FIGURE 6.9 The beautiful coat pattern of a tortoise-shell cat has its origin in X-chromosome activation. When such a cat is cloned, her clone will have a different coat pattern since X-chromosome inactivation is a random process.

that is difficult to recreate in a test tube. Members of the Missyplicity project together with scientists from Texas A&M University founded Genetic Savings & Clone and diversified their research to other animals, particularly cats. In 2001 the first cloned cat, aptly named CopyCat or CC, was born from a calico, or tortoiseshell, donor. Although CC's donor was orange, black, and white, CC was black and white and lacked the orange, despite being genetically identical. The reason for this is that the coat color of cats is determined epigenetically (Figure 6.9). Outwardly therefore, CC looked completely different from her twin donor. Although an interesting outcome scientifically, Hawthorne was disappointed by the result since the commercial success of pet cloning would largely depend on the pet showing a strong physical resemblance to the original animal. In 2006 Genetic Savings & Clone ceased to exist; a scientific success but a commercial failure.

A new company, called BioArts International Ltd., was set up by Hawthorne, independent of Texas A&M University. Collaboration was sought this time with the South Korean Sooam Biotech Research Foundation, with which Woo-Suk Hwang was associated as part of the research team. Missy had died in 2002, but some of her cells had been frozen and were flown to the Hwang team in Korea. In December 2007, Mira, the first clone of Missy, was born. Based on this success, an auction was organized where in a five-day period customers could bid for their dog to be cloned, but only the five highest bidders would be offered this opportunity. The auction raised approximately US$750,000 and indicated a potential market for dog cloning. All five dogs were successfully cloned. However, when BioArts International announced their Golden Clone Giveaway contest on May 30, 2008, where the prize was a free clone of the winner's dog, only 237 people signed up. This low number demonstrated the extremely limited market for a dog clone on demand, even when it was done for free, let alone when it would cost tens of thousands of dollars. BioArts International, therefore, decided to withdraw from dog cloning as a business.

To have your dog cloned, South Korea is still the place to be, however. South Korean Byeong-Chun Lee from Seoul National University, a former colleague of Hwang and part of the "Snuppy team," now works with the company RNL Bio. He has cloned a series of dogs, including a drug-sniffing dog because it was so outstanding and unique in its job. In 2008, the team presented its first commercially cloned Pitbull Terrier, "Booger," which had cost the owner an estimated US$50,000.

Despite the limited market, Sooam BRF announced on 22 March 2013 the first dog cloning competition for the United Kingdom. Just as the giveaway contest of 2008, applicants could send in a reason why their dog should be cloned. This time, however, the competition was restricted to residents of the United Kingdom. The winner has been promised that his or her dog will be cloned for free, provided that the winner agrees to participate in a documentary about the process.

Aside from any commercial application, dog cloning could help in providing new scientific knowledge. Several dog breeds suffer diseases similar to those in humans, and these dogs might be useful as animal models for human conditions. The availability of genetically identical dogs would eliminate the confounding factor of genetic diversity between individual animals in the disease studies. Moreover, dogs can be valuable for studying brain function and behavior, for instance. Again, cloning groups of dogs could circumvent the common problem of genetic variation normally present within dog groups. Pet cloning is controversial, however. Not only because dog cloning at Sooam is led by the once-disgraced Dr. Hwang, but, more importantly, because the costs are enormous, while pet shelters continue to become fuller each year.

6.4

THE CLONING OF DOLLY

Interview with Ian Wilmut

How did you come to clone a sheep?

Our objective was to be able to improve the health and productivity of farm animals. Cattle are by far the most important of farm animals, but they are extremely expensive. We had a great deal of experience of recovering and culturing sheep embryos and knew that embryo development in sheep is very similar to that in cattle. So we chose to work with sheep because they are cheap and we were confident that methods developed in sheep would be readily adaptable for the cow. That has, in fact, proved to be the case.

How did you feel when you realized Dolly's biological mother was pregnant?

Of course, we were very excited when we discovered that Dolly's mother was pregnant. She was approximately six weeks into pregnancy when ultrasound scanning first showed that she was carrying a lamb. However, previous experience had shown us that a considerable proportion of the fetuses that were present at that stage of development died during pregnancy or at the time of birth. So we were extremely cautious in our expectations from the time of the first scan until a few weeks after her birth when we became confident that Dolly was healthy and viable.

What did the birth of Dolly change in science?

The birth of Dolly provided important new understanding of the mechanisms that regulate mammalian development. Her birth showed that the mammary cell used in the cloning process contained all of the genetic information necessary to produce a viable offspring. Earlier, researchers had suggested that cells formed different tissues by losing segments of the chromosomal DNA that were not required for the functioning of a particular tissue. Clearly this was not the case (Figure B6.4.1).

The current hypothesis is that differentiation to form all the tissues of an adult is brought about by a sequence of changes in gene expression, which become progressively more specific for the final tissue. Before the birth of Dolly it was believed that the mechanisms that regulate these changes are so complex and so rigidly fixed that they cannot be reversed by the process of nuclear transfer. It was suggested that this was the reason why it had not previously been possible to produce a clone from an adult animal. The birth of Dolly

6.4 *(cont'd)*

clearly showed that this was not the case and has led to very important research to find methods that enable us to produce cells that closely resemble embryo stem cell from adult tissue. These so-called "induced pluripotent cells" will have a very profound effect in biomedical research. They are the most important outcome from the Dolly experiment.

FIGURE B6.4.1 Sir Ian Wilmut and Dolly. *Source: Roslin Institute, Royal (Dick) School of Veterinary Studies, University of Edinburgh, Edinburgh.*

What impact did Dolly have on you your life and career?

The project was led by Keith Campbell and I, and the birth of Dolly transformed our lives. We have both had greater opportunities to develop our careers following the birth of Dolly. Keith became a professor at the University of Nottingham and I became director of a research center in the University of Edinburgh. The other effect of the experiment is that we have both given an enormous number of interviews to radio, television, and newspapers.

6.3 JUST IMAGINE WHAT COULD BE

For the imaginative science fiction fan, cloning offers a plethora of interesting scenarios. Indeed many books and movies have already been made using cloning as a theme, most of them portraying the dark or absurd side of cloning (Figure 6.10). Cloning of endangered or extinct animals has also spurred the imagination of many science-fiction

FIGURE 6.10 Cloning has been an inspiring subject for both popular and scientific authors.

writers, movie-makers, and scientists alike. Is there a future for so-called conservation cloning? Several examples exist of endangered, even extinct, animal species that have been cloned, although the success of these cloning efforts in terms of efficiency is questionable. Cloning an endangered or extinct animal is in theory straightforward—transfer the nucleus of the animal to be cloned to an egg of the same or a similar, but more common, species—but the actual process itself is extremely complex. For animals already extinct, how and where to obtain (intact) donor cells and which egg cells are similar enough to the donor to accept the new nucleus are far from trivial questions. Then there is the difficulty of deciding which surrogate mothers would be best to implant the cloned embryo so that it can go on and develop in the uterus until birth. For endangered species, cells, eggs, and surrogate mothers would be rare; for extinct species they would be nonexistent.

The first critically endangered mammal to be cloned was the gaur (*Bos gaurus*), a species of formidably sized wild cattle that lives in Asia but is on the verge of extinction. Donor skin cells were taken from a five-year-old male gaur at autopsy, cultured for a few days, and subsequently frozen and stored. Eight years later the cells were thawed for cloning by SCNT. The eggs used for SCNT were not from the same species, but were instead isolated from abattoir ovaries of common domestic cows (*Bos taurus*). Forty cloned embryos were transferred to the uterus of recipient domestic cows. Only one calf was carried to term in 2001. Genetic analysis confirmed that the animal was a true gaur clone. Unfortunately the gaur clone died after only two days.

The Pyrenean ibex, or bucardo (*Capra pyrenaica pyrenaica)*, became extinct when the last surviving female died in 2000, allegedly crushed to death by a falling tree. Luckily, before her death, this animal had already been identified as the last remaining example of her species and skin samples had been taken and frozen (Figure 6.11). A research team of Spanish, French, and Belgian scientists took on the task of trying to bring this species back into existence. Since no bucardo eggs were available, the scientists again had to rely on eggs from a closely related animal, in this case eggs from domestic goats (*Capra aegagrus hircus*). As a recipient mother, a hybrid between a Spanish ibex (*Capra pyrenaica hispanica*) and a domestic goat was used. A total of 782 embryos were "reconstructed" of which 208 were transferred to surrogate mothers. In only one recipient did pregnancy continue normally. One of the difficulties of these kinds of cross-species techniques is the duration of pregnancy. The gestation times of animals that became extinct long ago are not known, but even when they are known, this period will most usually not match that of the species used as surrogate mother. Based on the gestation time of the Spanish ibex, a caesarean section was performed after 162 days of pregnancy on the goat carrying the cloned bucardo, and a baby bucardo was delivered that initially appeared normal. Disappointingly, this animal died within a few

FIGURE 6.11 When the first extinct mammal, a Pyrenean ibex, was cloned, the egg cells were retrieved from normal domestic goats.

minutes of birth. Nevertheless, a species that had been extinct had been resurrected for the first time, only rapidly to become extinct yet again. Many questions can be asked here. Was this another scientific breakthrough in cloning? A step forward in species conservation? Or was it ethically inappropriate and animal cruelty?

Even if the bucardo clone had been viable and healthy, it would have been virtually impossible to obtain a healthy population again since only cells from female animals had been rescued and frozen. The information on the bucardo's Y chromosome was lost, and any population generated later would have had to be generated from crossing the pure bucardo with males of the related ibex species. A new species would have been created, but the old species not rescued. Other endangered animals that have been successfully cloned by cross-species nuclear transfer include the African wildcat (*Felis silvestris lybica*) born from a domestic cat, coyote (*Canis latrans*) using oocytes from a domestic dog, and mouflon (*Ovis orientalis musimon*) using sheep (Figure 6.12).

One animal species that became extinct approximately 10,000 years B.C., which has been particularly avidly and imaginatively discussed as possibly "clonable" is the woolly mammoth (*Mammuthus primigenius*). The reasons for the interest in cloning a mammoth are obvious: in contrast to other long-extinct animals, the woolly mammoth is not known from fossilized imprints. Instead, rather complete and virtually intact animals are present in the Siberian and North American permafrost. This offers the possibility of retrieving cells with intact DNA from these carcasses. It has even been suggested that mammoths contained a kind of natural antifreeze that would help them to endure the extreme cold

FIGURE 6.12 The African wildcat (*Felis silvestris lybica*) from Namibia is one of the few "wild" animal species that have successfully been cloned. Domestic cats have been used as oocyte donor and embryo recipients.

the animals had to face and may have acted as a natural preservative for the cells. In 2013, Russian researchers reported finding a mammoth carcass at the Novosibirsk Islands that had muscle tissue with the color of fresh meat and even a brown liquid claimed to be blood, despite the subzero temperatures. Also, the mammoth has present-day close relatives, elephants, which could be used as egg donors and surrogate mothers. The remote possibility of bringing this epic animal back to life has nevertheless spurred the enthusiasm of a number of scientists, including those at Korea's Sooam.

In the meantime, Japanese scientists at the RIKEN Center in Kobe have cloned mice from cells that had been stored and frozen in a regular freezer for 16 years without using any specific reagents for protecting the cells against frost damage. In particular, brain cells were demonstrated to be suitable, possibly because of the high sugar levels in the brain, which could work as a natural antifreeze. The fact that around 80% of the mammoth genome has been sequenced from frozen hair shaft cells supports the idea that freezing under low temperature weather conditions might be sufficient to preserve cell viability for cloning purposes.

Other problems do need solving, however, before the mammoth can be resurrected: how would eggs from elephants be collected (based on size and phylogeny the Indian elephant *Elephantus maximus* seems the most appropriate), how would the womb of a surrogate elephant mother be prepared to enable the cloned mammoth embryo to implant and develop, and, finally, how would the cloned mammoth embryos be transferred into the elephant's womb (the distance from the vagina to the uterus in elephants is about 2.5 meters!) (Figure 6.13)? Of course,

FIGURE 6.13 Can the Indian elephant be used as a recipient mother for a cloned woolly mammoth embryo?

one could also question the value of the whole project; why clone a mammoth, except for ghoulish reasons and, perhaps, very understandable human curiosity?

6.4 CLONING DOMESTIC LIVESTOCK

The first mammal that was cloned from an adult cell was, of course, a sheep, but many other livestock have later been duplicated by SCNT, including species of agricultural importance such as cattle, pigs, and goats. The reason for cloning domestic livestock is not primarily to produce unlimited copies of, say, high milk production dairy cows, but rather to make genetic copies of a few elite animals to use as breeding stock (Figure 6.14).

Particularly in the cattle breeding industry, selection of stud bulls is a long and costly process. It usually takes several years before the progeny from potential stud bulls have been tested to see whether they are suitable for reproduction in high-end breeding. Cloning of bulls that had already been carefully selected would save considerable time. In principle, one could even make a clone of the clone to maintain an animal with the best characteristics, but this would not be of much commercial interest because of the constant need for genetic improvement.

In cattle, the birth weights of cloned calves are typically 40% greater than those of normal calves, but the length of the telomeres appears to be that of age-matched control calves. This means that the idea that

FIGURE 6.14 Extensive research has found no differences between meat and milk products from normal and cloned animals.

cloned animals might already be "old" at birth (i.e., just as old as the donor cell) is probably incorrect (see next section for more details). The fertility of cloned cattle appears to be similar to that of control animals, and offspring from cloned animals usually have normal birth weights. Although animals would not be cloned directly for meat or milk production, cloned breeding animals that have outlived their reproductive utility, as well as their offspring, may deliver both milk and meat to the food chain. Concern has been voiced over the safety of food products from clones and cloned progeny for human consumption. Both the U.S. Food and Drug Administration (FDA) and the European Food Safety Authority (EFSA) have made recommendations not only on the health and welfare of the clones themselves, but also on the safety of derived food products and the potential impact of animal clones on the environment. Both the FDA and the EFSA have advised that food derived from cloned cattle, swine, and goats or their progeny is as safe to eat as food derived from conventionally bred animals. However, despite the appearance of being normal, evidence has been accumulating that not only the cloned sheep Dolly was abnormal but all cloned animals show a higher incidence of developmental anomalies than animals resulting from, for example, artificial insemination. In a direct comparison of meat and milk characteristics of cloned animals and matched comparator animals, no differences were found in composition of milk. Also, of over 100 meat parameters analyzed, the large majority did not differ from those of control animals and all parameters were within the normal range of those detected in meat approved for human consumption. Although it is still unlikely that consumption of food products from cloned animals or their descendants has human health implications, the EFSA has recommended continued research in this area.

6.5 CLONING CHALLENGES

Animal cloning by somatic cell nuclear transfer is a highly inefficient process. Depending on the species being duplicated, less than 0.1% or up to maximally 5% of the attempts results in the birth of a viable clone. There are many known and unknown reasons for this inefficiency. The cloning procedure, for example, bypasses several critical events that take place after natural fertilization, because the egg is activated by a tiny electric shock or exposure to a chemical able to mimic the effects of fertilization and not by sperm entry. The age of the cell used for cloning is also important.

Another question that always arises when discussing clones is: how old is a cloned animal? Is it the same age as the cells from which it is derived, or is its age "reset" during the cloning procedure? One of the telltale signs of cellular age are the lengths of the so-called telomeres: these are the DNA caps at the end of the chromosomes that shrink with each cell division. The older the cells are, the shorter their telomeres. When the telomeres of Dolly the sheep were examined, it was found that they were shorter than normal: at one year of age the lengths of her telomeres were comparable to those of a six-year-old sheep, suggesting that Dolly might age unusually quickly. Later experiments established that in most cases telomeres in cloned animals are usually restored to a normal length, and the reason why Dolly was exceptional is not known. She was euthanized at six years of age because she suffered from a virally induced lung cancer and, unfortunately, there is no treatment for this disease. Sheep of Dolly's strain normally live up to 11 years. Interestingly, Dolly was actually cloned using a cell from a six-year-old animal so she might indeed have been effectively six years old at birth! However, the jury is still out on whether Dolly's diseases were the result of, or contributed to, premature aging, particularly since other aspects of her health appeared to be quite normal. Cloned mice, by contrast, seem to live as long as normal mice, and, once born, do not appear to be abnormal with the exception of tending to be obese. Despite this, both Rudolf Jaenisch, well known for his work cloning mice, and Sir Ian Wilmut, the "grandfather" of Dolly, have said "there will never be a completely normal clone."

One way to try to understand the biology of cloned animals is to make serial clones over generations: make a clone using a somatic cell from a previously cloned animal, and so forth. One could imagine, perhaps, that cloning would become more efficient over generations because of the selection of cells that are more easily reprogrammed. Conversely, one can also imagine the opposite: that recloning would become more inefficient because of the accumulation of genetic or epigenetic mistakes. Serial cloning has been attempted in pigs, cattle, and cats but with little success, only the third generation was reached. Japanese researchers have had more success with mouse cloning. Initial attempts at serial cloning were rather unsuccessful and, similar to that found with other species, the cloning efficiency, low as it was, decreased with every generation after which a point was reached were the animal seemed "unclonable." However, in 2013 these researchers succeeded in recloning mice for up to 25 generations. Importantly, these mice seemed completely normal, were fertile, and had life spans similar to those of normally conceived mice. The reprogramming efficiency did not change over generations. This "world record of cloning" has not been set to enter the *Guinness Book of World Records* but to provide us with valuable information about the biology underlying nuclear reprogramming.

6.5

IN MEMORIAM: KEITH H. CAMPBELL (1954–2012)

In October 2012, the Nobel Committee awarded the 2012 Nobel Prize in Physiology or Medicine to John Gurdon and Shinya Yamanaka for their work on the reprogramming of mammalian cells. In a bizarre twist of fate, Keith Campbell, who played a major role in creating Dolly the cloned sheep, had died a few days earlier. It was the cloning of Dolly that unequivocally demonstrated that differentiated adult cells can be reprogrammed to a pluripotent state.

Keith Campbell was born in the spring of 1954 in Birmingham in the United Kingdom. Having a keen interest in biological processes he went to study microbiology in London and obtained his bachelor's degree in 1978. He subsequently studied the development of frog eggs and yeast cell division at the University of Sussex and was awarded a PhD degree. In 1991, Keith joined the Roslin Institute near Edinburgh in Scotland, where he started working with Ian Wilmut. Together, they

FIGURE B6.5.1 Keith Campbell (23 May 1954–5 October 2012). *Source: András Dinnyés.*

studied cellular differentiation and mammalian cloning. Campbell realized that for successful cloning by somatic cell nuclear transfer, the donor cell and the egg cell had to be at the same state of the cell cycle. This could be achieved with a simple method: deprivation of the donor cells of important nutrients (called starvation), which left the cells in a dormant state very similar to the state of an unfertilized egg cell. Dolly was born in 1996 and our ideas on cell biology and differentiation were forever changed.

Keith Campbell left the Roslin Institute in 1997 to become head of embryology at PPL Therapeutics, a biopharmaceutical company, to produce cloned transgenic animals whose milk contained pharmaceutical products or whose organs could be used for transplantation. In 1999, however, Campbell returned to academia and became professor of animal development at Nottingham University where he stayed until his death. The importance of his work was acknowledged by the prestigious Shaw Prize in Life Science and Medicine awarded to Shinya Yamanaka, Ian Wilmut, and Keith Campbell in 2008 (Figure B6.5.1).

CHAPTER

7

Regenerative Medicine: Clinical Applications of Stem Cells

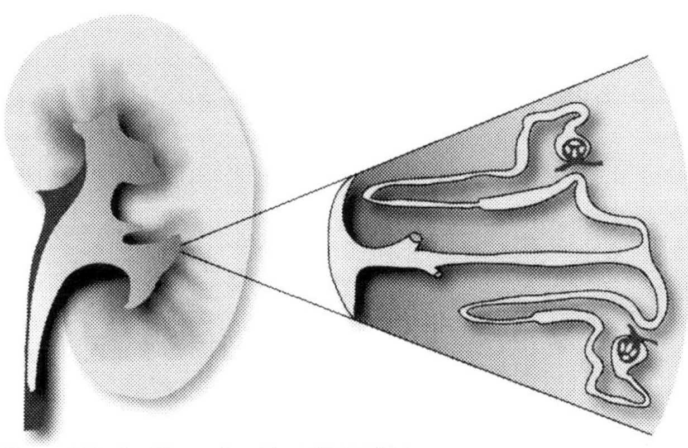

Bernard Roelen/Stamcellen *Veen Magazines*

OUTLINE

There is sometimes confusion about the difference between stem cell medicine and regenerative therapy in considering clinical applications of stem cells. There is a widely held perception that stem cell medicine is the same as (stem) cell therapy, which is actually not the case. While cell therapy (meaning a therapy with cells) is certainly a useful approach to treating some diseases, in others, especially those involving tissue and organs with an inherent regeneration potential, a preferable approach may, in fact, be to pharmacologically activate the natural regeneration process, mediated by adult stem or progenitor cells in the damaged organ itself. This "heal thyself" mechanism may even be part of the reason that some cell therapies work and others do not. To make matters more complicated, articles on the potential therapeutic use of stem cells in the popular press often do not explain which of the different stem cell types have been tested in a particular therapy, or, in the case of preclinical research, whether animal or human cells were used. Further, reporting and opinions are sometimes colored by the ethical debate on the use of embryonic versus adult stem cells.

For some diseases, recent developments in the stem cell field created such high expectations among patients and their families that their hope has developed into media hype. We explain in this chapter how the different mechanisms mentioned are benefitting (or in the future will benefit) patients with many different sorts of degenerative disease and injury. However, we view the field with a certain degree of caution, tempering the hype while at the same time reflecting with optimism on the potential that future generations may have to benefit from stem cell treatments. These may help them retain good health in later life, with "healthy aging" rather than endless youth as a realistic goal. Specific examples are used to illustrate the obstacles that will need to be overcome before new therapies can be tested clinically, and the reasons why stem cell biologists and clinicians believe that some diseases will be treatable relatively soon, but other diseases may require a much longer wait. Most importantly, translational stem cell medicine will rarely be achievable by a conventional team of researchers led by one chief clinician or scientist, because stem cell therapies are far too complicated for (even pioneering) individuals to succeed alone.

7.1

PROFESSOR ANTHONY HOLLANDER: A TISSUE ENGINEERING PIONEER

Professor Anthony Hollander, Arthritis Research Campaign Professor of Rheumatology and Tissue Engineering at the University of Bristol in the United Kingdom, has been working in the field of cartilage biology and arthritis research for nearly 20 years (Figure B7.1.1). His major research focus has been repair of knee injuries, but his most well-known "claim to fame" was the first tissue-engineered trachea (windpipe), utilizing the patient's own stem cells. In 2008, this engineered windpipe was successfully transplanted into a 30-year-old Spanish mother of two with a failing airway, following a severe case of tuberculosis. She was admitted to hospital with acute shortness of breath that made her unable to carry out normal everyday tasks, and she was slowly dying. The bioengineered trachea immediately provided the patient with a normally functioning airway, thereby saving her life (Figure B7.1.2).

A 7 cm tracheal segment came from a 51-year-old transplant donor who had died of cerebral hemorrhage. Spain has a policy of assumed consent for organ donation. Using a new technique developed in

7.1 *(cont'd)*

FIGURE B7.1.1 Professor Anthony Hollander, a cartilage tissue engineer, who cultured stem cells from Claudia Castillo to build a new windpipe (trachea). *Source: Melissa Temple-Smith.*

FIGURE B7.1.2 Claudia Castillo, who received a transplanted windpipe that was engineered from her own stem cells. *Source: Hospital Clinic of Barcelona, Spain.*

Padua University, the trachea was decellularized over a six-week period so that no donor cells remained. Stem cells were obtained from the recipient's own bone marrow, grown into a large population in Professor Martin Birchall's laboratory at the University of Bristol, and matured into cartilage cells (chondrocytes) using an adapted method originally devised for treating osteoarthritis by Professor Hollander.

The donor trachea was then seeded with chondrocytes on the outside, using a novel bioreactor developed at the Politecnico di Milano, Italy, which incubates the cells and allows them to migrate into the tissue under conditions ideal for each individual cell type. To replicate the lining of the trachea, epithelial cells were seeded onto the inside of the trachea using the same bioreactor. Four days after seeding, the graft was used to replace the patient's left main bronchus. The operation was performed in June 2008 at the Hospital Clinic, Barcelona, by Professor Paolo Macchiarini of the University of Barcelona.

These remarkable results show that it is possible to produce a tissue-engineered airway with mechanical properties that permit normal breathing and that is free from the risks of rejection seen with conventional transplanted organs. The patient did not develop antibodies to her graft, despite not taking any immunosuppressive drugs. Lung function tests performed two months after the operation were all at the better end of the normal range for a young woman.

Professor Macchiarini, lead author on the paper published in the top medical journal *The Lancet*, said: "We are terribly excited by these results. Just four days after transplantation, the graft was almost indistinguishable from adjacent normal bronchi. After one month, a biopsy elicited local bleeding, indicating that the blood vessels had already grown back successfully."

Professor Hollander concurred: "This successful treatment manifestly demonstrates the potential of adult stem cells to save lives." Since this pioneering procedure was carried out, a number of other patients have been treated in the same way. In addition, studies in Sweden have shown that it is actually possible to engineer the trachea itself without requiring a donor trachea at all. These completely artificial windpipes containing the patient's own cells have already been tested clinically with success.

Interview with Professor Anthony Hollander

You supervised the culture of the cells and played a vital role in engineering the trachea. How did you feel when the cells were ready to be transported to Spain?

We made the decision to go ahead and engineer a trachea in late March 2008, and by early June we had grown all the cells needed for the procedure. It was a relief to have completed that phase of the work and exciting to be ready for the final phase of engineering. We had to transport the cells to Spain and made arrangements for them to be taken on board a direct flight from Bristol to Barcelona by a young German thoracic surgeon working with Paolo Macchiarini.

7.1 *(cont'd)*

Unfortunately, on the day, the airline refused to allow the cells on board because they were being carried in a liquid and there was more than 100 ml volume, contravening antiterrorist measures. However, the German surgeon and my colleague Martin Birchall were able to rescue the situation by chartering a small private plane flown as a hobby by another thoracic surgeon living in Germany. They saved the day!

How did the public react to the report of the success of you and your team?

The public and press were completely supportive and I was astounded by the huge appetite for information about the science behind the clinical advance. I knew that the saving of a life would be of obvious human interest, but had not predicted the high level of interest in the stem cell biology. I felt at the time that we had made a shift from stem cell science to stem cell medicine, and was delighted that my comment on this aspect of the story at the press conference was picked up and used by the journalists. They did a great job in conveying the historical importance of our team's work and the public played their part too. That public interest continues today, and is reflected in the decision by the Science Museum in London to include a description of the project in their newly revised permanent exhibition. This high level of public interest in a stem cell breakthrough is a sign of a very healthy society that is open to new ideas and I am delighted by this enlightened attitude in twenty-first century Britain.

What would you think will be the next breakthrough in the area of cartilage tissue engineering?

Most cartilage tissue engineering work is focused on developing treatments for injured knee cartilage and for treating patients with early osteoarthritis. I expect that we will see a number of clinical trials using adult stem cells (most likely those isolated from the patients' own bone marrow). I doubt if the advances will be as rapid or as high-profile as the tracheal tissue engineering; the aim will be to improve quality of life rather than to save lives and so the outcomes will be rather less dramatic. But I am convinced that this area of stem cell biology/medicine can have a major positive impact for many thousands of patients, so long as we can find effective ways of integrating engineered cartilage implants with the natural tissues of the joint. At the same time, work will continue on tracheal tissue engineering and finding ways of making the procedure simpler and cheaper so that it can be used around the world.

7.1 THERAPEUTIC CELL TRANSPLANTATION

Most people are familiar with the concept of organ transplantation. Best known is probably kidney transplantation, where patients with diseased, nonfunctioning kidneys receive a healthy kidney from another person (Figure 7.1). If the transplantation procedure is successful, the patient can be spared the biweekly dialysis in the hospital, or at home, that artificially replaces renal function. Heart transplantation was among the first life-saving organ transplants that took place. This pioneering work of Dr. Christian Barnard in South Africa led the way to a better understanding of the immune system and immune rejection, so that multiple other organs can now be transplanted almost routinely: the lungs, liver, pancreas, skin, and cornea are all examples of tissues and organs that can be transferred from donor to patient.

Even though transplantation of tissues such as the cornea might not be life-saving, it can restore eyesight and therefore have a large impact on the quality of life. In most cases organs and tissues cannot be used from living donors but have to be donated by individuals or their families after death. As an exception, the kidney—since we have two kidneys—can be transplanted from a living donor; this happens especially within families where chances for good human leukocyte antigen (HLA) tissue matching (see section 7.10.1 "How the Body's Immune System Recognizes a Cell as Foreign") are best, but there are an increasing number of altruistic donors who donate their kidney anonymously. In general, however, the supply of organs for transplantation is much lower than the demand, so that waiting lists for organs worldwide are very long, with many hundreds of patients dying annually before an organ becomes available.

Cell transplantation has been advocated as a promising therapeutic alternative for the future. Instead of whole organ replacement, relevant cell types could, in principle, be transplanted into the damaged organ

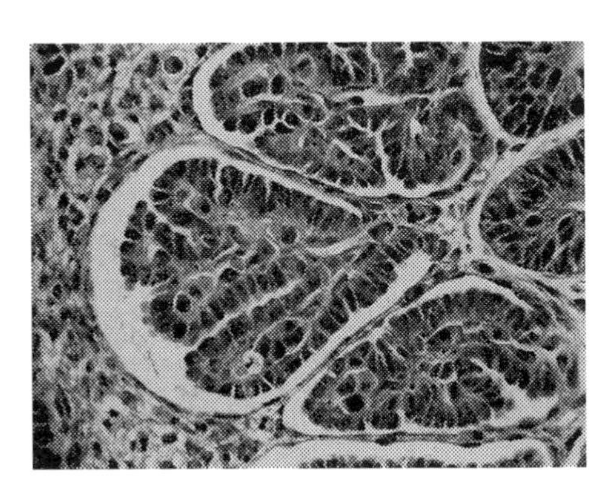

FIGURE 7.1 Complexity and diversity of the cell types in some organs are clearly visible under the microscope as in this tissue slide of a kidney from a horse embryo. *Source: Stamcellen Veen Magazines.*

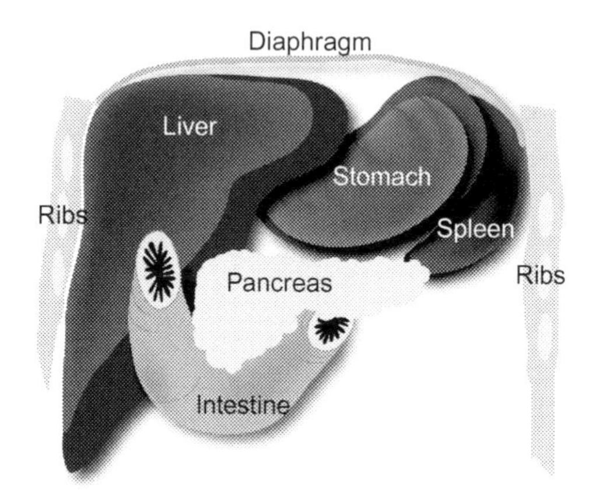

FIGURE 7.2 Today's medicine uses organs of deceased donors to replace damaged or diseased organs. Stem cells may offer an alternative to organ transplantation in the future. *Source: Stamcellen Veen Magazines.*

to replace cells that are dysfunctional or have been destroyed, and thus restore function. This is the principle underlying one aspect of regenerative medicine. These therapeutic cells could be delivered directly to the place in the body where they are required, for example, in the heart muscle. Alternatively, simple intravenous administration through a blood vessel would, in some cases, be sufficient, since many cells have an intrinsic capacity to find (or "home to") the tissue or organ to which they belong, especially if that tissue or organ is still attempting to repair the damage, as, for example, following a heart attack. Prerequisites for such cell-based treatments to become routine clinical practice (aside from evidence that they are safe and actually work) would be that the cells are available in sufficient quantities to actually treat the patients, and that they are available for repeated treatments in case the transplanted cells lose their function over time, for example, due to slow rejection by the immune system. Finally, the costs to the health-care system should be sustainable (Figure 7.2).

Depending on the disease application, the stem cells may have to be first differentiated to the specific cell type needed for tissue or organ repair. For example, for the brain, it would likely be necessary to turn stem cells into nerve cells of the right type before transplantation.

7.2

A BANK OF ARTIFICIAL HUMAN SKIN?

In the case of extensive third-degree burns covering a large part of the body, it is of utmost importance that the open wounds be covered as quickly as possible to prevent both acute loss of fluid from the wound itself, and also later infection. This is sometimes done by

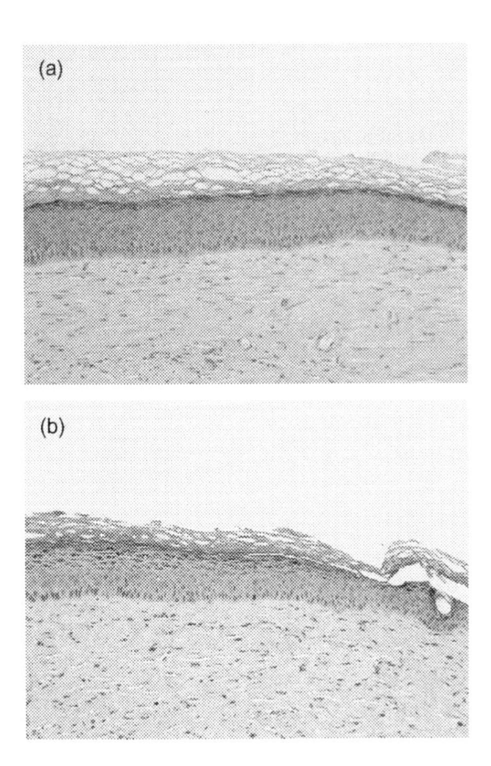

FIGURE B7.2.1 (a) Human epidermis from donor keratinocytes, grown as a xenograft on mice. (b) Human embryonic stem cells were differentiated to keratinocytes and the artificial skin implant was grafted on to mice, as in (a). The artificial skin implant derived from the stem cells is indistinguishable from the skin derived from normal human keratinocytes and very similar to normal human skin. *Source: Christine Baldeschi, I-Stem, Evry, France and Marcelo Del Rio, CIEMAT, Madrid, Spain. With thanks to Marc Peschanski, I-Stem, for input.*

submerging patients in a saline bath or transplanting skin from animals, usually pigs, or human donors on to the wound. However, saline baths cause much suffering and animal (or human) skin grafts are only temporary solutions, since the grafts are rejected by the immune system fairly rapidly. Growing cells from the patient's own residual skin and using it to create their own artificial skin by tissue engineering is a good alternative, but it is a slow solution (it takes a fair amount of time to grow enough cells) while the need is often acute.

A new option in development is to turn human pluripotent stem cells into artificial skin that can, in principle, be "banked" to provide a stock of skin ready for use after "matching" with the HLA type of

7.2 *(cont'd)*

the patient. Human pluripotent stem cells can now be turned into cells very like a skin cell called a *keratinocyte*, which is the main cell in the skin that gives rise to most of the other cellular components. By growing the cells at a culture-fluid–air interface, they can form multilayered structures very like human skin, which is rather different from that of most animals. These have already been successfully transplanted onto the skin of mice and survived well for several months. It has even been possible to produce cells called *melanocytes* from human embryonic stem cells. These are the cells that produce the brown pigment called melanin that is produced in response to sunlight and is transferred to the other skin cells where it protects them from damage by ultraviolet light. It may then be possible in the future to produce human skin grafts able to "tan" in the sun so that they are not acutely sensitive to sunlight.

The research institute I-Stem in France is a world leader in this area of translational research. Its ultimate intention is to produce skin on an almost industrial scale from adult skin cells from individuals with a wide range of HLA immunotypes, or even through appropriate differentiation of human induced pluripotent stem cell lines with similarly different HLA types, so that skin will be available for treatment of acute injury where the need for a rapid solution with a near human skin match is of utmost importance (Figure B7.2.1).

7.2 NUMBER OF CELLS NEEDED FOR CELL TRANSPLANTATION

Transplantation of cells for the purpose of regenerative medicine may require just a few million cells or even less, but in many cases (e.g., to repair large areas of tissue damage) a huge number, up to a few billion, of appropriately differentiated and well-characterized cells is likely to be necessary. For example, after a severe myocardial infarction, up to 10^8 to 10^9 cells may be lost and would need to be replaced to fully restore the ability of the heart to contract properly. Fewer cells might need to be transplanted if, for example, new blood vessel cells could be simultaneously transplanted to generate an improved blood supply. Also, fewer cells might be needed if they could somehow activate the natural repair or regeneration processes in the damaged tissue, even perhaps by recruiting stem cells already present to the site of injury.

The number of cells probably also depends on the particular cell type to be transplanted, the organ or tissue location, and the disease for which the patient is being treated. It is likely that fewer cells are needed when the transplantation is *autologous*, meaning the cells are derived from patients themselves, versus *allogenic*, when the stem cells are from a donor and would, even if immunologically "matched," still be prone to some rejection by the patient's immune system.

7.3

NEW SKIN FOR CLAUDIO

Dr. Michele De Luca from the University of Modena and Reggio Emilia in Italy studies the use of human epidermal stem cell cultures in the treatment of massive full-thickness burns and skin depigmentation. He is also investigating the use of epithelial stem cell mediated gene therapy for junctional epidermolysis bullosa, a serious genetic skin disease. Claudio, who is 42 years old, is one of his patients with this debilitating skin complaint, which is caused by a mutation in a gene called laminin 5. This gene codes for a protein important for sticking layers of the skin together. If it does not function properly, then the skin peels off in large, chronic blisters. Claudio has suffered all of his life from huge open wounds over the whole of his body. Although it is possible to make new skin from a patient's own epithelial skin stem cells, in Claudio's case this would not help, since he has a genetic defect and any new skin made from his own cells would have the same problem as Claudio's own skin.

Dr. De Luca has attempted to solve this by combining the epithelial skin stem cells with gene therapy. He has introduced a good copy of the laminin 5 gene into the defective skin stem cells and used these to make new skin by tissue engineering techniques, growing the cells on a carrier material. These sheets of genetically repaired skin have been transplanted to both of Claudio's upper legs and have attached well, giving him a large area of reconstructed skin. He has no pain and no open wounds in these grafts, which have remained stable for several years, without any sign of tumor formation or other negative side effects. Two technologies, gene therapy and stem cell technology, have been successfully combined in one treatment. Unfortunately for Claudio, EU legislation set in place to reduce the potential risks of stem cell therapies by categorizing them as "drugs" requires that many new safety criteria be addressed, and his further treatment has to wait

7.3 *(cont'd)*

until these new guidelines have been met. For him and other patients like him, however, there is at least hope on the horizon. Five years after the first transplantation, the skin remained as healthy as it had during the first year (Figure B7.3.1).

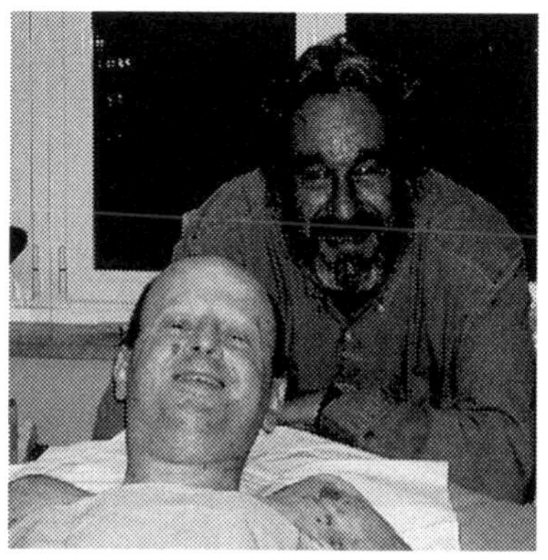

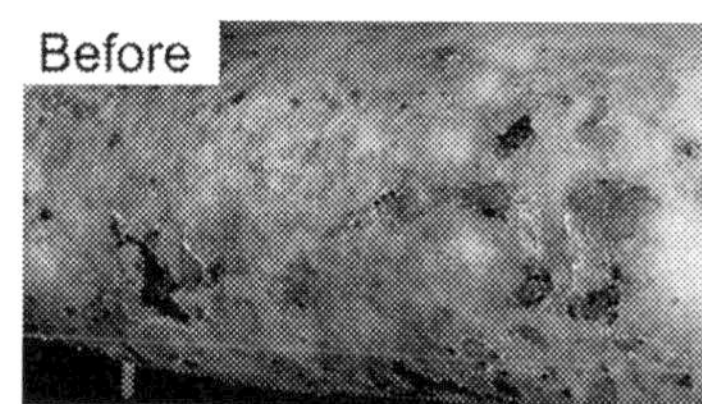

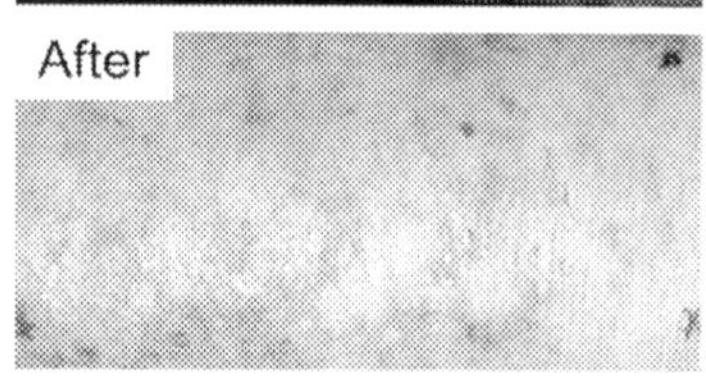

FIGURE B7.3.1 (a) Dr. De Luca (right) together with Claudio. (b) Skin at the top of Claudio's leg before and after transplantation with reconstructed skin. The cells used were Claudio's own but their genetic defect had been repaired by gene therapy. Five years later, the skin had remained healed. *Source: Michele De Luca, Centre for Regenerative Medicine "Stefano Ferrari", Modena, Italy.*

7.3 WHY SOME DISEASES WILL BE TREATABLE WITH STEM CELLS IN THE FUTURE AND OTHERS NOT

Not all diseases are equally well suited for stem cell treatment. The purpose of stem cell transplantation is to replace damaged or lost cells in an organ or tissue. Most organs contain their own small reservoir of stem cells, also called progenitor cells, as mentioned, which can be recruited to start dividing and replace cells that have died during normal aging, or to repair small areas of damage. Although small lesions can be rapidly and completely repaired by progenitor cells, these are usually present in insufficient numbers to compensate for more extensive damage. In the latter case, a significant number of cells are irreversibly lost and the function of the organ or tissue becomes compromised, a situation that can be rapidly life-threatening when an important organ, such as the heart, is affected. In such cases, under certain defined criteria, stem cell transplantation could potentially be life-saving. In the example of the heart, following a myocardial infarction, predominantly one cell type, the heart muscle cell, has died and new heart cells obtained by differentiating pluripotent stem cells or heart progenitor cells could conceivably replace the lost cells. This is discussed in more detail in Chapter 9, "Cardiomyocytes from Stem Cells: What Can We Do with Them?"

Generating a complete organ, made up of several cell types that work together in an intricate and often highly complex manner, would be much more difficult. For this reason, therapeutic applications of stem cells will, for the time being, be largely limited to those diseases requiring replacement of one, or maximally two, well-defined cell types that could be injected locally into the existing organ.

Thus, several criteria need to be fulfilled for a disease to be treatable using stem cells. These include identification of the cell type(s) lost, their numbers, and knowledge of the underlying cause of the disease. The greater the number of cell types, organs, or tissues involved in a disease, the more complex treatment by transplantation of cells becomes. In addition, accessibility of the organ and the risk of stem cell delivery causing extra damage need to be taken into account. It is not difficult to imagine that stem cell therapy for the skin and for the cornea of the eye might be relatively easy, while therapy for deeper-lying regions of the brain would be more difficult. Using such criteria, it also becomes possible to develop a rationale for distinguishing which diseases are likely to become treatable in the short term and those that may have to wait much longer for a solution. In this way, we come to a hierarchy of diseases that might be treatable, some high on the list, some lower.

7.4

MAKING THE BLIND SEE: REGENERATING THE CORNEA

The cornea is the thick transparent layer that covers and protects the eye. After damage, say from accidental burns, disease, or even violent acid attacks, the cornea can lose its transparency and become cloudy, so that light can no longer pass through it, dramatically reducing the patient's sight. Cornea transplants can solve the problem but, as for many diseases, there are too few human donors to meet the clinical need. Furthermore, corneal transplants are unsuitable in the presence of a destruction of the limbus, the narrow zone that surrounds the iris of the eye where the corneal stem cells necessary for the resurfacing

7.4 *(cont'd)*

of the donor cornea are located. Culturing limbal stem cells may be the solution.

Dr. Graziella Pellegrini from the University of Modena and Reggio Emilia in Italy is one of the small group of scientists who have managed to culture limbal stem cells in the laboratory. These cells can be collected from the "good eye" or from the damaged eye if a small area of the limbus has been spared in some cases of bilateral injuries. Once sufficient numbers of cells have been grown in the laboratory, the damaged cornea can be removed under anesthetic and the cultured limbal cells placed around the iris of the damaged eye on a scaffold material where they will start to regenerate the different layers that make up the cornea. Within months, the cornea can be completely reconstituted and the clear covering of the eye restored.

Dr. Pellegrini and Dr. Paolo Rama have patients who are able to see as well as before their accident, some even 15 years after the limbal cell transfer. Cultivation of limbal cells might offer an alternative to patients with unilateral lesion and a therapeutic chance to patients with severe bilateral corneal–epithelial loss (Figure B7.4.1).

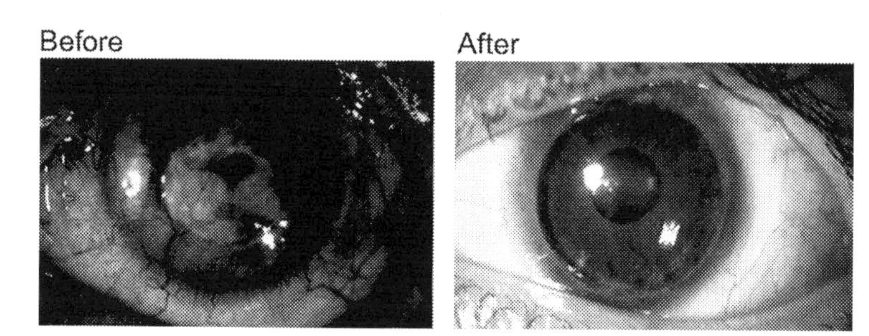

FIGURE B7.4.1 (a) A damaged cornea, the hard transparent layer covering the eye, can be reconstituted after damage that has made it opaque, by transplanting limbal cells. (b) The newly regenerated cornea is transparent and remains so for many years, restoring eye function. This type of blindness can thus be cured. *Source: Graziella Pelligrini, University of Modena and Reggio Emilia, Modena, Italy.*

7.3.1 Loss of One Cell Type or More Cell Types

What causes damage to, or loss of, cells in an organ? One cause is acute deprivation of the oxygen supply to a tissue; examples are myocardial or brain infarction, bleeding in the brain, or blockage of the blood supply to the kidney or even the spinal cord. This is known as *ischemic damage*. If it happens, multiple cell types at the affected site usually die rapidly. This creates a need for both the functional cells to be replaced and the tissue architecture to be rebuilt, like the functional connections between blood vessels and surrounding tissue. New vasculature may grow spontaneously from existing surviving blood vessels into the damaged area, provided that the right conditions are present, a process called *angiogenesis*, but most damaged heart muscle cells will not be renewed by the body itself. Here, stem cell transplantation to replace the lost cardiomyocytes seems a feasible option, although, in practice, introducing heart cells into the muscle and making them stay there while the heart continues to beat is far from trivial. In contrast, in the case of brain infarction, several different types of brain cells are lost and endogenous progenitor cells in the region of damage may not be able to recreate all of these cells. A stem cell based therapy would require production and injection at the right location of each of the different brain cell types, where they should integrate properly within surviving brain tissue.

Certain degenerative diseases, such as Parkinson's and Alzheimer's, also serve well to illustrate how the cause of the disease and the cell types involved in the disease may determine eligibility for stem cell therapy. In the case of Parkinson's disease, there is principally only one cell type lost: a neuronal cell type that produces the neurotransmitter molecule dopamine. This molecule is responsible for communication between nerve cells (Figure 7.3). Because Parkinson's disease is caused by progressive loss and malfunction of one single cell type it would, in principle, be a candidate disease for stem cell therapy. In fact, many researchers believe that this will be one of the first neurodegenerative diseases to be treated with stem cell therapy. It has been a challenge to obtain the right neural cell types from embryonic stem cells, but Lorenz Studer at the Sloan Kettering Institute in New York appears to have solved this and is now working toward studies in patients currently planned for 2017 (see Box 7.17, "Lorenz Studer: Pioneer Research on Stem Cell Therapy for Parkinson's Disease"). Other degenerative brain diseases have their origin in an abnormal accumulation of aberrantly folded proteins that interfere with the normal function of the cells and eventually leads to cell death. Such a mechanism plays a role in, for example, Alzheimer's disease. The etiology of Alzheimer's disease is far more complex than that of Parkinson's disease, and sticky plaques form on different nerve cells. This makes it much less suitable for treatments using stem cell therapy even if the right cells for transplantation could be made available.

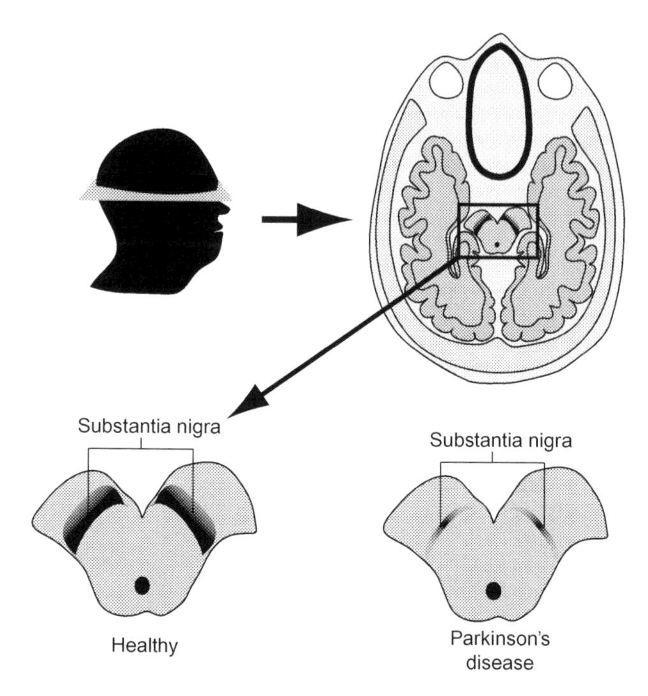

FIGURE 7.3 Parkinson's disease is caused by loss of specific cells in the brain. These cells produce dopamine, which is a protein important for brain function. Cells of the substantia nigra in the thalamus are those primarily lost in patients with Parkinson's disease.

7.3.2 Inherited and Acquired Diseases

If the cause of the organ cell dysfunction is not genetic, stem cell transplantation is likely to be only effective if, at the same time, the cause of the underlying disease is also treated, for example, by the administration of appropriate drugs. If the disease is under control, either allogenic or autologous cell transplantation might be an option.

If the disease is the direct consequence of a well-defined genetic defect, as is often the case in hereditary diseases, stem cell transplantation can potentially be used as therapy, provided that the transplanted cells are from a healthy donor. In such a case, autologous stem cells obtained from the patient might not be suitable or would only work for a short time because they would carry the same genetic abnormality. However recent advances in gene therapy offer a potential solution for autologous transplantation in these patients as well, since the stem cells may be genetically "engineered" to correct the defect in the DNA. This sort of genetic engineering used to work only in embryonic (pluripotent) stem cells, but now methods have improved so much that it will also work on adult stem cells. If these can be obtained from a patient, this may be a very good autologous cell therapy option in the future.

In some (genetic) diseases, it is not the cells with the mutated gene that are affected, but the neighboring cells. In amyotrophic lateral sclerosis (ALS), for example, a mutation in one specific cell type in the nervous system makes it kill its neighboring nerve cells. This causes the decline in motor function that is so debilitating and eventually fatal for ALS patients.

7.5

WHICH CELL TYPES ARE LOST IN WHICH DISEASE?

Parkinson's disease: Loss of a specific type of neuronal cells in different regions of the brain, especially neurons in the *substantia nigra*, which produce dopamine. Dopamine is necessary to control muscle movements correctly.

Demyelinating neurological diseases: Loss of a specific type of neuronal cell, the oligodendrocyte, which normally sustains the protective myelin layer around the extremely long neuronal extensions (axons) in the spinal cord. This loss can be caused by an accident, but also by certain chronic diseases such as multiple sclerosis.

Some forms of diabetes: Loss of insulin-producing cells from the islets of Langerhans in the pancreas. This is usually caused by immune cells attacking the body's own insulin-producing cells (an "autoimmune" reaction of the immune system against "self" cells) and killing them as a result. Alternatively, in some forms of diabetes there may be no cell loss but a genetic defect interferes with the function of the insulin-producing cells. Insulin is necessary for cells to take up the nutrient glucose (sugar). Diabetic patients control their blood glucose levels by injecting insulin.

Myocardial infarction: Acute loss of heart muscle cells (cardiomyocytes), usually in the left ventricle of the heart as a result of lack of oxygen (ischemia). Also referred to in laymen's terms as a heart attack.

Deafness: In some forms loss of "hair cells" in the inner ear.

Cancer: In certain forms of cancer (e.g., blood cancer, breast cancer), treatment with cytostatic (chemo) therapy causes loss of blood cells, as well as their precursor cells and blood stem cells. Cells of the immune system are also lost in severe combined immunodeficiency (SCID).

7.5 *(cont'd)*

Muscle diseases (e.g., Duchenne muscular dystrophy): Progressive loss of skeletal and sometimes heart muscle cells caused by a genetic defect in a protein called dystrophin, which is important for normal muscle cell contraction.
Some forms of infertility: Too few healthy oocytes (eggs) or sperm cells are produced (Figure B7.5.1).
Ulcerative colitis: Ulcerative colitis and other inflammatory conditions of the colon (called IBD, inflammatory bowel disease) are autoimmune diseases, characterized by severe inflammation of the intestinal wall, as a result of which patches of cells lining the intestine are lost.
Macular degeneration (age-related blindness): Retinal pigment epithelial cells at the back of the eye are damaged by new blood vessels that invade the retina.
Osteoporosis: Loss of bone cells associated with aging ("brittle bones").

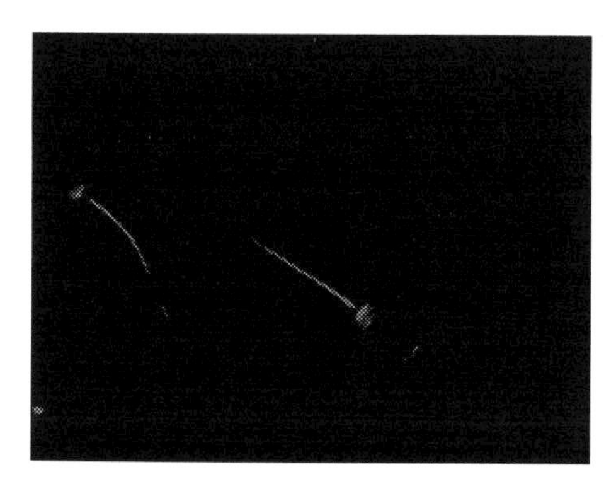

FIGURE B7.5.1 Just imagine stem cells differentiating to gametes ("sex" cells) like these sperm cells. They might be useful for treating certain forms of infertility. *Source: Stamcellen Veen Magazines.*

7.3.3 Diabetes Type 1: An Autoimmune Disease

In some diseases, the immune system of the body reacts against specific proteins on its own cells. These are known as *autoimmune* diseases. One relatively frequent type of diabetes, called type 1, is a good example of an autoimmune disease. In healthy people, specialized

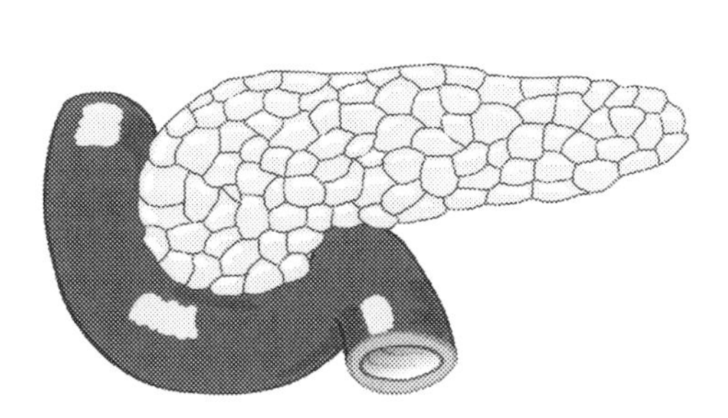

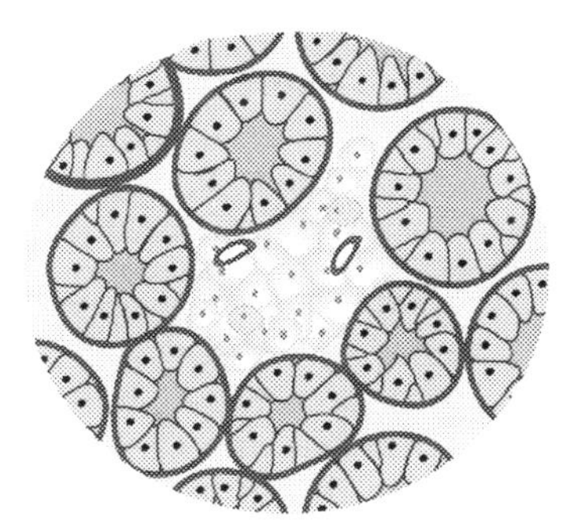

FIGURE 7.4 The pancreas as schematically drawn on the left is surrounded by the duodenal part of the intestines. In a cross section, a series of channels (green) can be seen, as shown in the diagram on the right. Digestive enzymes are produced that pass via the channels to the intestines. The islets of Langerhans (white) contain α-cells that produce glucagon and β-cells that are responsible for the production of insulin in response to glucose that is transported via blood vessels (red).

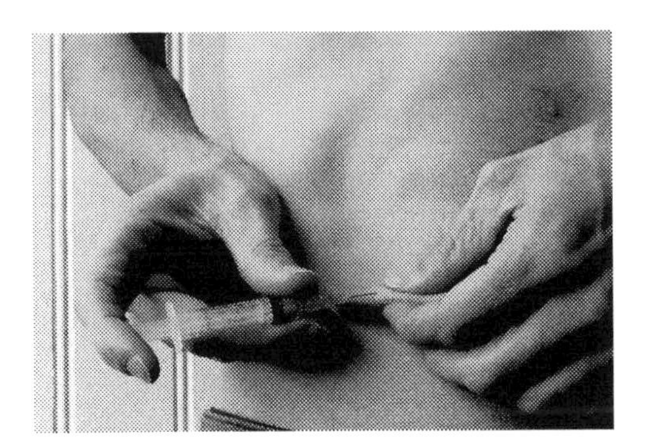

FIGURE 7.5 Many diabetes patients have to inject themselves regularly with insulin to maintain proper blood sugar levels.

cells, called beta cells, which reside in the islets of Langerhans in the pancreas, produce insulin in response to an increase in the concentration of glucose (sugar) in the blood (Figure 7.4). The insulin is secreted by the beta cells into the blood stream. Here, it induces rapid uptake of sugar by all cells of the body, which subsequently convert it to glycogen, a long-term energy store for cells. In patients with type 1 diabetes, abnormal activation of the immune system slowly leads to destruction of the insulin-producing beta cells. Only one specific cell type is lost. It is known that donor beta cells from pancreas can be transplanted successfully to patients with diabetes and actually survive and function for many years after transplantation in the liver. Some patients have been known to be free of the need to inject themselves with insulin for many years after receiving donor pancreas islets, suggesting that this disease is almost certainly a candidate for stem cell therapy (Figure 7.5). However, since the actual autoimmune cause of the disease cannot be effectively treated, it cannot be excluded that over time the transplanted cells might be killed by the same autoimmune disease process as the original beta cells, necessitating repeated transplantations with time. Ideally then, there should be another source of beta-islet cells other than donor pancreas, of which there is a shortage. Beta cells from pluripotent stem cells or adult progenitors in the pancreas could be the much sought-after renewable source of cells for treatment of diabetes.

7.3.4 Spinal Cord Injury and Multiple Sclerosis: Restoring Demyelination

Another opportunity for stem cell transplantation, which has undergone much research in animals and in some patients, could be a subgroup of patients with spinal cord injury. These patients are paralyzed below the level in the spine at which the damage has taken place. In most patients the nerve ends are severed. The body cannot generate new nerve cells in the spinal cord, and although nerve cells can be generated from stem cells, it would be very difficult for the cells to cross the extensive scar tissue at the site of damage and reconnect the nerve ends. Thus, stem cell therapy is not an obvious remedy for this type of injury, especially if it occurred some years previously. However, many patients have crush injuries of the spinal cord, in which the insulating layer of myelin surrounding the nerves has been lost, but the nerves themselves are still intact. For these patients a slightly different approach might work. Oligodendrocyte cells that produce myelin can be generated from stem cells; if these are injected into the site of the crush injury they could repair the myelin layer locally, and thus restore function and cure the paralysis (Figure 7.6). Of course, this is a very specific type of spinal cord injury. Treatment would only be possible in the acute phase, say, within a few weeks of the time of the injury.

Experiments in rats with this type of damage have been so promising that the stem cell company Geron in the United States was granted permission by the U.S. Food and Drug Administration (FDA) in 2008 to treat 10 patients with this type of spinal cord injury using stem cells. However, Geron itself put the study on hold after discovering unexplained cysts in the spines of some treated rats. Although probably not the feared teratoma that could arise from embryonic stem cell therapy, it was decided to undertake more research before initiating the first-in-man studies. The problem was resolved and five patients were treated as a safety study. While many secretly hoped for a cure for some of these patients, the permit given to Geron did not allow more cells to be injected than in the rats in the preclinical studies. It was perhaps destined to fail, and in 2011, Geron discontinued the trial,

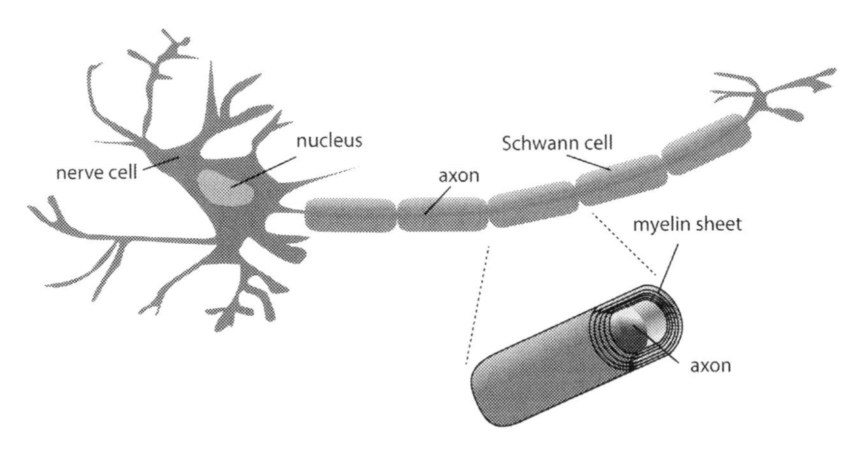

FIGURE 7.6 The axons of nerve cells (blue) are insulated by a myelin sheet (brown). The myelin that forms the sheet can be made by different cell types; the myelin sheet of motor and sensory neurons is produced by Schwann cells.

stopping their stem cell therapy program altogether. Was the trial a waste of time and money? Although the costs and effort were tremendous, the consensus is, probably not. A great deal was learnt about how to bring pluripotent stem cells into clinical use, what safety issues need to be addressed, and even how the cells behave in patients. This has benefitted efforts to bring pluripotent stem cells into the clinic for other diseases tremendously, with, for example, cell therapy for macular degeneration (age-related blindness) making the most advances, as will be discussed later in this chapter.

Interestingly, in patients with multiple sclerosis, nerve cells also lose their outer insulating layer of myelin, suggesting the option of a similar form of stem cell therapy. However, an unsolved problem as yet is how to inject the cells in such a manner that they spread throughout the central nervous system and restore all of the lost myelin around nerves.

For demyelinating diseases in general, fetal tissue can also be a source of myelinating cells and, although also controversial, various trials are ongoing. Results are eagerly awaited by many.

7.6

FUSION BETWEEN STEM CELLS AND DIFFERENTIATED CELLS

One curious property of stem cells is that if they are very close to differentiated cells they can spontaneously fuse with them, albeit rarely. In the past this has led to misinterpretation of some

7.6 (*cont'd*)

experimental results, and erroneous claims that adult stem cells may be "plastic" and able to form many cell types of the body. For example, when bone marrow stem cells, were marked with an extra gene that made them artificially produce a green fluorescent protein (GFP), they could be traced after injection into mice. If the mice had been exposed to X-rays before injection to destroy their own bone marrow cells, the GFP-marked bone marrow cells would form all cells of the bone marrow and eventually all blood cells.

The bone marrow of the mice is repopulated after bone marrow transfusion, much like bone marrow transplantation restores the blood and immune system in cancer patients treated by chemotherapy. Very surprisingly, in mice, after a few months, green fluorescent cells were not only found in the blood but also in other tissues and organs: some nerve cells and liver cells, for example, also became green. This suggested that the stem cell in bone marrow could *transdifferentiate* into nerve and liver as well as other tissues. However, this was incorrect. When scientists looked more carefully, they found that these differentiated cells had more than one nucleus, or twice the normal amount of DNA. They were *tetraploid*, indicating that the fluorescent bone marrow stem cells in the mouse had fused with differentiated body cells. This does not occur under normal circumstances in the body: blood stem cells in the bone marrow cannot turn into nerve, liver, heart, or any other type of cell except blood. It is unclear how often cell fusion occurs, and to what extent a cell with more nuclei can continue to function normally, but this cell-fusion effect would certainly not have any therapeutic effects because there is no increase in the number of differentiated cells.

Pluripotent stem cells can also fuse with differentiated cells in laboratory culture. This has provided useful experimental information about "reprogramming" of adult cells because the differentiated cells appeared to become pluripotent following fusion.

7.3.5 Macular Degeneration: The Ideal Disease

One could say that the eye as a target organ for stem cell therapy has a tick in all the right boxes. In the first place, because the eye is an immune-privileged site, transplanted cells that are not HLA matched are not rejected, and immunosuppressive medication is not required. It has been known for many years that corneas from eyes of deceased donors can be transplanted into patients, and, more recently, that cells derived from a donor cornea can also be used for this purpose. Secondly, the eye is a "closed" unit: any stem cells injected would find it difficult to misbehave and escape to other parts of the body. And thirdly, because we have two eyes, it may be possible to treat one eye without the risk of losing vision altogether—of course, on the premise that the other eye still functions to some extent. Macular degeneration means that the part of the retina that is crucial for central vision is lost. It is also called "age-related blindness" because onset is commonly in old people and loss of the macular or retinal pigment epithelial cells causes loss of vision in the affected area—in the center of the field of vision, which then appears black. This is a very common disease affecting millions of individuals, increasing in incidence as the population ages.

Human embryonic and pluripotent stem cells can easily differentiate into retinal pigment epithelial cells, the cells that need replacing. Experiments in rats, mice, and, more recently, pigs have shown that if these cells are cultured on supporting scaffolds, they can be surgically inserted into the retina. There they pick up signals from light and indeed transmit images to the brain. While generation of the right cell type in numbers suitable for therapy is quite feasible in this special case, the surgical procedure to insert the transplant effectively in the retina of patients is still challenging. Nevertheless, the interest and investment of commercial companies and the relatively straightforward nature of the disease—with just one cell type affected—has placed macular degeneration high on the list of diseases potentially treatable with stem cells. As a result, half a dozen or so biotechnology companies have been established in this area, and even the larger pharmaceutical companies and funding agencies such as the California Institute of Regenerative Medicine (CIRM) in the United States and the Medical Research Council (MRC) in the United Kingdom, among others, have started to support clinical studies. More than 20 patients have now been treated with no adverse effects but to date only very few cells have been used, the same cautious start as the Geron trial for spinal cord injury (as previously mentioned). Time will tell whether increased doses of retinal cells will actually restore sight.

In Japan, home of the discoverer of induced pluripotent stem cells, the focus is on these stem cells for treating macular degeneration. Whether embryonic or induced pluripotent stem cells prove best remains to be seen. Although not a life-threatening condition, macular degeneration severely impacts the quality of life of those elderly people affected. A solution, so that they can read a book without a magnifying glass or watch TV normally, would therefore be much welcomed by them and their families.

7.3.6 Deafness

Some forms of deafness are caused by loss of a particular cell type in the inner ear, called the *hair cell*. As understanding of how the ear develops in humans has grown, so also has knowledge of the cells and signals controlling the processes. Using this knowledge, several researchers have shown that they can generate inner ear hair cells from pluripotent stem cells, both from mice and humans. When the stem cell derived inner ear hair cells from mice are transplanted to the ears of mice in which deafness was induced by the same chemotherapy reagents as used to treat cancer, their hearing was restored. This equals hope on the horizon for patients with chemotherapy-induced deafness and any other form in which hair cells have been destroyed.

7.7

ETHICAL ISSUES OF DERIVING AND USING HUMAN EMBRYONIC STEM CELLS

Embryonic stem cells are derived from a small group of cells present in early embryos. If these cells are removed from the embryo during the process of creating an embryonic stem cell line, then the embryo is destroyed. If the embryos are human, then there are inevitably ethical issues around destroying a living entity that has the potential to become a new human being, even if the embryo would otherwise not be used for that purpose. This is the case with embryos that are left over after a successful *in vitro* fertilization (IVF) procedure, or when (only in some countries) embryos have been specially created for the purpose of deriving stem cell lines.

The ethical discussion largely focuses on the moment when human life begins, and what the moral and legal status of a human embryo is in this very early stage of development. Does the embryo have the same right of protection of life as a born individual, or is the right to protection something that grows as its development proceeds? For example, does the embryo have more right to protection after it has implanted in the uterus than it did when it was just a fertilized egg in a laboratory test tube? If so, how should we then regard abortion that occurs after implantation? Some societies equate the moral status of an egg immediately after fertilization with a more developed fetus, so that stem cell and abortion debates become intertwined. By contrast, other societies judge the moral status of an IVF embryo in the laboratory as less than that of a later-stage fetus developing in the uterus, even though it still has the potential to become a new individual. In their opinion, the IVF embryo may be used to isolate stem cells for research on cell therapies for serious diseases, as long as it is no longer required for reproduction and thus destined for destruction, and has been freely donated by both of the parents (sperm and egg cell donors).

However, what to think of human embryonic stem cell lines that will not be directly used for development of a therapy but instead for drug discovery by pharmaceutical companies or for screening for toxic effects of novel drugs? Is this a morally acceptable use of embryonic stem cells that might originally have been derived solely for the purpose of developing stem cell therapies? And is the issue circumvented by deriving an embryonic stem cell line for a single blastomere at the eight-cell stage, leaving the remaining seven cells to go on to become a healthy embryo? Was the one blastomere a second "life"? These are technically challenging moral debates that often only take place once something morally challenging has taken place. Should we have IVF developed on such a massive scale before we had decided as a society what to do with the embryos that would inevitably be left over?

In this context, it is of note that not all religions consider the moral status and right to protection of the embryo in the same way. The Christian belief is that life begins at fertilization or first cleavage of the embryo. The Jewish and Muslim religions consider the beginning of life as the moment when "the soul enters the body," which may be 40 days after conception. From any of these perspectives, several topics for ethical discussion, such as two identical twins stemming from one fertilized egg (see Box 7.10, "The Twin Paradox"), intrauterine devices (a contraceptive method whereby a copper wire placed in the uterus prevents implantation of the embryo, which is thus lost), and spontaneous early miscarriages, complicate some of the arguments on the right to protection. In some countries, the law is based on the prevalent religion; in other countries, it is not. It seems there are almost as many different perspectives on the moral status of the embryo as there are different societies. In countries where stem cell derivation from human embryos is allowed by law, the consensus is usually that embryos have a small but finite moral status and right, which increases as gestation advances, finally becoming equivalent to a living person by the end of pregnancy and the time of birth.

In practice, there are now more than 1000 disease-bearing and normal human embryonic stem cell lines worldwide (see the hESReg database *www.hESReg.eu*) and relatively few stem cell labs are actively deriving new embryonic stem cell lines. The exceptions are those attempting to derive clinical grade human embryonic cell lines that would be suitable for cell therapy in the future.

7.7 (*cont'd*)

An immediate burning question is whether developments with respect to induced pluripotent stem cell lines as a novel source of pluripotent stem cells will supersede the need for new human embryonic stem cell lines—particularly as disease models—or whether human embryonic stem cell lines will remain the gold standard. Only time and research will tell.

An additional dilemma involves cloning. As described elsewhere, in the procedure to produce an embryo by cloning, fertilization does not take place. The Christian definition of the start of "life" then covers cloning, as it is considered to begin at the first cleavage division. Cloning can be divided into two categories based on purpose: nonreproductive cloning, where cells are derived from cloned embryos either for treatment of specific diseases or for generation of disease-bearing cell lines for research purposes, or reproductive cloning, where a new individual is created. The latter is prohibited in most countries and considered ethically unacceptable—a poorly defined threshold that, however, should not be crossed. Aside from the *Boys from Brazil* or *The Island* types of science fiction, the question of how it would feel to be a clone of someone else is an important ethical issue. Moreover, it is most unlikely that any cloned human beings would be completely normal: most would expect to have unpredictable medical problems. In some countries (e.g., the United Kingdom, Australia, and Belgium), nonreproductive therapeutic cloning is allowed by law, as it only results in a blastocyst-like embryonic stage, with limited rights of protection—comparable to those of leftover embryos from an *in vitro* fertilization procedure.

For therapeutic cloning, unfertilized oocytes are needed. To obtain these, hormone treatment in combination with a limited surgical procedure is required, which involves healthy women volunteers. After collection of oocytes, as for IVF, DNA of the oocyte is replaced by DNA from a normal, differentiated body cell, and sperm entry is mimicked by a tiny electric shock, which initiates the growth of the embryo. A cloned blastocyst embryo will thus develop that has DNA identical to that of the body cell. Stem cells derived from this embryo will be genetically identical to the donor of the differentiated cell, and any tissues derived from the stem cells would not be rejected by this donor's immune system. Alternative sources of oocytes for therapeutic cloning include oocytes derived from existing human embryonic stem cell lines, or animal oocytes; cloned embryos that result from these sources would nevertheless be considered human embryos and fall under existing embryo and stem cell legislation.

In discussing this sensitive topic, it will clearly not be easy, in fact almost impossible, to reach a consensus on the moral status of the embryo, its right to protection, and what that degree of protection would be. The issue is certainly driving research to search for alternatives and the generation of induced pluripotent stem cells is one positive outcome of this. It has virtually eliminated most research on therapeutic cloning with the exception of a few research groups studying how reprogramming by an unfertilized egg cell actually works.

7.4 THE BEST STEM CELLS FOR TRANSPLANTATION

The type of stem cell to use to treat a particular disease is predominantly determined by the type of disease itself. If the disease is acute and life-threatening, only cells "off the shelf" and essentially ready to go could be used. In general, these will be allogenic and derived from a living (e.g., in the case of bone marrow) or deceased (e.g., insulin-producing beta cells) donor. In the future these off-the-shelf options could include embryonic or induced pluripotent stem cells (see Chapter 3, "What Are Stem Cells?" and Chapter 4, "Of Mice and Men: The History of Embryonic Stem Cells"). On the other hand, for chronic diseases there is more time to select, collect, culture, expand, and, if necessary, differentiate cells for transplantation; here the use of autologous stem cells should, in principle, be possible.

Myocardial infarction is an example of an acute and life-threatening condition. In many patients, ischemia that accompanies an infarct caused by a temporary lack of blood leads to a large loss of heart muscle cells. This is repaired, more or less, as a normal wound but the heart cells are, over time, replaced by scar tissue. Since scar tissue contains no heart cells, it does not contract and the heart is no longer able to pump blood efficiently around the body. The remaining healthy heart muscle tissue works harder to compensate but finally starts to become weak and fail. The result is chronic heart disease that is fatal for many patients (up to 65%) within five years. Replacement of the damaged heart muscle cells could be a therapeutic option, but a large number of stem cell derived cardiomyocytes would be needed for transplantation. However, an interesting recent option is *direct reprogramming*. Direct reprogramming is a process in which selected genes encoding certain proteins (known as transcription factors) are introduced into somatic cells such as fibroblasts. These genes, or gene combinations, are used selectively by some differentiated cell types and determine their identity. They are so powerful that they are able to change the identity of the fibroblasts into the differentiated cell type from which the genes were derived. Examples include skeletal muscle, blood cells, and, more recently, nerve and heart cells. Fibroblasts can be turned into these cell types simply by expressing these so-called master genes. In the future, it may be possible to inject genes into the heart or brain and turn the fibroblasts there into heart cells or nerve cells. So far, this has only been shown in organs in mice (although it has worked on human cells in culture), but it is an example of what we have learned from stem cell medicine that might be useful in regeneration therapy.

Whatever the solution, for patients with acute heart failure it should be available at short notice. Even if heart cells are grown from the patient's own progenitor cells in the heart, it would take at least a few months to generate sufficient numbers of cells for a transplantation. In this case, heart muscle cells off the shelf from, for example, pluripotent stem cells deposited in a certified stem cell bank could, in the future, be an option to treat these patients.

Another example of acute tissue loss is a complicated bone fracture, caused perhaps by an accident. This is usually not immediately life-threatening and there is usually no underlying disease that needs to be treated first. In these cases, *osteoblasts* (bone-building cells) or their progenitor cells could be collected from the patient, expanded in culture, and then inserted at the fracture site to facilitate and speed up the healing process. Mesenchymal stem cells from the patient's blood, fat, or bone marrow can be turned into bone, cartilage, or fat cells in culture. They are already being studied in the clinic for their ability to help in the healing of large bone wounds, particularly in combination with three-dimensional scaffolds (*tissue engineering*) that mimic the bone structure and shape. Perhaps in the future these will be derived from genetically matched induced pluripotent stem cells from a stem cell bank, which may be an off-the-shelf option. It is not thought at present that induced pluripotent stem cells will be an autologous option because of the prohibitively high costs of such individualized stem cell production for routine health-care use.

Diabetes is another interesting example that illustrates that even for one disease, the stem cell choice can depend on the etiology of the disease. As mentioned earlier, type 1 diabetes is caused by a hyperactive immune system, which slowly destroys the body's own insulin-producing beta cells. The insulin-producing cells themselves are not the cause of the disease, so replacing them with new (stem cell-derived) insulin-producing cells could cure the disease, if they responded correctly to the blood glucose level after transplantation and remained functional for a long period of time. In this disease, it does not, in principle, matter much whether the stem cells are derived from the patient or from an external (donor) source, although in the latter case drugs would be needed to suppress the patient's immune response. This is associated with multiple side effects, including susceptibility to serious infections and even certain types of cancer. For this reason, autologous stem cells are the most attractive therapeutic option. They could be derived from, for instance, residual progenitor cells in the patient's own pancreas or, as already discussed, induced pluripotent stem cells generated from the patient and differentiated to beta cells. A renewable source of insulin-producing beta cells, such as those derived from a stem cell line, would have the advantage that the transplantation could be easily repeated if necessary. This is relevant because if there is an autoimmune mechanism killing the patient's own beta cells then transplanted cells would, over time, probably undergo the same fate.

In rare inherited forms of diabetes the cause of the disease lies in the beta cells themselves, the result of a genetic defect that interferes with the proper insulin-producing response of the cells to the blood sugar concentration. In such a patient, transplantation with genetically normal insulin-producing cells would be a likely cure. Transplantation of cultured adult stem cells or progenitor cells from pancreatic tissue derived from a (deceased) donor is clinically feasible but is still limited by the shortage of donors. Genetically (HLA) matched pluripotent stem cells, or (progenitor) stem cells derived from the patient in which the defective gene is corrected in the laboratory may present interesting new opportunities to treat these (largely) young diabetic patients in the future. Again, however, the potentially high costs will determine if this actually becomes widely used clinical practice.

7.8

BONE MARROW TRANSPLANTATION

One of the most frequently used clinical applications of stem cells is the transplantation of blood-forming, hematopoietic, stem cells from bone marrow for the treatment of a number of malignant blood diseases (e.g., leukemia and lymph node cancer), but also for the rescue of bone marrow function after treatment of breast cancer with aggressive chemotherapy. In all of these diseases, the patient's own bone marrow is destroyed by chemotherapy, which kills dividing cells. In the case of blood cancer, the aim of chemotherapy is to kill the malignant cells present in blood and bone marrow, while with breast cancer the loss of bone marrow cells is not the purpose, but an unfortunate and undesirable toxic side effect of chemotherapy. Blood stem cells normally supply all the different cell types to the blood: white blood cells of the immune system, red blood cells that transport oxygen, and platelets

7.8 *(cont'd)*

for blood coagulation. These blood stem cells are often used for an autologous transplantation, which means that cells from the patient's own body are used for the transplantation. However allogenic transplantation of blood stem cells, where a carefully selected donor provides the patient with cells, is also possible. For this, as with organ transplantation, it is of the utmost importance to have a good immunological HLA match between donor and patient to avoid either rejection of the transplant by the patient, or a graft-versus-host reaction where the transplanted cells attack the tissues of the patient. If possible, autologous transplantation is preferred, but this can only be done if sufficient healthy bone marrow can be obtained from the patient before the chemotherapy treatment, or a technique to free the bone marrow of residual tumor cells can be used.

To obtain the blood stem cells from the bone marrow, bone marrow aspiration from the pelvic bone is usually carried out under anesthesia. More recently, it has become common practice to collect stem cells directly from the blood after giving certain growth factors (e.g., granulocyte-macrophage-colony stimulating factor or GM-CSF) to the patients. These growth factors temporarily recruit stem cells from the bone marrow into the blood. This is called *stem cell mobilization*. The stem cells thus obtained can be stored until needed. Following collection by either procedure, the patient is treated with cytostatic drugs (chemotherapy) or undergoes irradiation. The blood stem cells can subsequently be returned to the blood of the patient at any chosen time point through a simple intravenous infusion into the blood stream. They will find their own way back "home" to the bone marrow, where they will settle and start producing new blood cells, at no risk of rejection since they are the patient's own cells. If healthy blood stem cells from a patient are not available, for example, when malignant leukemia cells are still present in the bone marrow, an attempt will be made to find a suitable transplant donor. This is a high-risk treatment because immune rejection of the transplant can be life-threatening, and a graft-versus-host reaction is a much-feared complication. To avoid this, the HLA match between donor and patient should be as good as possible.

Worldwide, more than eight million donors are registered as bone marrow donors, of whom a little bone marrow is taken for HLA typing. This information is registered in a data file, and only when that particular bone marrow type is required, is the donor called on to donate. One liter of bone marrow is removed and properly stored for the transplantation to follow.

7.5 COMBINING GENE THERAPY WITH STEM CELL TRANSPLANTATION

As discussed for diabetes, not all diseases are caused by loss of functional cells. Instead diseases may also be caused by the intrinsic inability of cells to produce the right proteins in the right amounts. These include several, relatively rare, inherited diseases, such as hemophilia A. These hemophilic patients cannot produce sufficient amounts of blood clotting protein factor VIII in their blood due to a DNA mutation in the gene for factor VIII. As a consequence, blood clotting is impaired and patients bleed very easily, which can obviously have serious consequences. The disease is typically treated by regular intravenous injections of the missing protein but, apart from being inconvenient, this is often prohibitively expensive. A combination of stem cell and gene therapy may in the near future offer a better solution. Blood stem cells would be collected from bone marrow of the patient, a normal factor VIII gene introduced into the stem cells in the laboratory, and the cells returned to the patient to secrete the missing coagulation protein into the blood stream. The patient would, however, not be completely cured, mostly because the production of proteins in the body is complex and tightly regulated to make just the right amount. This intricate regulation cannot be performed by a genetically modified stem cell. Fortunately for patients with hemophilia, sustained low levels of factor VIII suffice to protect patients from major bleeding.

In diabetic patients, the situation is quite different. It is essential for them to have precisely the right level of insulin in the blood at any particular moment, and this can vary depending on whether they are eating, running, or sleeping. This is a major hurdle for gene therapy. Too much insulin secreted in the blood can be fatal since it removes sugar from the blood too quickly and the resulting low blood sugar level or *hypoglycemia* damages the patient's brain cells; this is immediately life threatening. Therefore, in the case of diabetes, just introducing a gene for insulin into a blood stem cell would not work, since insulin production by these cells would not be regulated in response to sugar levels in the blood. A better approach would be first to repair the genetic defect in the insulin gene in stem cells from the patient using *homologous recombination*, then culture and differentiate them to insulin-producing cells that now would be capable of sensing and reacting properly to changes in glucose concentration. Induced pluripotent stem cells might be a future stem cell source for this approach since the high costs would be set off by a "lifetime treatment" option. Progenitor cells from the pancreas of the patients might be a shorter-term option as it is now feasible to carry out gene repair in stem cells that are not necessarily pluripotent. This illustrates the rapid advances in this area over the last several years.

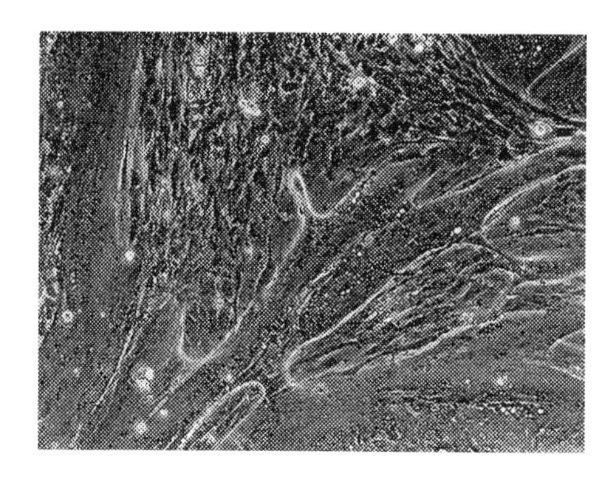

FIGURE 7.7 Stem cells can differentiate *in vitro* into skeletal muscle cells. Differentiated skeletal muscle cells can be identified morphologically by the presence of multiple nuclei in each cell instead of one. This results from fusion of numerous cells to form muscle fibers.

Another disease that might be treatable by a combination of gene therapy and stem cell transplantation is Duchenne muscular dystrophy, a genetic disease caused by mutations in the dystrophin gene. The disease is characterized by progressive weakness of all muscles, often including the heart, respiratory muscles, and skeletal muscles, and is usually fatal by 30 years of age (Figure 7.7). If muscle stem cells (or myoblasts) from these patients could be genetically repaired, differentiated into muscle cells, and injected back into the muscle, this could mean the muscle would be repaired. A very particular type of gene therapy has recently been developed for this disease called *exon skipping*. A special gene construct is introduced into the cells which, when the DNA is "read," allows the polymerase enzyme to jump over ("skip") the mutation while producing the RNA strand. From the resulting RNA molecule, a normal protein is made that is fully functional, but just a tiny bit shorter than the native protein. The muscle cells can then behave normally. Each of the very many different mutations that can cause Duchenne muscular dystrophy requires its own exon skipping construct. These need to be tested on human cells, since the mouse dystrophin gene differs from that in humans and it is therefore not an option to test these in mice. Here, induced pluripotent stem cell lines generated from patients and differentiated into (skeletal) muscle could be very helpful in identifying the right exon skipping construct.

7.9

EMBRYOID BODY: DIFFERENTIATION OF EMBRYONIC STEM CELLS IN CELL CULTURE

There are many ways to induce pluripotent stem cells to differentiate to more specialized cell types. One of the first methods developed was the generation of *embryoid bodies*. These are small aggregates of

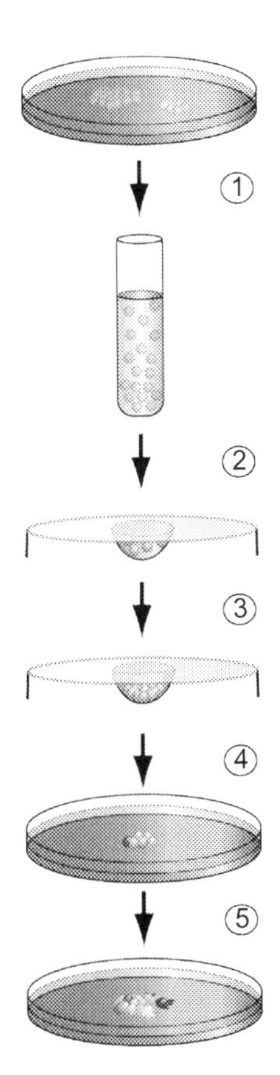

FIGURE B7.9.1 Undifferentiated pluripotent stem cells are added to a tube with culture medium (1). A droplet of fluid containing cells is removed and put on to the lid of a culture dish. The lid is inverted over the dish, so that the drop "hangs" (2). Stem cells in the hanging drop start aggregating (3) and form a single clump of cells in each drop. This structure resembles the inner cell mass of early blastocyst stage embryos to some extent and interactions between cells induce differentiation, just as in an embryo. Derivatives of all three "germ layers" can form. Cells are later plated on a culture dish (4), where differentiation becomes even more extensive and the cells become morphologically distinct. *Source: Stamcellen Veen Magazines.*

cells that are usually grown in very small drops of culture fluid, which are deposited on the lids of a petri dish. When the lids are inverted, the cells "hang" in the droplets in such as way that the stem cells stick together forming cells on the inside and the outside, which creates a signal for differentiation into the three basic cell lineages—ectoderm, endoderm, and mesoderm—similar to the differentiation that takes

7.9 (*cont'd*)

place in a real embryo. Embryoid bodies can be retrieved from the droplet after a few days and placed in a cell culture plate in culture medium where they will attach, grow, and differentiate further to cell types such as beating cardiac muscle or nerve cells with long axonal extensions. Although this method is considered somewhat old-fashioned, it has been used to make complex structures from other stem cell types, such as those from hair follicles that grow hair when grown as a cell aggregates or organoid cultures, which form well-organized tissues in culture (Figure B7.9.1).

This method does not work very well for human pluripotent stem cells because they do not stick together as well as mouse embryonic stem cells. For human cells, alternatives have been developed in which the cells are forced to stick together, for example, by centrifugation. These are called *spin embryoid bodies*.

7.6 WHERE TO TRANSPLANT STEM CELLS AND THEIR EFFECT

The site in the body to which stem cells should be delivered depends on which disease they are being used to treat. Some cell types will only function within the context of the organ in which they belong, while others function just as well at sites entirely separate from those in which they would normally be found.

When heart muscle cells are lost from the heart after a myocardial infarction, new cardiac cells would obviously be needed in the heart. New cells would have to be transplanted directly into the heart muscle. Cells that were not actually heart muscle but might be of benefit to the heart could be administered intravenously into the blood stream where they would need to find their own way into the damaged heart. This is called *homing* and is actually more plausible than one might think. Following any tissue damage, a local inflammatory reaction occurs around the damaged spot; here immune cells secrete special inflammatory protein molecules, called *cytokines*, that can attract the injected stem cells to the right location in the heart muscle. While the relatively homogenous cellular architecture of the heart might be favorable for local stem cell therapy, other organs such as the kidneys, have a more complicated structure. The glomeruli in the kidney that filter the blood are connected to an intricate series of tubular structures, each with different functions, together forming a large number of tiny suborgans.

In each of these, highly specialized cell structures work together to concentrate the urine, remove any molecules that might still be useful to the body, and add waste molecules. This finally leads to the production of urine in the bladder. Much like heart muscle cells, replacing kidney cells with stem cell derivatives will need the transplanted cells to exert their function within the kidney itself. The expectation is that in such a complex organ structure, with so many different collaborating cell types, it will be extremely difficult to replace nonfunctioning cells exactly at the right location.

In contrast to the heart and kidney, insulin-producing cells of the pancreas could actually act at a distance. The large number of islets of Langerhans in the pancreas function quite independently of one another and only need a good blood supply to be able to sense the levels of sugar and respond by secreting insulin into the blood stream. Any replacement beta cells, from stem cells or pancreas progenitors, for example, could just as well be transplanted at another site in the body and do not necessarily have to become part of the pancreas. The portal vein of the liver, to which the cells can be delivered simply under local anesthetic, is, for this reason, already being used as a transplantation site for beta cells from donors.

One of the challenges in treating Parkinson's disease is actually where to put the dopaminergic nerve cells in the brain. They exert their effect through the neurites, which are at a site distant from where the cell body "sits" in the brain. In mice and rats the neurites from these neurons can easily grow out over that distance, but in the much larger human brain, this can be several centimeters and is very difficult. Preclinical experiments in nonhuman primates (monkeys) are expected to show whether the neurons will also survive at the site in the brain where they are supposed to act (which is rather distant from the site where they normally "live"). In this way, the neurons do not have to find their way through the maze of brain cells to carry out their function at the right place in the brain.

7.10

THE TWIN PARADOX

A central question regarding the ethical discussion on the use of human blastocyst stage embryos to generate human embryonic stem cell lines, is whether such a procedure leads to the deliberate destruction of human life. When is an embryo equivalent to a human being? This question can be answered in several ways, not least dependent on

7.10 (*cont'd*)

religious arguments. Some religions, including parts of the Catholic church, consider oocytes and sperm cells as new life and are, therefore, against the use of contraceptives. Others consider the egg immediately after fertilization as the beginning of life, and thus as a new individual with the right of complete protection. Yet others consider life to begin after implantation of the embryo in the womb, or when the embryo has developed somewhat further to form the three germ layers of differentiated cells, or when the brain develops in the fetus. It seems that every society, religious group, or even individual has a different definition of what life is and when it begins. However, if the definitions are considered in a context relevant to hESCs, there are some paradoxes that are unresolved, especially if we consider the blastocyst stage of human development as being morally equivalent to an adult person.

The paradox of twins is a very interesting example. There are two types of twins: identical or monozygotic twins, arising from one fertilized egg, and fraternal or dizygotic twins, arising from two eggs usually released into the oviduct in one menstrual cycle and each fertilized by a different sperm cell. About 1–40 per 1000 births (depending on race) results in the birth of dizygotic twins, while chances of a monozygotic twin are somewhat smaller, around 1 in 400. Monozygotic twins develop when an early embryo—it can happen at a variety of stages—divides into two. In about one-third of cases, each of the fetuses has its own placenta. This means that the embryo would have split in two before the blastocyst stage on day 5 after conception. Nearly all other monozygotic twins share one placenta, although each of the fetuses has its own inner amniotic membrane. In this case, the embryo would have split into two between days 5 and 9 after conception, during the blastocyst stage, either just before or just after implanting in the womb. A very small number of twins arise from the embryo splitting into two at a later stage, after day 9. These embryos have a common placenta and develop within one amnion sac. If the splitting of the embryo is incomplete or the two fetuses later fuse, the twins will be conjoined (Siamese twins). (See Figure B7.10.1.)

Back to the paradox: if a blastocyst is considered as a living human being, how should one interpret the division of one blastocyst into two embryos? No one would consider monozygotic twins as just one individual. Has one individual "died" and been replaced by two new individuals? Or has individual A originated from individual B, or the other way around for that matter? And how should we regard the

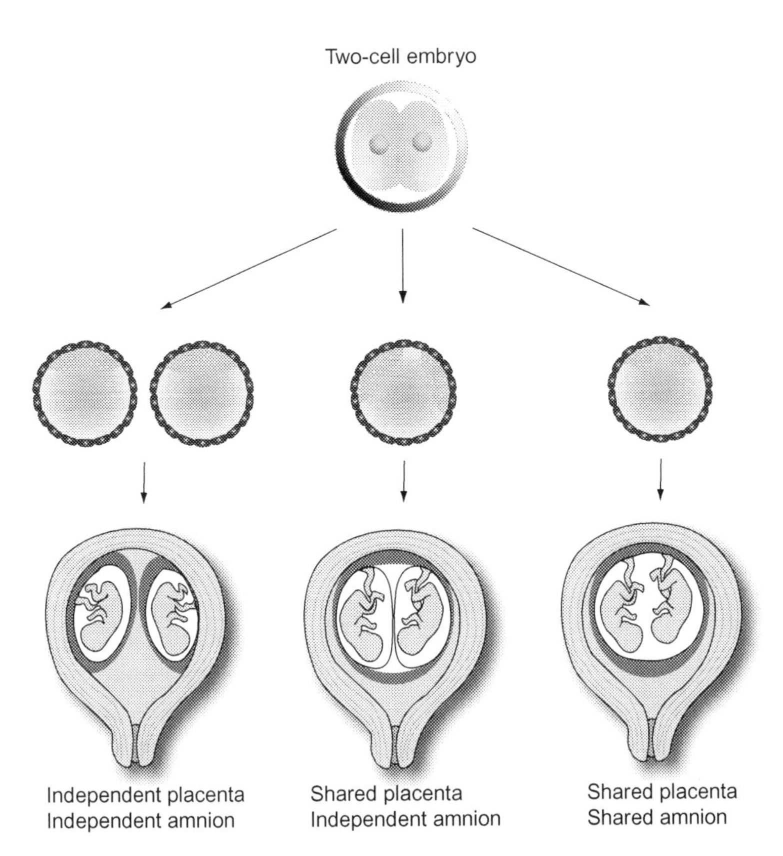

FIGURE B7.10.1 With monozygotic twins, the division of the fetal membranes and placenta depends on when the original embryo splits into two. In most cases this happens at the early blastocyst stage, and both fetuses develop their own amniotic cavity and membrane but have a common (shared) placenta (middle). In some cases the embryo is already divided at the two-cell embryo stage, resulting in two separate blastocysts, each developing its own placenta (left). Only very rarely does the embryo split at a later stage: here the two new embryos share a placenta and amniotic membrane (right). *Source: Stamcellen Veen Magazines.*

moral status of conjoined twins, which, depending on the embryonic stage that division occurred, may have a common spinal cord and thus a partly shared nervous system, and thereby senses. Are these two individuals or should we consider them as one single person? Nearly everyone would agree that twins, including conjoined twins, are two individuals, even though they originate from one embryo and may share part of the nervous system (Figure B7.10.2).

7.10 *(cont'd)*

FIGURE B7.10.2 Monozygotic twins originating from one embryo are genetically identical: essentially clones of the same age. The twin on the left is one of the authors. *Source: Jan Roelen/Stamcellen Veen Magazines.*

Another "freak of nature" is the formation of a natural chimera. Although an extremely rare event, it can happen that dizygotic twins at a very early stage of development "fuse" to form one developing embryo containing two separate DNA blueprints. This probably happens at the two-cell stage and in most cases will go completely unnoticed, with an entirely normal baby apparently being born, although in fact the baby is made up of cells as genetically distinct as two brothers or two sisters are. Usually such a condition would remain unnoticed, but in certain situations it is discovered, for instance, when a patient has to be prepared for organ transplantation and undergoes histocompatibility (HLA) testing. However, it may be that one embryo is female and the other male. In this case, the baby born will have around half of its cells carrying the female XX-genotype and the other half the male XY-genotype, which can lead to serious gender problems. The paradox: is the resulting individual one person or two? Nobody in practice would consider this person as two individuals with two "souls" and identities. Thus the moral status of a blastocyst is not as simple as it might seem at first sight.

7.7 CELL TYPES AVAILABLE FOR CELL TRANSPLANTATION

Cell replacement therapy is unlikely to be developed for all cell types and all organs. There are a number of practical issues, such as limitations in producing and expanding the right types of cells, irrespective of stem cell source, that will remain obstacles for the foreseeable future.

7.7.1 Adult Stem Cells

Adult stem cells can usually only form the cell types present in the organ from which they were derived. Differentiation to other cell types does not appear to occur naturally in the body and can at present only be carried out artificially, by changing the DNA in the cells using genetic manipulation techniques in the laboratory (a process known as direct reprogramming, as described earlier).

As discussed in Chapter 5, "Origins and Types of Stem Cells: What's in a Name?" the number of organs from which sufficient stem or progenitor cells can be isolated, expanded in the laboratory, and differentiated in the right direction is limited. On the other hand, a small number of stem cells can be derived from a surprisingly large collection of body tissues and organs, such as lungs, large and small intestine, thymus, kidney, liver, platelets, and even teeth. A few years ago, it was believed that normal adult stem cells had only a limited life span and could not be maintained in culture over extended periods (i.e., in contrast to pluripotent stem cells, adult stem cells were not immortal) but it is now clear that their potential to grow is much greater than expected. This is one of the most spectacular breakthroughs of the last several years.

Adult stem cells were first isolated from the large intestine of mice in which the stem cells of the intestine were marked with a fluorescent protein (see Chapter 10, "Adult Stem Cells: Generation of Self-Organizing Mini-Organs in a Dish"). It turns out that these cells can be cultured over long periods, and under certain conditions will self-organize into "organoids" or "mini-guts." Moreover, it also appears that these stem cells are present in many organs of the gastrointestinal tract, skin, and perhaps even other organs, and that some of them, at least, are also present in human tissues. There are already advanced plans to use these cells, for example, to treat intestinal conditions in which the intestinal epithelial layer is damaged, such as ulcers, with mini-guts. In mice, at least, the organoids can be introduced through the anus and will settle down to functionally fill in the gap in the intestinal wall (see Chapter 10, "Adult Stem Cells: Generation of Self-Organizing Mini-Organs in a Dish"). This could have far-reaching implications for cell therapies for other organs in humans. The outcome of new studies is eagerly awaited, particularly regarding how much of this work in mice translates to humans.

Another case in point with respect to adult stem cells are the hematopoietic stem cells obtained from umbilical cord blood (Chapter 5, "Origins and Types of Stem Cells: What's in a Name?"). There are too few hematopoietic stem cells in one cord blood sample to enable reconstitution of the blood cell population of a patient after, for example, destructive chemotherapy. Generally, two or more matched cord blood samples are required for one adult patient. Being able to expand the stem cells in one sample would therefore be fantastic. However, despite many attempts and development of multiple novel culturing techniques, this research has only advanced slowly over the last several decades. However, one study in a top scientific journal reported that the culture of cord blood cells together with mesenchymal stem cells allowed the hematopoietic stem cells to expand, such that just one sample was needed and the time to repopulate the patient's bone marrow. Restoration of the immune system was shortened by several days, which meant that the patient spent a shorter period being vulnerable to infections that could be life-threatening because, at that point, they no longer have a functional immune system.

Mesenchymal stem (or stromal) cells can be derived from the bone marrow as well as from adipose (fat) tissue. Mesenchymal cells are kinds of fibroblast cells, thought to play a role in building the three-dimensional architecture of bones and organs and in repairing small amounts of tissue damage. These cells appear to be flexible and able to differentiate into at least cartilage, bone, and fat cells. They are also thought to affect the immune system by repressing its activity, although it is not clear how this works. For this reason, many clinical trials have been carried out, are ongoing, or about to be started, using these cells because of their apparent immunosuppressive activity. Some diseases for which such clinical trials have been designed are heart failure, bone and cartilage defects, kidney disease (lupus nephritis), fistulizing Crohn's disease, and graft-versus-host disease. In addition, there are investigations ongoing to see whether their immunosuppressive activity could be used to reduce the amount of immunosuppressive drugs needed after the transplant of a donor organ, such as a kidney. An up-to-date list with planned and ongoing clinical trials can be found at the website of the U.S. National Institutes of Health (NIH).[1]

[1]http://clinicaltrials.gov.

7.11

UMBILICAL CORD BLOOD AND HURLER'S DISEASE

Hurler's disease, also called mucopolysaccharidosis I or MPS I, is a hereditary metabolic disease. The disease is caused by deficiency in the enzyme alpha-L-iduronidase, responsible for the breakdown of specific sugar molecules in cells. If the enzyme is absent, the sugars accumulate in the cells and interfere with their normal function. Clinical symptoms become apparent in the first year of life, but because the disease is so rare the right diagnosis is often missed. Symptoms include stiff joints, typical facial features, reduced growth, and mental retardation. Later in life, heart and lung problems arise, and most children die before the age of seven. Clinical trials have shown benefit from transplantation with bone marrow or umbilical cord blood from a healthy donor. Rather like bone marrow, umbilical cord blood also contains stem cells that are easily harvested after birth and stored for later use in stem cell transplantation.

The underlying mechanism of the therapeutic effect of stem cell transplantation in Hurler's disease is not yet clear, but one attractive explanation is that the bone marrow or cord blood stem cells distribute themselves through the body and the alpha-L-iduronidase enzyme they produce may be sufficient to compensate for the deficiency in the other cells. However, cells that have already been damaged are not repaired; reason enough to start stem cell transplantation as early as possible!

7.7.2 Embryonic and Induced Pluripotent Stem Cells

Embryonic and induced pluripotent stem cells are, in principle, able to differentiate into around 200 cell types present in the body, provided the right culture conditions can be established. Until recently, this was very difficult, certainly if pure populations of one cell type are required. Since an increasing number of scientists are working in the field after the discovery of induced pluripotent stem cells, the number of good growth and differentiation methods, including the availability of necessary reagents, has increased enormously. This has made research much easier so that indeed the field has moved forward significantly. Use of these cells as a reproducible source for transplantation has become a reality. Although difficulties such as obtaining pure populations of the desired cell type, the unknown risk of formation of tumors (called teratomas), proper integration of the cells with the host tissue after transplantation, and immune rejection of the transplanted cells are still present, the problems are no longer considered impossible to solve. It is important to realize, however, that the required quality control and high level of hygiene together with use of special clinical grade culture reagents produce the high costs of cell production.

Embryonic and induced pluripotent stem cells are most easily induced to differentiate to ectodermal derivatives such as neural cells (e.g., cells of the brain and retinal epithelial cells), and there are efficient and defined protocols to induce differentiation to these cells types. Mesoderm derivatives such as cardiomyocytes, endothelial and vascular smooth muscle cells that make up the blood vessels, and, to a lesser extent, cartilage and bone can also be made fairly easily. For reasons unknown, however, the efficiency with which each of these cell types can be produced still varies between available cell lines, and now lies for cardiomyocytes between 20% and 90% of the cultured cells, although the uniformity is increasing with the use of commercial culture reagents. Around the beginning of the twenty-first century, efficiencies ranged from 0.1% to maximally 5%, which illustrates the progress that has been made. Cells of endodermal origin proved much tougher to produce; examples are the insulin-producing cells that are so much needed for diabetic patients, and lung cells. However, this has also now been successful and has been extended to thyroid and thymus cells.

Many cell types are only formed after undifferentiated stem cells have been forced to proceed through a complex sequence of differentiation steps that mimic those taking place during normal development in the uterus. Knowledge of these steps in themselves is still incomplete, with the result that steering cells along the right track is far from trivial. Just consider the many highly specialized cell types that together make up the kidney, and it will be clear that we are still some way from producing all the individual cell types that would be necessary for producing a functional organ replacement. That said, there is no need for pessimism; even scientists are frequently surprised by the huge and unexpected advances in research that can appear from one day to the next. It turned out, for example, that mouse embryonic stem (ES) cells that were only partly differentiated into a particular direction (e.g., neurons) quite unexpectedly specialized and matured in the organ itself (in this case the brain) after transplantation, under the influence of the local cell environment. The same was true for early beta cells which also matured when transplanted in mice into the pancreas. It is thus quite reasonable to expect that transplanted stem or progenitor cells may be stimulated by their new environment to further differentiate to individual cell types in the organ, which may result in restoration of function.

7.12

DUCHENNE MUSCULAR DYSTROPHY

Duchenne muscular dystrophy (DMD) is a disease caused by a mutation in the gene for dystrophin. Dystrophin is the largest and probably most complex gene in the human genome. While in the simplest cases in biology, one gene codes for one protein, the dystrophin gene can code up to 18 protein variations. This intriguing gene is located on the X chromosome, which explains why the disease mostly occurs in boys, and that girls can be carriers without symptoms. This is because females have two X chromosomes, and therefore have two copies (alleles) of the dystrophin gene. If one copy does not function properly, the remaining one still guarantees production of a sufficient amount of the protein. However, males only possess one X chromosome and thus one gene copy. As already implicit in the name *muscular dystrophy*, the disease primarily involves skeletal muscle, heart, and respiration muscles, but may affect also other organs, such as brain, eyes, kidneys, and liver. Patients with this slowly progressive disease usually do not live beyond the age of 20. The most common causes of death are respiratory problems and cardiac failure. Because DMD is not uncommon (1 in 3500 births), considerable research has been dedicated to elucidate the molecular mechanism causing the muscle dysfunction. Unfortunately, patients are often severely ill and their muscle tissue cannot be used for research. Clinical scientists therefore frequently use animal models as an alternative source of research tissue: dogs, cats, even zebrafish, with a mutation in their dystrophin gene have been studied. The mouse remains the most popular animal model for this disease, however, even though progress of the disease seems to be very different from that in human patients. The most representative animal model is the Golden Retriever dog but its use is limited by ethical concerns and high costs.

Recent research efforts are attempting to develop human model systems for DMD based on human pluripotent stem cells. The advantage of this approach is that the affected cells are human and the stem cell origin guarantees an almost endless source of cells to support reproducible experiments. A culture model based on human induced pluripotent stem cells is very suitable to test large numbers of drugs for efficacy, most importantly because the two cells types affected by DMD, skeletal and cardiac muscle, can both be derived from the stem cells of the same patient. In some DMD patients, the skeletal muscle is affected; in others, the heart; and in some, both tissues. In addition to

7.12 (*cont'd*)

drug screening, these types of cells could provide answers to scientific questions arising from the ongoing work on animal models. The expectation is that although this type of stem cell research will probably never fully replace animal research, it does have the promise of significantly adding value to currently available methods.

7.8 TRANSPLANTATION OF STEM CELLS: WHERE WE STAND

Permission to carry out a clinical trial in human volunteers or patients always requires substantial experimental evidence of how the therapy affects experimental animals both in terms of safety and effectiveness. The animals used for these experiments preferably have symptoms of the disease or condition to be treated. The most commonly used experimental animals are mice, because there are many different strains of mice with mutations in their DNA that are similar to those causing human disease (Figure 7.8). They are also fairly easy to handle and are relatively inexpensive to keep. Sufficient experience is available with respect to humane treatment of the animals, even when they do have particular diseases. In fact, some diseases in mice have very mild symptoms compared to, say, humans or even larger animals such as dogs. This type of research in animals is called *preclinical research*. Many stem cell therapies in development are already at this stage, which shows promise with respect to future clinical application. In this section we provide examples of the status of some of these preclinical studies, the outcome of which essentially determines the chance of real clinical applications.

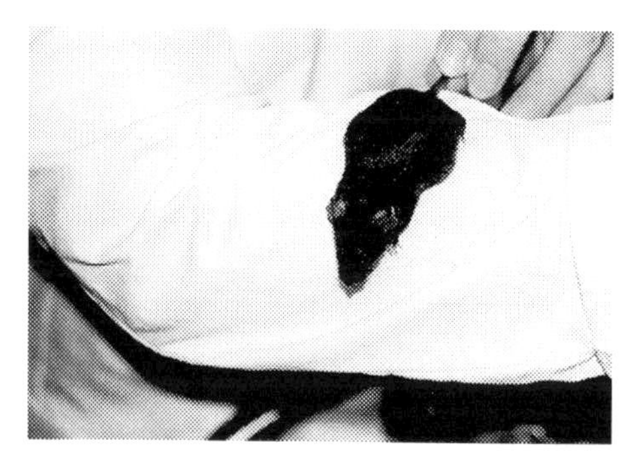

FIGURE 7.8 Diseases caused by the effects of interplay between several different cell types are difficult to model in cell culture systems. By using special mice, often genetically modified to simulate disease symptoms, the disease can be investigated more precisely and the results have a better chance of being applicable to humans. *Source: Stamcellen Veen Magazines.*

To date, most preclinical research with human pluripotent stem cells has been performed in mice. In addition, rats and pigs have also been used, with the advantage that these larger animals more closely reflect human physiology. One of the most interesting types of preclinical experiments has been the transplantation of differentiated stem cells to rats that were paralyzed and unable to walk because of an experimentally induced spinal cord lesion. Under anesthesia, the nerves of the spinal cord were either completely severed or crushed (Figure 7.9). After injection of embryonic stem cells that had been differentiated in the laboratory to oligodendrocytes, the rats regained much of the ability to walk. Spectacular as films of these animals are to see, extrapolation to humans can only be made to a limited extent in terms of effectiveness. This is because in these rodents the spinal cord not only contains nerve extensions, as in humans, but also complete nerve cells. This suggests that electrical signaling from the brain with associated movement of the legs in these animals is organized differently than in humans, implying that the therapeutic requirements for restoring nerves are also likely to be different. Moreover, the mechanism by which paralysis in the animals improved was unclear from the experiments: did the transplanted cells replace damaged cells, or did they just deliver a factor that somehow induced repair of the damaged resident cells? Despite this reservation, expectations were understandably high. After extensive discussions with the regulatory authorities, and 20,000 pages of paperwork, the biotech company Geron was finally granted permission to start "first-in-man" studies in a small group of human patients. However, as we mentioned earlier, this research in humans was later stopped.

Parkinson's disease, as also mentioned earlier in this chapter, is another potential candidate for stem cell therapy. Clinical experience has already been acquired in patients through transplantation of cells from the brains of aborted human fetuses into the brains of patients.

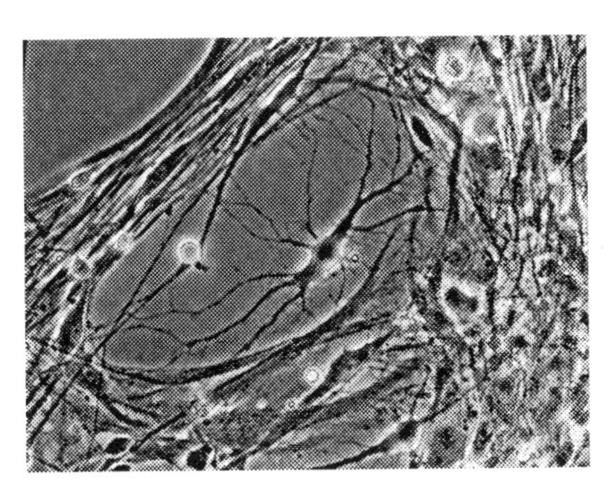

FIGURE 7.9 Stem cells can differentiate *in vitro* to cells that resemble oligodendrocytes. *Source: Leonie du Puy, Utrecht University.*

Clinical results from a Swedish trial were initially promising, and symptoms such as uncontrolled movements improved significantly in a number of patients. Unfortunately, during a later clinical trial in the United States, patients were observed where there seemed to be a cell "overdose," leading to a disastrous increase in uncontrolled movements and inevitably the halting of the trial. This event emphasized the extreme care that is required when introducing brain cell therapy in the clinic, but also highlighted the ethical issues associated with using human fetal abortion material. At one point, wealthy American patients were traveling to China to receive fetal brain cell transplants without evidence that the abortions were entirely voluntary, while it could also not be excluded that pregnancies were established "for sale." This issue would, of course, be redundant if established pluripotent stem cell lines were the source of cells for transplantation.

Cells from these stem cell lines can be differentiated efficiently to nerve cells producing the molecule dopamine, which is produced in insufficient amounts by brain cells of patients with Parkinson's disease. Transplantation of these cells into mice with a form of Parkinson's, indeed, did alleviate some of the symptoms and the cells survived well in the mouse brain. In rats, similar positive results were obtained. Most strikingly, nerve reflexes were partially restored so that rats could pick something up with a front paw after touching it with their whiskers. However, when the brains of these rats were analyzed, few surviving cells were detected (Figure 7.10). This again raises questions about the mechanism underlying the improvement in paw reaction: could it be possible that the transplanted cells indirectly helped restore function, for example, by secreting one or more specific factors that stimulated the regenerative capability of the resident brain cells?

FIGURE 7.10 Micrograph of stem cells that have differentiated to astrocytes. Astrocytes are supporting cells for neurons. These cells have typical cytoplasmic extensions, here visible in red. The nuclei of the cells are stained in blue. *Source: Leonie du Puy, Utrecht University.*

One of the most important discoveries of late has been the realization that in the first patient studies, the right type of neurons might not have been transplanted, which caused a side effect known as *kinesis*. This is continuous and exhausting involuntary movement. In fact, the earlier prestudies on pluripotent stem cells tested in mice and rats may also not have used the right neurons. Evidence suggests that this issue may now have been solved, and research toward a clinical trial has been financed in New York, with a view to starting in 2017 (see Box 7.17 "Lorenz Studer: Pioneer Research on Stem Cell Therapy for Parkinson's Disease").

We anticipate that the experiments described here are only the beginning of an era in which the benefit of stem cells will be explored for many more diseases. Clinical trials for stem cell treatment of deafness, macular degeneration, and diabetes are all close to commencement or have begun, as described earlier.

7.13

STEM CELLS FROM UMBILICAL CORD BLOOD

If there is no HLA-matched bone marrow donor in any database available for a patient, for example, someone with leukemia, stem cells derived from umbilical cord blood may offer an alternative. This is in part because cord blood contains large numbers of stem cells; for ethnically complex backgrounds (say children of an African mother and Chinese father) it is often easier to find a match; and, finally, cord blood can be quite easily obtained after delivery of the baby from umbilical cord and placenta—of course with permission of the mother. The blood is subsequently processed and frozen in liquid nitrogen (−196°C), where it can be stored for long periods of time.

To have stem cells with different HLA types available for transplant use, many countries all over the world have established cord blood stem cell banks to store these cells. Via hospital networks, these stem cells are available on a noncommercial basis for any patient in need of such transplantation. The cord blood banks are required by law to work according to well-defined safety and hygiene criteria for derivation and handling, transport, and storage of the blood. Part of the procedure is testing of the blood from the mother and the cord blood for the presence of viruses and bacteria, determination of the blood group, and HLA type. A disadvantage of cord blood is its relatively small volume, limited to 50–200 ml, just enough to treat a child. Methods for

7.13 (*cont'd*)

culture of stem cells from cord blood are still under development. Thus, an adult requires two cord blood samples for successful transplantation, although in the end only one of these will finally repopulate the bone marrow of the patient and provide all of the new blood cells. How one cord blood batch "wins" over the other is unknown but is being investigated. It may be possible one day to expand human cord blood stem cells in culture so that just one sample will be sufficient to transplant to an adult patient. An exciting development is that culture of cord blood stem cells on mesenchymal stromal cells has now been shown to work, and in patients treated with culture-expanded cord blood stem cells, the time to engraftment is shortened, meaning that the blood stem cell graft "kicks in" earlier. The period that the patient is more or less without an immune system and thus very vulnerable to infection, is therefore reduced.

Unfortunately, the therapeutic potential of cord blood is also being exploited commercially and not always with the best intentions (see also Chapter 9, "Stem Cell Tourism"). Various companies, especially in the United States, now offer to collect and freeze cord blood for future private use by the child from whom it is derived, often at considerable cost (up to around $2000). Apart from the collection costs, an additional storage fee per year is often required (on average around $110). Colorful flyers advertise with slogans like "the gift of life" to try to encourage pregnant women (or future grandparents) to invest in storing cord blood after delivery of the baby as a kind of "insurance": if an organ or tissue would fail later in the child's life as a result of disease, the deep-frozen stem cells would be available to repair the organ and restore its function. A correct claim is that these cells would not be "foreign" to the child's body and thus would not be rejected. However, the disadvantages are never mentioned in these brochures or internet pages with photos of happy smiling babies. Cord blood has so far only provided a potential treatment for a limited number of diseases. The chances that the baby will acquire one of these diseases are small. Also, the disease may be genetic so the cord blood would carry the same (defective) genes, and, in the case of leukemia, the cord blood itself may already contain tumor cells at birth. A study in the United Kingdom also showed that 30% of cord blood samples collected and stored by commercial companies were infected with bacteria (from the delivery room in general) so would not actually be suitable for transplantation anyway.

7.9 RISKS ASSOCIATED WITH A STEM CELL TRANSPLANTATION

The most serious potential complication of transplantation with pluripotent stem cells is probably the development of a benign, or even worse, a malignant tumor. This risk is anticipated to be real, but fortunately much less for adult stem cells than for pluripotent cells, and would be related only to any mutations introduced during culture and not to the intrinsic nature of the cells. Extensively cultured cells may acquire new DNA mutations that promote cell growth, creating a risk that the cells become cancerous. Therefore, for adult cells such as mesenchymal stromal cells, only short culture periods are used prior to clinical application.

The risk of tumor development is inherent in undifferentiated pluripotent stem cells that form so-called *teratomas* when introduced experimentally under the skin of mice (see Chapter 2, "Embryonic Development"). This is also likely to happen in humans. Such a tumor could also become malignant, a *teratocarcinoma*. For this reason, cell transplants containing undifferentiated pluripotent cells would be unacceptable in humans, and it is the highest priority for the regulatory authorities to make sure these cells have been removed in advance. Methods to do this are still being developed but include both selecting only the differentiated cells of interest, or killing any residual undifferentiated cells. It is possible, for example, to sort the cells that are specifically needed for transplantation by labeling them with an antibody that binds to the cell membrane. A fluorescent molecule could be coupled to this antibody, enabling separation between fluorescent and nonfluorescent cells using a flow cytometer or a *fluorescent activated cell sorter* (FACS) (see Chapter 9, "Cardiomyocytes from Stem Cells: What Can We Do with Them?"). Alternatively, the antibody could be bound to magnetic beads, which could then be used to pull the required differentiated and labeled cells from the cell mixture using a magnet. One could also temporarily introduce a gene construct which makes the cell of interest resistant to antibiotics. If grown in the presence of antibiotics, all other cells would die, leaving a pure population of the cells needed for transplantation. Finally, reagents have been developed that specifically kill undifferentiated cells. Cells for transplantation could be treated with these reagents to ensure the undifferentiated cells are gone by the time the transplantation takes place. Another way of doing this after transplantation into patients, to remove any undifferentiated stem cells, is to introduce a *suicide gene* into the cells before transplantation. The suicide gene would be activated in response to a specific drug; this would cause any residual undifferentiated cells in the transplant to die automatically after the transplantation, after administration of that drug to the patient. The obvious disadvantage here, and of other transgenic methods, is that foreign DNA is introduced in the cells. Regulatory authorities such as the FDA are generally reluctant to allow genetically transformed cells to be used as treatment for patients.

In addition to the tumor risk, for many stem cells not obtained from the patient (i.e., not autologous), immune reactions can be the cause of serious side effects and rejection of the transplant.

Aside from direct stem cell associated risks, the transplantation itself may also pose a significant risk. In many cases the cells will be introduced into some of the body's most vital organs—the heart, brain, spinal cord, lungs, and kidney—and these are not trivial procedures.

It is important to realize that careful risk–benefit analyses will need to be made prior to any clinical introduction of stem cell based therapies. In general, the more serious the disease and the fewer the treatment options, the higher the acceptable clinical risk. Disasters in commercial stem cell clinics have usually been the result of the (surgical) procedure to introduce the cells and not the cells themselves (see Chapter 11, "Stem Cell Tourism"), although severe immune reactions have been known.

7.14

STEM CELL THERAPY FOR MACULAR DEGENERATION

Macular degeneration is a common cause of partial blindness in the elderly. They lose their central field of vision because of degeneration of the retina cells in the macular region of the eye. Although some vision remains, reading, for example, becomes nearly impossible for these patients, except with a magnifying glass. Therapy options are still very limited. Macular degeneration is often referred to as age-related blindness.

In March 2010, the FDA announced that it had granted "orphan status" to a human embryonic stem cell therapy directed at providing a cure for a special rare form of macular degeneration in younger patients, called Stargardt's macular dystrophy (SMD). This means that, because the disease is rare, it can be given what is called an "orphan" (or exceptional) status and can be developed for clinical application faster. The therapy first started being developed by a commercial company called Advanced Cell Technology (ACT) and was already documented by several research groups to work in mouse and rat animal models of macular degeneration. Human embryonic stem cells differentiated to retinal pigment epithelial cells were placed by a surgical procedure underneath the retinal cell layer in the eye, and took over the function of the lost retinal cells in supporting the survival of the photoreceptors that lie on top of the epithelium. It requires exceptional surgical skill to place new epithelium under the photoreceptors.

In addition to ACT, several other commercial companies, such as Pfizer in the United States and United Kingdom and CellCure in Israel, and research organizations such as the California Institute of Regenerative Medicine (CIRM), are developing therapies based on retinal pigment epithelium derived from human embryonic stem cells.

In 2012, a report was published in *The Lancet*, a medical journal, of the first clinical treatment of two patients with macular degeneration; one had the Stargardt's form of macular degeneration and the other had the age-related form. The two patients also received drugs to prevent rejection of the transplanted cells. Although the goal of the study was to investigate the safety of the treatment and very few retinal cells were actually delivered, the study raised hopes that once larger numbers of cells can be tested, vision may indeed improve. Within the short, four-month follow-up time, no side effects were detected and, most importantly, no signs of teratoma (tumor) formation were found (Figure B7.14.1).

By 2013, 30 patients with the "dry" form of macular degeneration or SMD had been treated in one eye with a low cell dose as part of a first safety trial. Three of these had improved visual acuity (they could once more count an investigator's fingers and see letters, something they had lost the ability to do during the progress of the disease) in

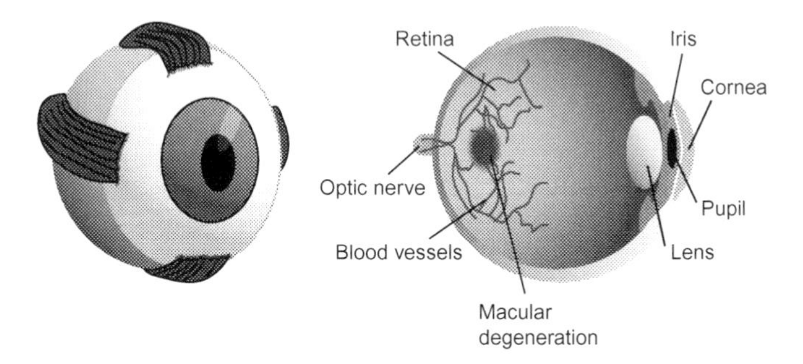

FIGURE B7.14.1 Schematic presentation of the area at the back of the eye in the retina that is damaged in patients with macular degeneration (age-related blindness).

7.14 (*cont'd*)

the treated eye, while the other eye had continued to deteriorate. Japan has concentrated on the use of human induced pluripotent stem cells for the same purpose, and in 2013 obtained permission from the regulatory authorities to treat the first patients. The next developments will include using stem cells to replace the photoreceptors that have been irreversibly lost, to extend the treatment to a broader group. Many millions of elderly people with macular degeneration eagerly await the outcome of these early trials for this distressing condition. The "wet" form of the disease, however, remains untreatable (Figure B7.14.2).

FIGURE B7.14.2 Patients with macular degeneration lose their central field of vision, so what they actually see looks somewhat similar to the picture on the right. The left-hand image shows the same picture as seen through the eyes of healthy people.

7.10 STEM CELLS REJECTED AFTER TRANSPLANTATION

The risk that transplanted cells or tissues will be rejected by the immune system plays an important role in choosing which stem cell type to use to treat a disease. Stem cells derived from tissue of the patient's own body or a monozygotic twin, such as bone marrow cells or induced pluripotent stem cells, are not likely to be rejected by the immune system. An autologous stem cell source would, therefore, in principle, be preferable, except when the disease of the patient is caused by a genetic defect or is necessary for an acute condition in an emergency. In this section we will discuss how the body recognizes something as "foreign."

7.10.1 How the Body's Immune System Recognizes a Cell as Foreign

Our immune system has developed during evolution to kill and eliminate unwanted invaders. These may range from macroscopic splinters to microscopic bacteria or malignant cells, but more recently, of course, also include transplanted cells or tissues. To reject foreign objects, the immune system has first to be able to distinguish very efficiently between invaders and the body's own cells. This recognition process is mostly based on differences in protein content, although some other types of molecules may also play a role. Bacteria obviously have protein repertoires that differ from those of humans and are thus relatively easy to scout as invaders. However, differences also exist between the cells of two different human beings. Particularly relevant are the proteins that belong to the human leukocyte antigen family (HLA proteins). Hundreds of different genes code for these types of proteins, which are incorporated into the outer part of the cell membrane of all cells in the body.

The blood contains a variety of blood cell types, many of which take part in the immune response against transplanted "foreign" (blood or bone marrow) stem cells. This attack is organized in several ways (Figure 7.11).

The immune response of our body to "invaders" consists firstly of recognizing pathogens (micro-organisms such as bacteria or viruses) or other foreign substances, then rendering them harmless by destroying them. The immune response has an innate (inherited) and adaptive (acquired) component, with a clear distinction between the two. The adaptive response carries a memory of the invader and is consequently able to eliminate it in a very specific, but rather slow, manner. The innate response, on the other hand, is a kind of first-line defense, acting rapidly but not very precisely against all kinds of infections and abnormalities. The innate immune response is, in evolutionary terms, much older and is present in various animals, including relatively primitive species. The activities of both immune systems are closely intertwined, and multiple cell types play a role in the overall immune response. With the innate immune response, *phagocytosing* cells such as macrophages "eat" the enemy: they bind to the invading microorganism, engulf it, and efficiently inactivate and destroy it. With the adaptive immune response, lymphocytes recognize the invader, whether inside or outside the cell. Lymphocytes are divided into B-lymphocytes, which produce antibodies against all kinds of pathogens outside the cells (e.g., bacteria), and T-lymphocytes, which are required for the defense against intruders residing inside cells, such

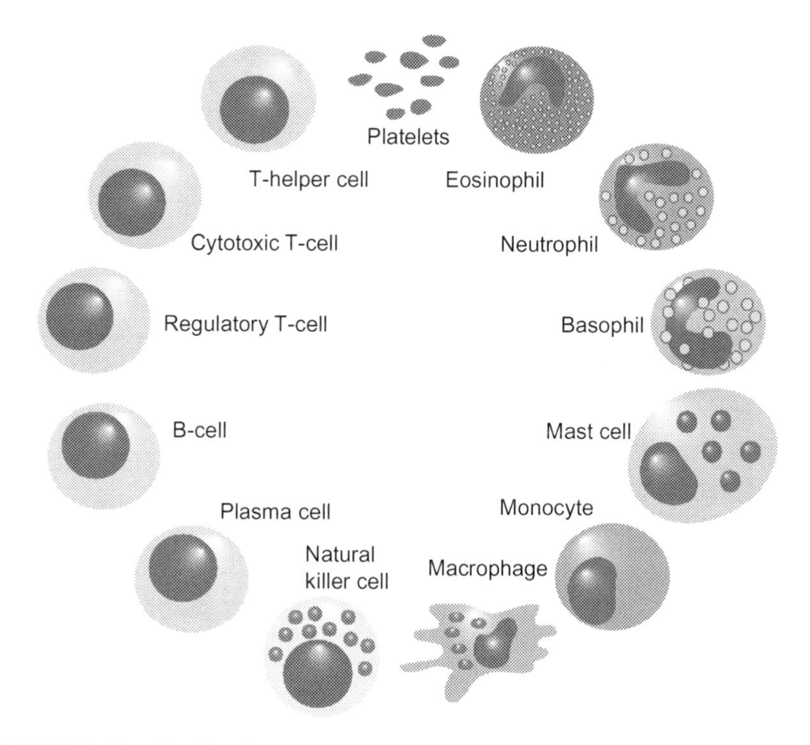

FIGURE 7.11 The blood contains a variety of different types of blood cells, many of which take part in the immune response against transplanted "foreign" stem cells. *Source: Stamcellen Veen Magazines.*

as viruses, and also malignant cells or foreign cells from a transplant. The T-lymphocytes, in turn, can also be subdivided into several categories:

1. *T-helper lymphocytes*, which provide support to other T-cells, but also help B-cells and macrophages
2. *Cytotoxic T-cells*, which directly kill infected, malignant, or foreign cells
3. *Regulatory T-cells*, which are important to balance the activity of the immune response
4. *Natural killer T-lymphocytes*, which recognize abnormal (transplanted, infected, or malignant) cells that do not carry the appropriate HLA proteins on their membrane, and kill those

Within the HLA protein family, two classes are recognized: class I (A, B, and C) and class II (DP, DQ, and DR). These proteins form complex three-dimensional structures in defined combinations on the

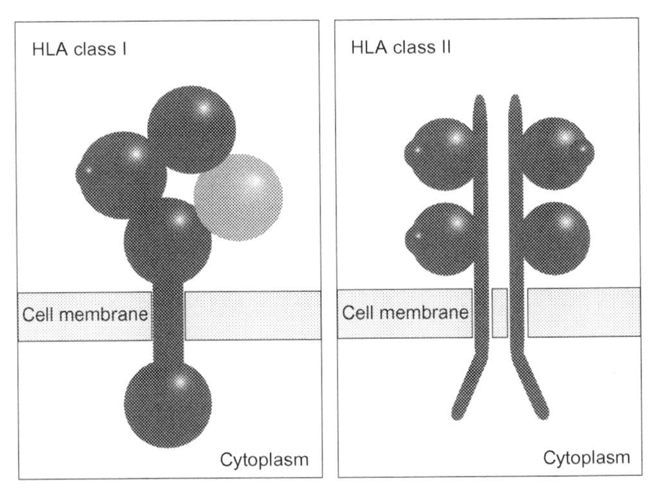

FIGURE 7.12 HLA proteins consist of a glycosylated polypeptide (blue), coupled to beta2-microglobulin (green). The protein contains a carbohydrate group (red). HLA protein class II consist of two noncovalently bound proteins. The shortest protein molecule has one carbohydrate group; the longest carries two carbohydrate groups. *Source: Stamcellen Veen Magazines.*

outside of cells in such a way that they can expose small pieces of other proteins produced by the cell on its surface. HLA class II proteins are present on the cell membrane of a variety of immune cells, for example, monocytes and macrophages (Figure 7.12). These membrane HLA proteins expose pieces of proteins that the cells have acquired through ingestion of, for example, proteins or bacteria to lymphocytes; with membrane exposure comes the message to produce antibodies against the protein and destroy all cells carrying these protein markers. These immune cells are called *antigen-presenting cells*. In contrast, HLA class I proteins are located on all body cells and present small pieces of the cell's own proteins on the cell surface.

Within the human population, many variations exist in the genes coding for the HLA proteins. For this reason there are a large number of different variants of each HLA protein and, as a consequence, the overall profile of these proteins on the cell is rarely the same between two individuals. Only monozygotic twins share identical combinations of protein variants. All the cells that make up an individual carry the same unique combination of HLA-proteins. In addition to this aspect of the immune system, there are *ABO-histocompatibility antigens*, the well-known blood group antigens. Besides blood cells, most other cell types also have one of these three protein variants on their cell surfaces.

Even before birth, this variable and personal protein pattern is imprinted in the memory of the developing immune system of the fetus. From that point onward, the lymphocytes of the immune system will recognize their own HLA proteins as "self" and belonging to the body, and leave all cells carrying these proteins alone. Cells with another HLA profile, however, will be recognized as foreign and consequently killed. Transplantation of cells with the wrong HLA protein combination on the cell membrane, or the wrong blood group type, usually leads to the death of the foreign cells and rejection of the transplant. In the case of embryonic stem cells, HLA type I proteins are only minimally present and blood group proteins not at all. This would seem to be favorable for transplantation, although after differentiation the immune proteins do become present on the cell's surface. In addition, little is known as yet about the reaction after transplantation into a real patient. One would expect to find more or less similar effects with induced pluripotent stem cells.

An extremely interesting phenomenon is that a few places in the body are relatively protected against any activity of the immune system; so-called *immune privileged* sites. Examples are the brain and spinal cord (of the central nervous system), but also the eyes and testicles. Rejection of transplanted cells would, as a result, be less complicated when treating a disorder such as Parkinson's disease or macular degeneration, where the stem cells are injected into the brain compartment or eye and are therefore less likely to be attacked by the immune system—although this remains to be formally established.

7.10.2 Reverse Rejection: Graft-Versus-Host

A graft-versus-host reaction, where cells from the patient are attacked by the transplanted cells, only occurs when the transplant contains antigen-presenting cells and lymphocytes. These cells have the task of recognizing foreign proteins and cells, which, in the case of cell transplantation, are obviously abundant in the patient. It is a well-known and dangerous complication in patients who have received bone marrow transplants that contain antigen-presenting cells, for example, to recover from chemotherapy treatment against leukemia. In contrast, a limited and well-controlled graft-versus-host reaction can also be used to the benefit of the patient with leukemia because the responsible cells will also kill remaining tumor cells. Pluripotent stem cells do not appear to spontaneously differentiate to antigen-presenting cells, and so this is not expected to occur with transplantations involving these stem cell sources (Figure 7.13).

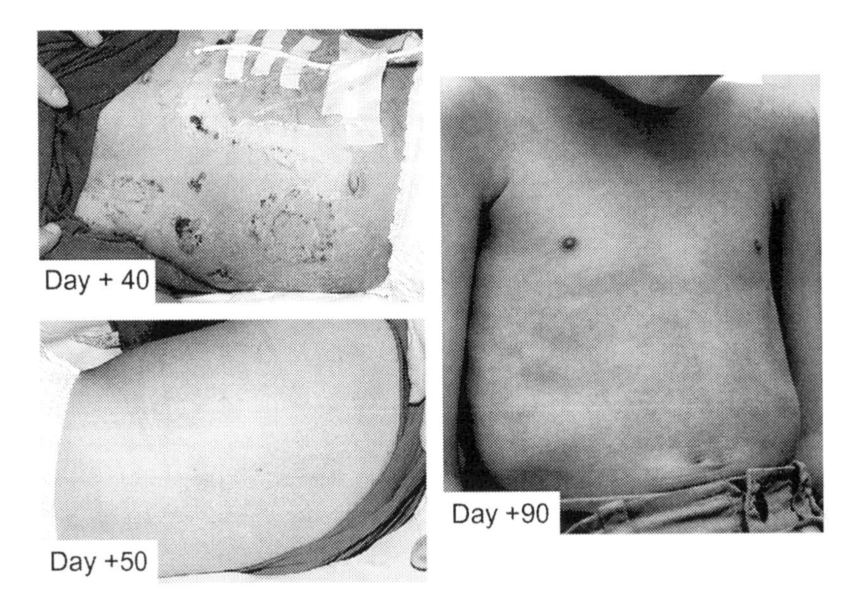

FIGURE 7.13 A boy with leukemia was given chemotherapy to destroy the blood cancer cells. To replace the blood stem cells that had also been (accidentally) killed by the chemotherapy, chemotherapy was followed by transplantation with immunologically matched bone marrow from a donor. The boy unfortunately developed an adverse reaction to the transplanted bone marrow cells, called graft-versus-host disease. This resulted in many open sores on his skin, which are clearly visible in the picture. Treatment with mesenchymal stem (or stromal) cells reduced the adverse reaction, and within three months the skin had healed completely. *Source: Lynne M. Ball, Departments of Pediatrics, Immunology and Hematology (Professor W. Fibbe), Leiden University Medical center, Netherlands.*

7.15

GENE THERAPY

One of the first experimental applications of gene therapy was the treatment of patients with a genetic disease called *ornithine transcarbamylase (OTC) deficiency*. Patients born with this disease lack a specific enzyme that is essential for removal of nitrogen, a waste product of protein metabolism, from the blood. As a consequence, these patients have a high concentration of ammonia in the blood stream, which may ultimately lead to severe brain damage and coma. In less severe cases, the disease can be suppressed by a diet with reduced protein content in combination with special drugs, or by transplantation of a healthy liver from a donor. Scientists from the University of Pennsylvania in Philadelphia in the United States have tried to cure this disease by

7.15 (*cont'd*)

means of gene therapy. This entailed introducing DNA coding for the missing enzyme into the liver cells, which then allows the cell to produce the missing protein. The coding DNA was packed in special viruses (the *vector*), which were injected into the portal vein of the liver of the patients. Mice and monkeys with the same enzyme defect had already been successfully treated by this approach in preclinical studies. Sixteen human patients with the disease volunteered to take part in a clinical trial to investigate whether the treatment was safe. In seven patients the injected DNA was indeed found expressed in liver cells but, unfortunately, no improvement was found in the patients' condition. Then disaster struck: an 18-year old patient developed an extremely serious immune reaction to the injection and died, retrospectively because of an error in the dose of virus given. The trial was immediately stopped.

Severe combined immunodeficiency (SCID) is a collective name for a number of inherited diseases characterized by a defective immune response that leads to seriously compromised resistance to infections. In the absence of effective treatment, children with SCID in general do not live much beyond four years of age. Allogenic bone marrow transplantation is currently the only therapeutic option, despite there being an appropriate (HLA-matching) bone marrow donor available for only one in three patients. Children with the disease for whom there is no donor match are, as a last resort, treated with a bone marrow transplant that is not optimally HLA-matched, reducing their chance of survival to only around 75%.

SCID has also played a role in the history of gene therapy. At the end of the last century, clinical tests were performed in hospitals in Paris, London, and Milan in which bone marrow was first collected from children with SCID and the genetic defect corrected using a specially constructed vector that contained DNA coding for the missing protein. The vector used in these studies was an inactivated retrovirus. The cells were infected with this retroviral vector and the genome in the cells was checked for the presence of the new DNA. Then the cells were returned through intravenous infusion to the patient. In total, over 14 patients were treated in this way. At first it seemed a revolutionary success: the treated children showed the appropriate immune response following vaccination and for the first time in their life they were allowed outside their normally sterile environment. Only several years later did a second gene therapy disaster surface: 3 of the 10 patients treated in Paris developed leukemia (blood cancer). The exact cause of this is still not known. The most likely explanation, however, is that the virus, containing the new DNA, randomly inserted itself in the genome of the blood cells as is usually the case, but by accident ended up in the middle of another gene, probably one that protects blood cells against development of leukemia. This insertion interrupted the code in the gene for the protective protein, which, as a consequence, could no longer be made by the cell, probably causing leukemia in these patients.

These disastrous events had a dramatic effect on progress with gene therapy in the clinic. It took over a decade before confidence in this therapeutic approach was restored. The U.S. Food and Drug Administration (FDA), which is the most powerful regulatory authority for approval of clinical treatments, decided to temporarily restrict the development of gene therapy that made use of viral vectors to introduce new DNA into cells. Research in this promising field is now slowly picking up speed again, although few clinical applications for gene therapy have yet been officially approved. However, in 2008 a U.K.-based clinical trial (UCL Institute of Ophthalmology and Moorfields Eye Hospital NIHR Biomedical Research Centre) was performed to treat a form of blindness (Leber's congenital amaurosis). This disease develops soon after birth and is caused by a DNA mutation that destroys a specific gene, called RPE65. The missing gene was introduced into the retinal cells surgically, and the outcome showed that this treatment was safe, as well as creating some improvement in vision. Another success occurred in patients with an advanced stage of melanoma (a highly malignant form of skin cancer). Lymphocytes from these patients were genetically modified in the lab to enable these white blood cells to recognize, migrate toward, and efficiently attack melanoma cells. Several patients for whom no other treatment options remained showed impressive improvement and survival. Many other gene therapies are currently being worked on, in principle, always first in animal models. In cases of positive effects, clinical trials on patients will be started. Despite initial optimism about this approach, progress remains extremely slow.

7.10.3 How to Prevent Rejection of a Cell Transplant

Several approaches can be used to prevent or limit rejection of transplanted cells. Of course, specific drugs can be administered to the patient to suppress the immune reaction and limit the process of rejection; this is usually done in the case of allogenic (from an unrelated donor) bone marrow transplantation and transplantation of whole organs. This carries with it definite disadvantages because, at the same time, the natural resistance to infections as well as to certain malignant cells is reduced. It would be best if the immune cells would leave the transplanted cells alone while, at the same time, performing their important surveillance role in the body normally.

Since immune rejection is mostly caused by differences in HLA proteins between the cells, the problem would be solved if the transplanted cells and the cells from the patient carry identical HLA complexes. It is for this reason that for organ transplants (e.g., of a kidney) and bone marrow transplants, searches are generally made to find a donor or donor-organ carrying HLA and blood group molecules that have the highest similarity to those of the recipient patient, although recent developments in immunosuppressive drugs are making this less essential for some organs. Nevertheless, an HLA-matched tissue will be the least perceived as foreign. The donor search has been professionally organized in Eurotransplant (*www.eurotransplant.nl*) and covers an international registry of available donors. Defining the right blood group for this purpose is relatively easy, while, in addition, blood group O is acceptable for all patients. However, because a perfect match between the complete HLA profiles of two individuals is extremely rare, some medication to suppress the immune system usually remains necessary, even if the best donor has been identified.

By analogy, for stem cell transplantation, the best match among a number of pluripotent stem cell lines could be chosen from a stem cell bank containing cell lines with the most prevalent combinations of HLA and blood group genes. For human embryonic stem cell lines and, more recently, induced pluripotent stem cells, an effort in this direction has already been made with a stem cell bank in the United Kingdom (*www.ukstemcellbank.org.uk*) and in Spain. More recently, a similar bank has been planned for induced pluripotent stem cells. In the United States, the national stem cell bank contains information on availability and characteristics of both embryonic and induced pluripotent stem cell lines, however, the cell lines are only available for research purposes, not for transplantation (*www.nationalstemcellbank.org*). Whether embryonic stem cells or induced pluripotent stem cells become the cells of choice in the future is presently unclear: each has advantages and disadvantages. One can nevertheless envision building large public collections ("libraries") of pluripotent stem cell lines, derived from an ever-increasing range of different individuals and covering most HLA gene combinations in the world's populations. Such an approach would ultimately enable tight matching of the HLA profile between the stem cell line and the patient, and thus would be expected to prevent most rejection reactions—just like the standard protocol for matching an organ donor to the transplant recipient. Initiatives to establish commercial stem cell banks for transplantation purposes are an obvious alternative scenario. Banks of diseased pluripotent stem cell lines are of particular interest to the pharmaceutical industry (see Chapter 13, "Human Stem Cells for Organs-on-Chips: Clinical Trials without Patients" and Chapter 14, "Stem Cells for Discovery of Effective and Safe New Drugs").

Even in case of the optimal HLA match, other proteins than those encoding the HLA antigens are produced by cells and these may still show minor compatibility differences due to variations in the population, so some degree of rejection may be unavoidable. Of these proteins, the *minor histocompatibility antigens* (proteins) of males, for which the coding genes lie on the uniquely male Y chromosome, play a relevant role in rejection by immune cells from female patients—they will not recognize these proteins as "self." To solve this, it might be preferable to select female stem cell lines which lack a Y chromosome and its associated proteins.

An alternative to a perfect HLA match is "re-education" of the immune system of the patient to the extent that the transplanted cells are no longer perceived as foreign cells. Over the past decades, experience with this approach has been built mainly in the area of bone marrow transplantation. For pluripotent stem cell transplantation this could be done by differentiating stem cells from a pluripotent stem cell line into blood stem cells, and subsequently introducing these cells into the patient's bone marrow, where they can exist next to the patient's own blood stem cells: a chimeric bone marrow has been created. This procedure will induce "tolerance" for the transplanted cells, and needs to be performed prior to the therapeutic cell transplantation. If successful, such an approach will allow recognition of the transplanted cells as self instead of foreign.

In some cases, transplantation at immune-privileged sites might offer a way to avoid rejection. In the case of Parkinson's disease, spinal cord injury, or macular degeneration, this happens anyway, since the cells would be introduced into the central nervous system or eye. For other diseases, this usually only presents an option if the transplanted cells can function independently and outside of their normal organ context. It might be a feasible option, for example, to treat patients with diabetes with insulin-producing cells, on condition that the cells in their new location can properly sense blood glucose levels and, as a response, secrete appropriate amounts of insulin into the blood. An alternative would be to encapsulate the transplanted cells so that they are inaccessible to the large molecules and cells of the immune system, while smaller molecules, for example, insulin, can still be secreted into the blood as required.

7.16

NON FEDERAL GOVERNMENT STEM CELL RESEARCH INSTITUTES IN THE UNITED STATES

Wealthy businesspeople Bob Klein in California and Susan Solomon in New York both have children with diabetes (Figure B7.16.1). Their wealth however, cannot cure their children. All they are able to do is visit the doctor with their children for check-ups and insulin, just as any parent from any walk of life. But they had a vision of doing more. Both have set up large research initiatives to see whether stem cells can be of help, not only in the treatment of diabetes but also in developing therapies based on stem cells for many other types of chronic disease.

Bob Klein was the driving force behind Proposition 71 (or Prop 71 as it is sometimes known) in California. This is part of a voter referendum passed in 2004, in which the state of California authorized the sale of bonds for 10 to 14 years for the sum of $3 billion. This was to be distributed by a new institute, the Californian Institute of Regeneration (CIRM), to scientists at research centers around the state. Susan

FIGURE B7.16.1 Susan Solomon. *Source: David McKeon, NYSCF, New York, NY, U.S.A.*

Solomon likewise invested enormous personal effort in raising funds to found the New York Stem Cell Foundation in 2005. Both institutes fund basic and applied research to develop therapies based on stem cells for a multitude of chronic diseases. Most importantly, at a time when President George W. Bush had banned most research using National Institutes of Health (NIH) funds to develop therapies based on human embryonic stem cells, these private institutes actively supported research on human embryonic stem cell lines outside of those on the so-called "presidential list" (the cell lines eligible for NIH funding). Researchers supported by these institutes could use any human embryonic stem cell line that was most suitable for their research (Figure B7.16.2).

Harvard University, likewise and with the same motivation, established the Harvard Stem Cell Institute (HSCI). The HSCI appointed Douglas Melton and David Scadden as its codirectors and supported Melton's research group in the isolation of more than 70 new stem cell lines for basic and applied research. Significant financial support for the creation of those new lines came from another private entity, the Howard Hughes Medical Institute, a private philanthropic foundation. Many other U.S. universities followed suit, so there is now a large network of privately funded stem cell institutes throughout the United States, all able to carry out research on stem cell lines not supported by NIH funds at the time. These institutes largely prevented the United

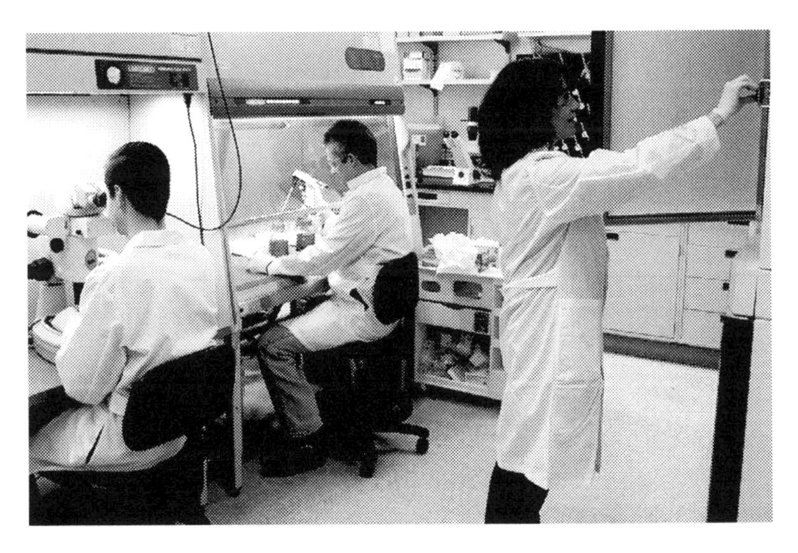

FIGURE B7.16.2 NYCSF (New York Stem Cell Foundation) scientists at work. *Source: David McKeon, NYSCF, New York, NY, U.S.A.*

7.16 (*cont'd*)

States from lagging behind research carried out in some of the rest of the world where legislation and/or funding were less restrictive. This allowed U.S. researchers to quickly take advantage of the new political climate when President Obama lifted the restrictions and NIH began funding research with new cell lines in December 2009.

7.11 TISSUE ENGINEERING

With the donor organ shortage as a powerful incentive, doctors, biologists, and tissue engineers are trying to create whole organs or parts of organs in the laboratory to replace and repair diseased tissues; this is called *tissue engineering* and forms part of the field called *regenerative medicine*. Constructing three-dimensional tissues from two-dimensional cell layers outside the body is a major challenge and requires very specific knowledge and skills. First, the right cells are needed. These can be derived from stem cells but the stem cells first need to differentiate into the right direction and become cells of the tissue to be repaired. In addition, tissues are made up of different cell types, so that often more than one cell type is required. The heart, for example, is made up of only 30% cardiomyocytes (contracting heart muscle cells); the rest are blood vessel cells and myofibroblasts, a kind of connective tissue cell specific to the heart. Second, the cells need a structure or scaffolding on which to attach, grow, and carry out their specialist function, such as producing bone or cartilage proteins. Last, but not least, the three-dimensional tissue-engineered organ part should have an adequate supply of oxygen and cell nutrients, and cellular waste products need to be disposed of. In addition, for therapeutic transplantation, the three-dimensional substrate on which the cells grow needs to be "immunologically inert," which means that it does not activate the immune system in the body of the patient being treated. Depending on the particular disease application, it may also be useful for the implanted scaffolding to be biologically degradable and resorbed with time so that in the end it is fully replaced by the body's own support tissue (Figure 7.14).

Taking some of these requirements into consideration, it may seem a small miracle that the tissue engineering approach is actually successful at all. In some cases it has not only been possible in a tissue engineering laboratory to mimic the complex process of organ development that normally takes place inside the body, but it has also been possible to transplant the organ parts successfully to patients. Bioartificial

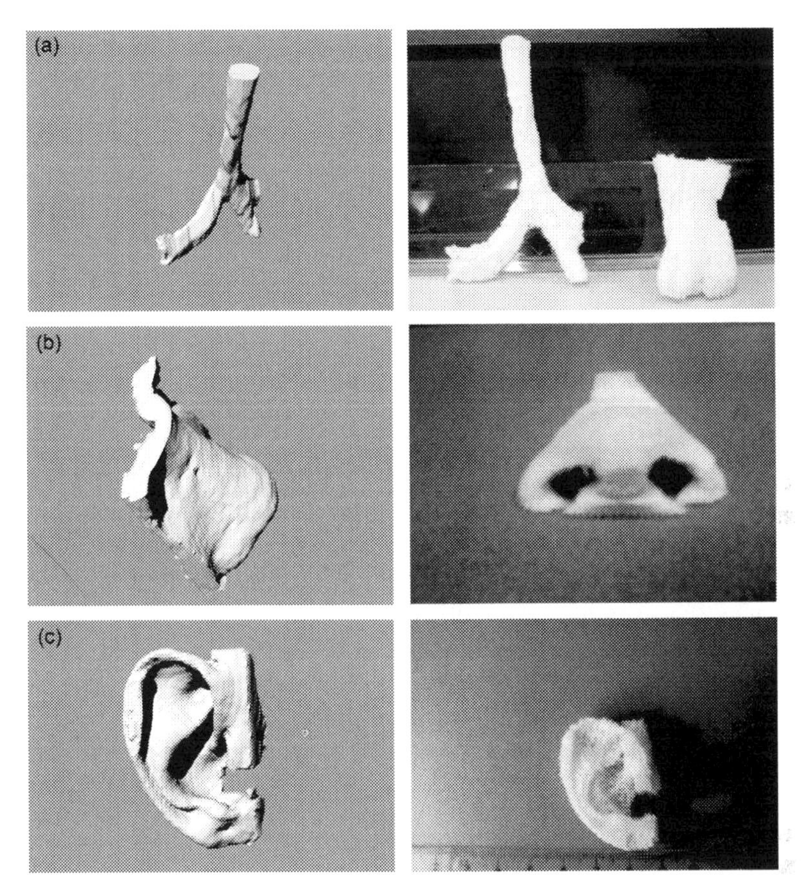

FIGURE 7.14 Three-dimensional scaffolds can be fabricated by *fused deposition modeling* (a rapid phototyping technology) in different anatomical shapes through the acquisition of computed tomography or magnetic resonance imaging datasets. Here, examples for (a) trachea, (b) nose, and (c) ear are shown. *Source: Adapted with permission from Moroni L., et al., J Biomater Sci Polym Ed 2008; 19(5):543–72.*

bladders for patients who were incontinent because of spina bifida and a tissue-engineered trachea (windpipe) for a young woman almost dying of the damage caused by tuberculosis (see Box 7.1, "Professor Anthony Hollander: A Tissue Engineering Pioneer") are recent success stories of the tissue engineering field. On the other hand, for many tissues, it is still extremely difficult to imitate the normal process of tissue formation in the laboratory. Within an organism, cells are brought together in the right format and kept in exactly the right position by complex intercellular interactions and interactions with a protein substrate, built of proteins from the cells themselves. This protein layer is called the *extracellular matrix*. Specific signal molecules are secreted by cells and bound to other cells, to transmit information into the cell on how fast the cells need to divide, where they have to migrate within the developing organ, and in which direction they have to differentiate. In addition, an extensive network of small blood vessels develops throughout the organ architecture, which takes care of appropriate delivery of oxygen, nutrients, and removal of waste. Considering the huge differences between various tissues (e.g., the hardness of bone compared with cartilage or soft skin), it may not be so surprising that tissue engineering for many organ parts will still require a substantial research investment before it can become clinically applicable.

The rapid development of the stem cell field has, however, proved to be a powerful new motor for tissue engineering research. Successful production of tissues and organs in the lab requires the availability of sufficient numbers of healthy specialized cells; a reliable stem cell source is then an excellent solution for this purpose. Also, three-dimensional structures need to be developed to which the cells can easily attach, grow, and further differentiate when required. "Bioreactors" need to be designed, rather like the huge vessels used to produce yeast or beer, to imitate the growth conditions present in the body suitable for large-scale cell production (Figure 7.15).

To grow living tissue outside the human body, *in vitro*, as this is called, cells are first allowed to attach to an artificial substrate, which resembles as closely as possible the structure of the tissue that needs to be produced (in terms of hardness, composition, and shape), and also at the right time releases the necessary chemical signals to guide cell growth and differentiation, and build a three-dimensional tissue structure. These substrates are preferably made up of porous, biologically degradable materials. Different components can be used, both natural and artificial polymers created in the chemistry laboratory, each with its characteristic advantages and disadvantages. Substrates obtained from biological sources, for example, protein components purified from animal sources such as collagen (often from rat tails) and a complex extracellular matrix extract marketed as Matrigel® (obtained from mouse sarcoma cells), have the advantage that they are easily degraded and remodeled by the cells. On the other hand, these proteins are difficult to purify, may be rejected by the immune system because they are recognized by the body as foreign, and may contain pathogens (virus particles or prions) that can cause disease in the patient. Plant extracts, such as alginate from seaweed or algae, are alternatives that are of increasing interest because they circumvent some of these problems. Interesting other options are synthetic polymers, such as polyglycolide, polyactide, and polylactide-coglycolide (Figure 7.16). Because synthetic substrates only contain components that have never been in contact with reagents derived from animals, they are free of the

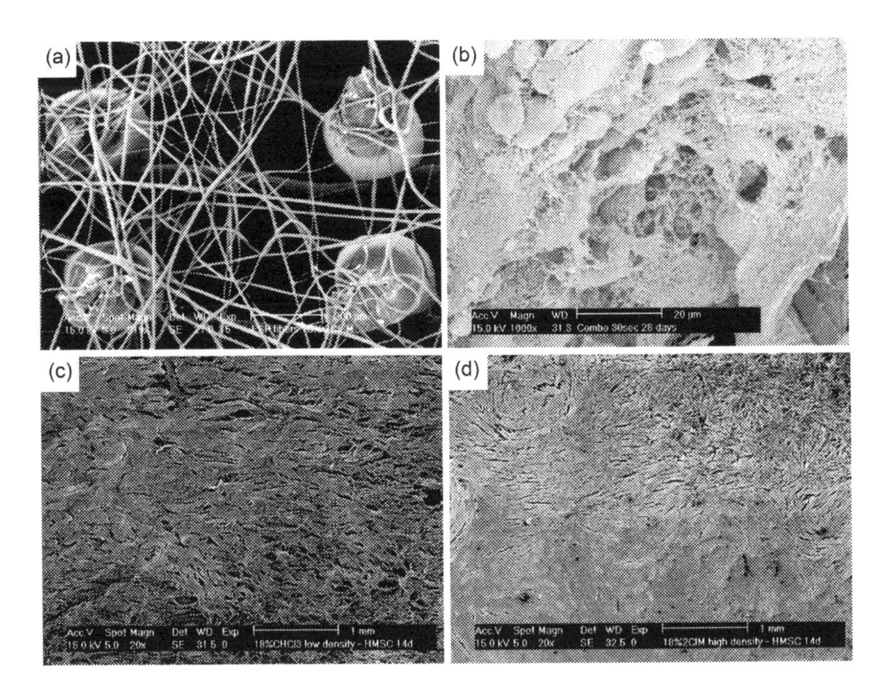

FIGURE 7.15 Three-dimensional scaffolds can be used alone or in combination with other technologies such as (a) electrospinning, enabling the generation of multiscale porous matrices with fiber dimensions mimicking those of natural extracellular matrix (ECM). (c) and (d) Cells adhere and proliferate at a high rate in these ECM-like structures producing (b) a densely packed ECM after 4 weeks of culture. *Source: Lorenzo Moroni and Clemens van Blitterswijk, Tissue Regeneration Department, University of Twente, Netherlands.*

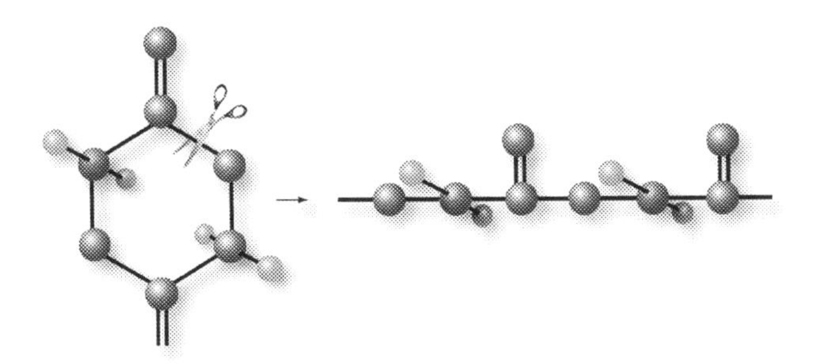

FIGURE 7.16 Schematic representation of a polyglycolide. Polyglycolide is a polymer structure built out of ring-shaped molecules. *Source: Stamcellen Veen Magazines.*

risk of transmitting infectious diseases. This is a very important advantage over natural sources, and is the main reason that these materials are most promising for the future. In addition, they can be treated to release specific chemical and physical signals at controlled rates and concentrations, which may allow better simulation of natural biological materials than natural materials themselves.

The creation of off-the-shelf fully functional organs and organ parts that can be directly transplanted into patients is something we can presently only dream about. However, for the future there is every reason to believe that at least some potential applications will become routine clinical practice. Tremendous progress has already been made in creating relatively uncomplicated tissues, such as bone and cartilage. Three-dimensional cultured artificial bone and cartilage tissue has already been in clinical use for several years, to treat patients in which the amount of skeletal tissue lost as a result of trauma or disease is so large that it would be impossible to repair the defect by traditional surgery. Large bone loss because of removal of a tumor mass is one example. Another that has been successful is the replacement of skull bones lost as a result of a shotgun accident. The construct is often seeded with bone marrow stem cells derived from the patient to be treated. While often successful, for reasons unknown, it does not work for all patients, and chances of success decrease significantly with age. Much ongoing research is devoted to addressing these problems (Figure 7.17).

Tissue engineering is also being used successfully in clinical practice to treat wounds that do not heal spontaneously such as can occur in diabetic patients. Fibroblast cells are first obtained from the foreskins

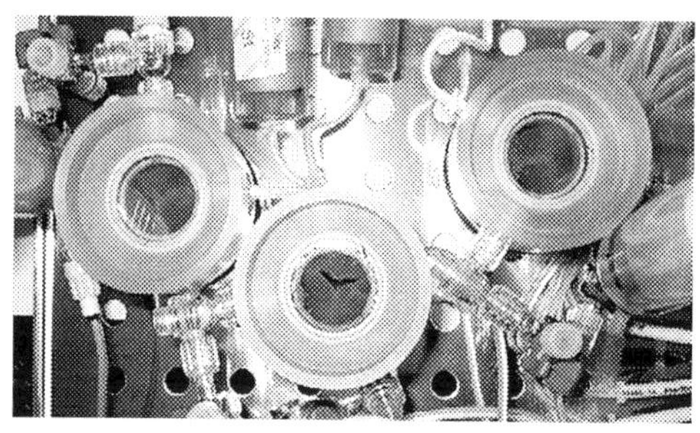

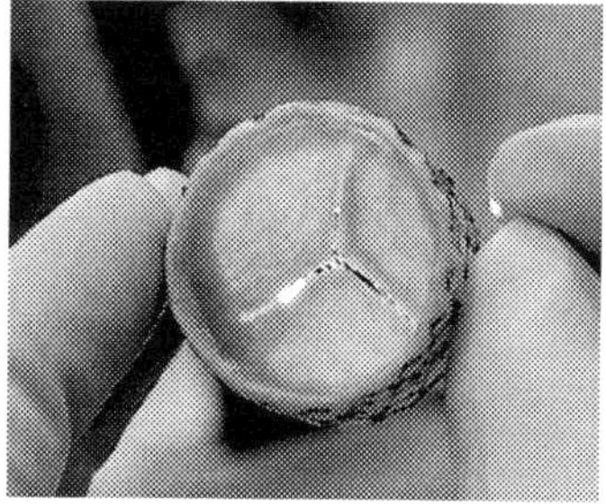

FIGURE 7.17 Tissue engineered heart valves. A structure in the form of a heart valve is engineered from a structural extracellular matrix protein, collagen, seeded with stem cells and placed in a "bioreactor," which mimics the flow of blood. Three are shown in (a). The valve is thus conditioned and remodeled before being transplanted into a real heart. The valve ready for transplantation is shown in (b). The triangles in the middle are artificial leaflets, which open and close with each heart contraction to prevent blood flowing back into the heart. At this stage, the research is preclinical, the valves being transplanted to animals such as sheep, but the expectation is that they will eventually replace valves collected from animals, to become "living valves" that even grow. This would be very useful in treating children where the heart is still growing and the valves become too small every few years and need to be replaced to stop them leaking. *Source: Petra Dijkman, Anita Driessen-Mol, Carlijn Bouten, Frank Baaijens. Soft Tissue Biomechanics and Tissue Engineering, Department of Biomedical Engineering, Eindhoven University of Technology, Netherlands.*

of circumcised baby boys, although sometimes they may come from the skin of the patients themselves. These cells are then cultured in the laboratory and grown on a substrate carrier made of polylactide-coglycolide until a tissue layer is formed that resembles the inner layer of a normal skin. This product can then be used for transplantation. Another artificial skin product consists of both this inner layer and the outer visible layer of the skin. Patients with extensive skin defects as a consequence of burns, for example, are already being treated with such tissues. This process is still considered experimental because, just like bone, the success of the treatment is variable for every patient, even when the cells are derived from healthy areas of their own skin. Other areas of interest include a three-dimensional synthetic structure for a heart valve, seeded with the patient's own cells and placed for "conditioning and remodeling" in a bioreactor that imitates the stress of blood flow on the engineered valve. It is hoped that in the future, heart valves can be made that are not only functional, but also grow and adapt to a growing heart. This would be applicable in children where growth of the heart, as the child ages, means that inert valves become too small and need replacing every few years until the child is an adult.

One of the practical restrictions for culturing tissues *in vitro* in the lab is the relatively short distance over which oxygen, the essential fuel for the cell, can diffuse and penetrate a layer of cells—at best a quarter to a half millimeter. A potential solution could be to add blood vessel cells to the cells of interest, and culturing the cell mixture on substrates that release chemical signals or growth factors in a controlled way to stimulate the growth of new blood vessels. Obviously for *in vitro* culture the active circulation of blood cells through these blood vessels is another difficulty.

7.17

LORENZ STUDER: PIONEER RESEARCH ON STEM CELL THERAPY FOR PARKINSON'S DISEASE

People with Parkinson's disease do not have enough dopamine. Dopamine is produced by a special kind of nerve cell, or neuron, in a part of the brain called the *substantia nigra*. It sends messages to parts of the brain that control movement. The disease kills dopamine-producing cells and as these cells die and the dopamine levels decline, patients develop tremors and rigidity, their movements slow down, and they can also lose their sense of smell or suffer from sleep

7.17 (*cont'd*)

disorders, depression, constipation, and sometimes dementia in the later stages of the disease.

Lorenz Studer was part of the first team that, in 1995, attempted to treat Parkinson's patients with fetal tissue. Although first results were encouraging, there were problems, not least being the difficulty of obtaining enough fetal tissue regularly to treat multiple patients and the ethical sensitivity of the cell source. Studer turned to stem and progenitor cells and, while working at the National Institutes of Health in the United States soon after, pioneered techniques that allow the generation of dopamine cells in culture from dividing precursor cells (Figure B7.17.1). In 1998, he demonstrated that after transplantation, dopamine-producing neurons generated in culture improved clinical symptoms in rats with Parkinson's disease. Studer then went on to show that virtually unlimited numbers of cultured mouse dopamine cells could be obtained from embryonic stem cells. More recently he has succeeded in making highly efficient dopamine-producing neurons from human embryonic stem cells and has transplanted them into the brains of rats and mice with Parkinson's disease. The cells did not multiply abnormally (form teratomas) and improved some symptoms.

The breakthrough was that he could finally make the right type of neurons, called mid-brain neurons, at the right stage of development to replace those lost in the disease. When he transplanted the neurons into monkeys he was also able to show that they would survive and function, even in these larger animals. This was not obvious, because the neurons have to send long extensions from the place in the brain where they are formed to the part of the brain where they have

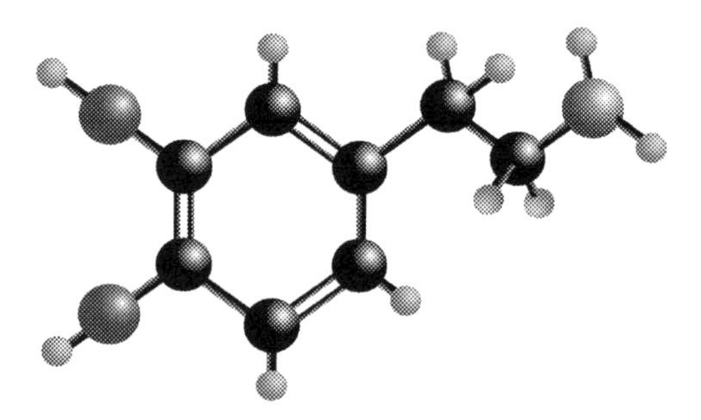

FIGURE B7.17.1 Dopamine. A reduction of dopamine production by the cells in the substantia nigra in the brain can cause Parkinson's disease.

their effect. This is easy in small rodent brains, but not so easy in larger animals and humans.

Much work is still needed before tests can begin on human patients: the neurons need to be made in sufficient numbers to be effective and be produced in a way that ensures the cells are safe. In addition, scientists are still investigating whether the dopamine producing neurons could actually be transplanted at their site of action rather than in the substantia nigra, saving them the challenging journey through the brain tissue. The scientists have received funding to move toward therapeutic applications and hope that early clinical trials may start in 2017 in New York. They are already applying for approval from the FDA for this purpose.

CHAPTER

8

Stem Cells in Veterinary Medicine

OUTLINE

Animals have many of the same ailments and injuries as people so stem cells could, in principle, also be used to treat animals with damaged tissues and organs. In fact, the potential effectiveness of most stem cell therapies is actually pretested in animals, and these studies can therefore be useful for obtaining more fundamental information on animal development, differentiation of animal cells to different tissue types, and development of disease. It is hoped that this will lead to the development of new drugs and therapies for veterinary medicine, particularly for valuable show or breeding stock. In addition, pet animals such as dogs and rabbits can be extremely important as preclinical models for human medicine, in some cases actually serving this purpose better than standard laboratory animals (e.g., mice and rats). Human medical conditions, such as heart disease, are often recapitulated more accurately in larger species than in small rodents. In light of this, one might expect that progress in the veterinary field would be substantially advanced than in human medicine, but actually the opposite is true.

The development and use of stem cells as therapy in the veterinary field has been slowed down by a number of factors. Firstly, really good embryonic stem cell lines meeting the stringent requirements of stable pluripotency have not been generated for any animals other than rodents and primates. This means that embryonic stem cell lines from, say, the pig, horse, cat, or dog have not (yet) been generated. Secondly, the motivation for applying regenerative techniques to companion animals has been rather low because of the expected high costs involved for something that in the first instance would be only experimental. For farm animals, veterinary medicine has changed from trying to cure diseased or injured animals with costly therapies toward management that prevents diseases occurring in these animal groups. A hard economic fact is that where there is no demand, there is no development. Thirdly, the therapies that have been tried have been done on a rather ad hoc basis, with treatment of only single animals, poorly described protocols, and little or no documented follow-up. Often these "one off" treatments come with a high price tag and regrettably little rationale. Finally, stem cell research in large animals lacks many of the molecular biological and immunological "tools" (e.g., antibodies and gene sequences) and this has hampered proper research. In this respect, we know much more about mice and ways to treat them than we know about diseases and possible treatments for pets, horses, and farm animals.

As importantly, large animals are also excellent preclinical models for human medicine, not only because of their size and physiology, but also because they do not necessarily require an injury to be induced. Rather, injuries and diseases that occur in these animals naturally are very comparable to those that also occur in humans. Of the ~400 genetic diseases known in dogs, approximately 58% resemble specific human disorders caused by mutations in the homologous gene. In parallel, over 250 genetic diseases have been described in cats, of which almost half have a high homology with human genetic diseases. The recent complete sequencing of dog and cat genomes will further aid in the understanding of gene function in these animals, and this will benefit both veterinary and human medicine. This fact is often not recognized, however. In addition, companion animals such as dogs usually share the same environment as humans, and this makes them reliable as models for environmental diseases. Furthermore, companion animals that are kept until they are very old suffer from diseases similar to age-related diseases in the elderly human population and these may be treatable by tissue regeneration. Such conditions include chronic organ failure and tissue degeneration.

8.1 TREATMENT OF FAMILY PETS

The use of stem cells in veterinary medicine is currently mainly focused on tissue regeneration of the musculoskeletal system. In particular, tendon injuries and osteoarthritis in horses and various bone and cartilage defects in dogs could be treatable using stem cells. Importantly, there is a growing market for such treatments as patient owners are willing to pay significant amounts of money to treat their (prize-winning) animals. We have to be cautious here, however, so that the same unproven claims of cure plaguing human medicine do not contaminate the veterinary field for the same reasons of disproportionate economic gain. Pet owners can be almost as vulnerable as desperate patients and their families.

A big difference with the use of stem cells in human medicine is that regulatory authorities such as the U.S. Food and Drug Administration (FDA), U.S. Department of Agriculture (USDA), and

FIGURE 8.1 Horses are particularly prone to tendon injury.

comparable organizations have not so far formulated any regulations on the use of stem cells in veterinary medicine. As a result, veterinarians in private practices can and are offering "therapies" using stem cells of various origins for the treatment of diseased horses, dogs, and the occasional cat without proper control studies, reports of efficacy, and follow-up. In general, there is only anecdotal evidence for the efficacy of many of these procedures; well-controlled studies with enough animals included for proper statistical comparisons by, for instance, veterinary colleges hardly exist. Obviously, patient owners who have paid between 1500€ and 2300€ (~US$2,000–3,000) for a treatment are likely to be biased toward seeing a benefit and believe that their treasured animal has improved. Just as in any commercial enterprise, there will be differences in pricing and quality between the various companies. Proper, double blind, randomized placebo-controlled studies still need to be performed to determine the efficacy and safety of the different stem cell treatments that are commercially available for veterinary use.

Currently, the main application for stem cells in veterinary medicine are for treatment of tendon, ligament, or cartilage tissue in horses and dogs using mesenchymal stem (or, rather, stromal) cells.

8.1.1 Tendon in the Horse

In horses, a large tendon known as superficial digital flexor tendon (SDFT) is the equivalent of the Achilles tendon in humans (Figure 8.1). Especially in competitive performance horses used for racing or jumping, these tendons are prone to injury. It has, for instance, been estimated that about a quarter of the horses that are in training for hurdle

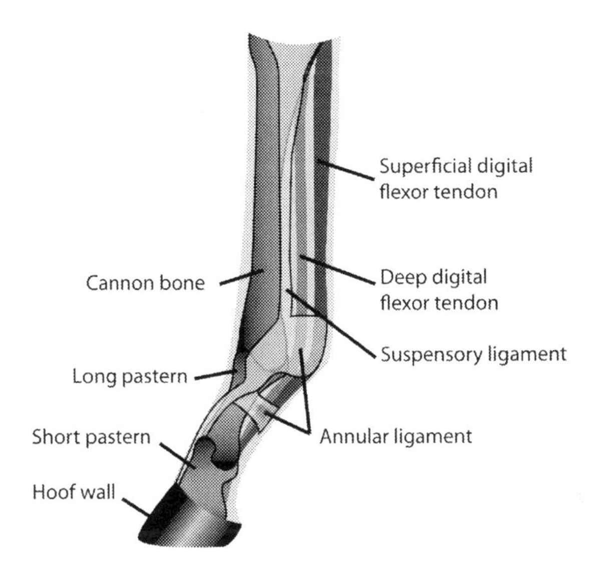

FIGURE 8.2 Schematic picture of a horse leg showing the major tendons.

and steeplechase racing suffer from tendon injuries at some time during their careers. The injury rate in horses that are used for flat races tends to be somewhat lower, but still averages at about 1 in 10. The injury can be chronic and progressive (e.g., inflammation) due to intense training or more acute, such as in overloading or overstretching or, in the most severe cases, complete tendon rupture. Tendon injuries, in general, take a long time to heal (more than a year), and they may not heal at all, partly as a result of the poor blood supply to tendon tissues. Natural healing often results in the formation of scar tissue, which leads to inferior biomechanical properties of the repaired tendon (Figure 8.2). After recovery, the horse often has poorer functionality than prior to damage and is predisposed to reinjury. It has been proposed that stem cell therapy can be used to encourage and facilitate tendon regeneration.

In many countries, because treatment using stem cells in veterinary medicine is less restricted by regulations than human medicine, there is large variety in the amount of detail in the documentation of efficacy and also on follow-up of animal patients. This does, however, provide an interesting window of opportunity to investigate an equivalent to the human conditions.

In stem cell treatments of injured tendons in horses, a decision first has to be made on which types of stem cells are to be used. Overall, there is a tendency to use bone marrow stem cells that can be isolated from the same horse; these are autologous and will therefore not be rejected by the horse's immune system. Using a relatively simple procedure, bone marrow cells can be isolated from the sternum of the animal under local anesthetic. The cells are then generally cultured *in vitro* to allow them to divide a number of times to increase cell number. Rather than differentiating these cells *in vitro* to tendon tissue, the cells are often introduced into the animals directly while undifferentiated. The idea is, then, that the cells would form tendon tissue in the animal or aid in the formation of tendon by other cells. There is some question about how the cells would best be delivered into the animal: systemically by injection into the blood stream from where they could " home" to the site of damage, or locally by injection at the site of injury. The latter option has the advantage that more cells can be directed specifically to the injury site; with systemic injection, cells might end up in places in the animal where they could cause damage, for example, by blocking small capillary vessels.

Cells introduced at the site of tendon injury can be done so in a kind of scaffold matrix such as a collagen sponge. However, it has been observed that cells within a scaffold matrix have a tendency to specialize. This specialization is not necessarily toward cells that make up tendon (tenocytes) but may also include, for instance, ectopic bone spicules. Obviously, bone tissue within a tendon structure would not favor normal tendon function but, rather, would inhibit its ability to behave normally. For this reason most scientists and veterinarians choose to inject undifferentiated cells directly at the lesion site, without the presence of a scaffold.

Generally, between 2 and 10 million cells resuspended in blood serum are injected at the lesion site. Such treatments do indeed seem to result in improved tendon recovery. Most horses treated this way can have a shortened time to functional recovery and, importantly, the reinjury rates seem to be lower in horses treated with stem cells compared with horses treated in other ways. The exact working mechanism of improved recovery is unknown. Whether the injected cells themselves synthesize new tissue, or whether these cells or the injected blood serum provoke the resident tendon cell population to form new tendon tissue through, for instance, secretion of bioactive trophic factors remains to be determined. Alternatively, the anti-inflammatory function of bone marrow stromal cells could enhance endogenous repair mechanisms. Although no significant adverse effects after injection of bone marrow cells have been observed, a better understanding of the mechanism of action is essential. This would require well-documented experiments with proper numbers of both experimental and control animals and detailed follow-up.

8.1.2 Osteoarthritis in Dogs

In large dogs in particular, the joints can suffer from a progressive degeneration of articular cartilage and subchondral (adjacent to the cartilage) bone, a condition known as osteoarthritis. The degeneration of cartilage leads to loss of function due to pain, and clinical symptoms include lameness and refusal of the animal to participate in strenuous activities. This condition is currently mostly treated using nonsteroid anti-inflammatory drugs, but stem cell therapy could possibly lead to better results. For this reason, veterinary stem cell companies have made stem cell therapy for treatment of osteoarthritis commercially available. Both mesenchymal stem (stromal) cells from bone marrow and from fat tissue are used for these treatments, but, overall, cells from adipose tissue, so-called adipocytes, are preferred because of the ease of access to subcutaneous fat tissue.

First, cells are harvested from the fat then about 4 to 5 million cells are injected directly into the affected joint. Clinical follow-up by trained veterinarians is usually for a limited time period (e.g., three months) and in these three months in many cases there is indeed improvement of the dogs clinical symptoms: the animals show improvement in lameness and fewer pain responses after manipulation. However, much as in human medicine, outcomes in veterinary medicine can also be biased by something of a placebo effect. Not so much because the "patients" believe they have been treated, of course, but because the patient owners and even skilled veterinarians convey positive reactions. Dogs that are injected with a simple salt solution, for instance, instead of fat or bone marrow cells also showed improvement in walking when analyzed by veterinarians, providing that they were unaware of the type of treatment at the time of scoring. This placebo affect is even more pronounced if coming from the patients' owners. In experiments where dogs were treated with either stem cells or a salt solution in a blinded fashion (patient owners did not know to which group their dog belonged), the same level of improvement was observed by the owner, independent of the treatment. Whether or not the animals suffered from any unwanted side effects longer times after the treatments has so far not been properly documented.

What could be the mechanism of the improved condition in dogs with osteoarthritis? It is very unlikely that the fat cells, or the stem cells residing in those cell populations, would form osteochondral tissue. Rather, the cells are thought to secrete factors that downregulate the immune system, much like nonsteroidal anti-inflammatory painkillers such as carprofen, etodolac, and deracoxib do. In addition, the injected cells could activate the endogenous repair mechanism to stimulate endogenous cells to form cartilage.

8.1

HAMBURGERS FROM STEM CELLS

Regenerative medicine is one promising application of stem cells. There may be an unexpected application beyond regenerative medicine: meat from stem cells! This is because one of the tissues that at least some stem cells could make is skeletal muscle, leading to a form of extreme engineering: using stem cell-derived skeletal muscle from pig or cow for the generation of food.

For many, this is a bizarre thought at first. Traditionally, man has domesticated and kept animals such as cattle, pigs, goats, and sheep, to be fattened, slaughtered, and eaten. Indeed, meat is the most important source of protein for much of the developed world. In addition, this is also a valuable source of iron, zinc, and especially vitamin B12 as this is not present in plant products. Why would one want to change this?

For some people, keeping animals just to eat is cruel or incompatible with their religious beliefs; these are the present-day vegetarians and vegans. In addition, meat from domesticated animals comes at the expense of the environment. With a growing global population this price will continue to increase over the next decades. The environmental impact of farm animals is also high: the emission of anthropogenic greenhouse gases from farm animals (particularly methane production by cows) has been calculated as being even larger than that from the whole worldwide transport sector: cars, trucks, trains, ships, and planes combined. In addition, an estimated 30% of (ice-free) land on the planet is used for grazing livestock or growing animal feed and about 8% of global fresh water use is due to livestock. While impressive, these numbers are not yet cause for concern, although a search for alternatives would seem reasonable just on these grounds. However, the world population is continuing to increase and, in parallel, so is meat consumption. It has been estimated that the global meat consumption will increase from 228 million tons in 2002 to close to 465 millions tons in 2050. Earth by then will be inhabited by around 8 to 10 billion people. How are we going to feed the world with meat without damaging the planet?

One solution being investigated is tissue engineering meat from stem cells. Meat is a mixture of skeletal muscle, fat, cartilage, blood, and vessels. Although mouse stem cells could, in principle, be used most easily for this because they are easy to grow in the laboratory, consumers would probably not like the idea at all, even though it is simply a way of converting amino acids into protein. Stem cells from traditional farm animals such as cattle and pigs would be more acceptable, but these are more challenging and how exactly to culture and instruct them to become muscle cells is unclear. In addition, knowledge about stem cells from farm animals would not be enough to make large amount of meat substitute from stem cells. Knowledge on scaling up cell cultures to large quantities, bioreactors, and food processing would also be needed before edible products could be made on an industrial scale (Figure B8.1.1).

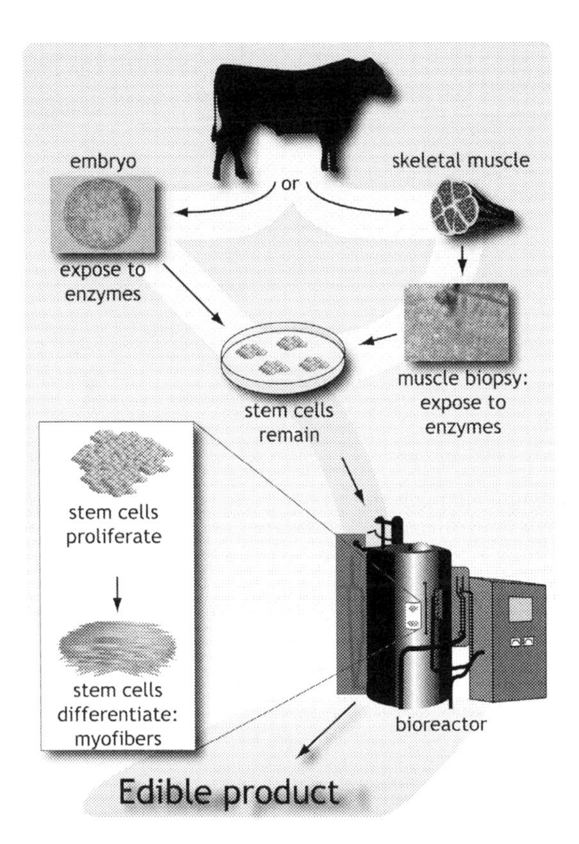

FIGURE B8.1.1 Scheme showing how meat could be made from stem cells. Stem cells isolated that are either pluripotent or have been isolated from adult tissue such as muscle (adult stem cells) can be grown in a bioreactor to generate large numbers of cells. When there are enough, different factors can be added to turn them into skeletal muscle cells. These can be combined with "ingredients" such as connective tissue protein to make an edible product.

8.1 *(cont'd)*

How close would "cultured meat" be to the real thing? Realistically it would not be much different from, for instance, factory-made cheese or beer. Both are made by combining natural ingredients in a bioreactor and after some time the products are formed. Cultured meat, however, could drastically reduce greenhouse gas emissions and land and water use. Two other ingredients would also be important for consumer acceptance: food safety and price. Cultured meat would be as safe as regular meat and, in fact, could reduce the outbreaks of diseases such as common foot and mouth disease that have led to worldwide preventive culling of millions of animals in the past decades. For most consumers, cost will be very important and, aside from the development costs, a stem-cell burger should not cost more than a real burger.

In the summer of 2013, a beef burger from cultured cow muscle cells was presented during a spectacular publicity stunt in London, pan-fried by a professional chef and tasted by a culinary writer and a nutritionist. This "proof of concept" patty had cost around 250,000€ to produce. Taste is not debatable: the panel decided that the burger definitely had a meat flavor and bite, was a bit dry, and could do with some more fat than the oil it was baked in. It is expected that it will take another 10 years at least before cultured meat can be found in the grocery store next to real meat and plant-based meat substitutes. However, the well-orchestrated publicity stunt of 2013 may attract investment and make the concept familiar by the time the consumer has an opportunity to buy.

8.1.3 Induced Pluripotent Stem Cells

Well-characterized pluripotent embryonic stem cells have only been generated from rodents and primates (human and nonhuman). Why the generation of embryonic stem cell lines from other mammalian species such as cats, dogs, cows, or pigs has proven to be so much more difficult is perhaps related to differences in early embryological development (and the natural prevalence of tumors called teratocarcinomas; see Chapter 2 "Embryonic Development"). Pig, cow, and dog induced pluripotent stem cell lines have been described, but their maintenance required continued forced expression of the four pluripotency genes for reasons unknown. This makes them presently unsuitable for therapeutic applications, although they are exciting material for understanding why some species are resistant to reprogramming. The development of induced pluripotent stem cells has opened up avenues of generating pluripotent stem cell lines in humans without being dependent on preimplantation embryos. Whether this will be the case in other animal species remains to be seen, but there is certainly interest, since these species are excellent models for human disease, as we have seen in this section.

CHAPTER

9

Cardiomyocytes from Stem Cells: What Can We Do with Them?

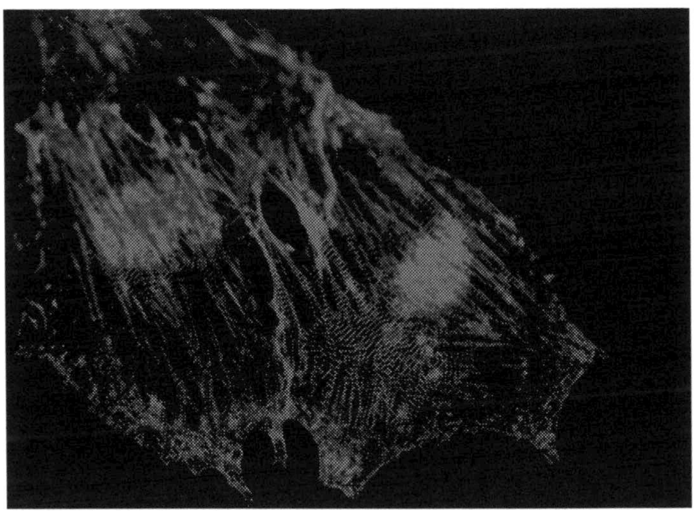

Susana Chuva de Sousa Lopes

OUTLINE

Cardiomyocytes are the cells in the heart that make it contract. There are several different kinds of cardiomyocytes in the heart. By investigating how they form (or differentiate) in the early embryo, it has been possible to develop ways of making them from some types of stem cell. In this chapter we discuss how they might be used in biomedical, clinical, and pharmaceutical research. Although the focus is on cardiomyocytes, the same principles of how we find out how to make them from stem cells and what we can do with them are applicable to many other types of cells: neurons from the brain and central nervous system, or, for instance, cells of the liver, lungs, or kidney. Differentiation of nearly all of these cell types from stem cells was greatly facilitated by basic research on their formation in embryonic development. Their application in biomedical, clinical, and pharmaceutical research, likewise, has many parallels both in the potential applications and challenges that need to be made for clinical translation in tissue repair. We thus use the heart as an example, and give counterparts in other tissues.

9.1

PRECLINICAL STUDIES ON TRANSPLANTATION OF HEART CELLS IN ANIMALS

Several laboratories around the world have transplanted beating heart cells derived from human pluripotent stem cells into the hearts of animals, usually after inducing a myocardial infarction artificially. This is done while the animal is under anesthesia by tying off one of the heart's blood vessels. Rats or mice that lack a normal immune system have mostly been used for these studies, as they do not reject transplanted human cells, but pigs have also been used. This is because pigs' hearts are similar to humans in many ways and they require drugs for immune suppression, just like humans. The cells are injected using a needle coupled to a syringe containing either a suspension of single heart cells or small clumps of cells into the damaged part of the heart. The cellular composition of the heart muscle is relatively simple (cardiomyocytes, cardiac fibroblasts, and blood vessel cells) and new heart muscle cells are thus injected to end up among other more-or-less similar cells.

It was expected that integration and formation of proper electrical connections between the transplanted and endogenous heart cells would take place. After cell transplantation, heart function in the animals was carefully monitored to see whether the animals that had received the new heart cells had heart function better than the control animals without the new cells. After a few weeks or months, during which the animals made a rapid recovery, the hearts were examined to see whether the transplanted cells were still present. Various methods have been used to distinguish transplanted cells from those of the host, including fluorescent markers that tag the cells before transplantation, or antibodies that recognize human cells lying in-between the cardiomyocytes of the animal (Figure B9.1.1).

The results of these complex experiments have, in some ways, been disappointing. Although some of the transplanted cells could indeed be found back in the heart (and in some studies quite a large number when special cell survival factors were added), many cells had somehow been lost. Even more importantly, the function of the heart only improved marginally and only temporarily in most studies. For the cells to contribute to the contractile function of the heart, full integration within the remaining viable heart tissue would be required. For this, the cells should be properly aligned and connected to one another.

The conclusion of these experiments was that, while some cells survived for many months, they probably did not integrate in the host tissue as hoped for, resulting in limited and short-lived improvement in heart function. Part of the problem was that the stem cell derived heart cells were from pluripotent stem cells of a different species (human to

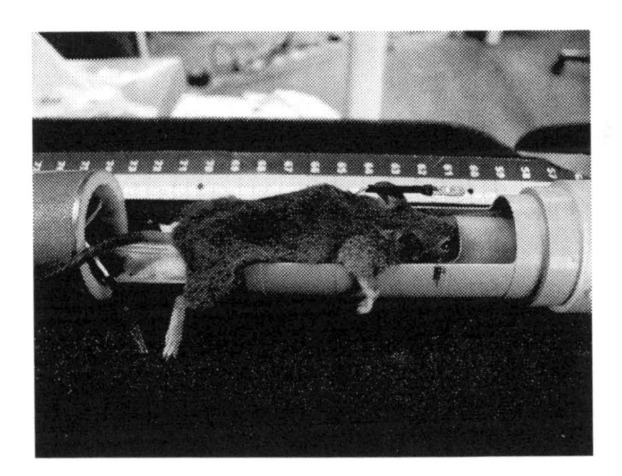

FIGURE B9.1.1 Magnetic resonance imaging (MRI) can be used to make a scan of a mouse heart in much the same way as scans are made of patients, but with the exception that the mouse is given an anesthetic prior to the procedure. In this way, the effects of treatment with stem cells can be studied in a mouse that has had a myocardial infarction. It is possible to investigate whether the function of the heart has improved as a result of the treatment. *Source: Cees van Echteld, University Medical Center, Utrecht, Netherlands/Stamcellen Veen Magazines.*

9.1 (*cont'd*)

mouse, rat, or pig). This *xenografting* is a problem. The only solution is to use donor cardiomyocytes from the same species. Pluripotent cells have only been derived reliably from mouse, nonhuman primates (monkeys), and humans. The most informative experiments for humans would be the nonhuman primates, but these experiments are not ethically acceptable in many countries. The United States does allow these experiments in exceptional cases. The first results from these studies are encouraging: engraftment seems greatly improved above xenografts and there is also functional improvement in the heart (Figures B9.1.2 and B9.1.3).

How, then, should these results be transferred later to human patients and what have they taught us about the challenges ahead? We need to make sure that, when introducing beating cells into the

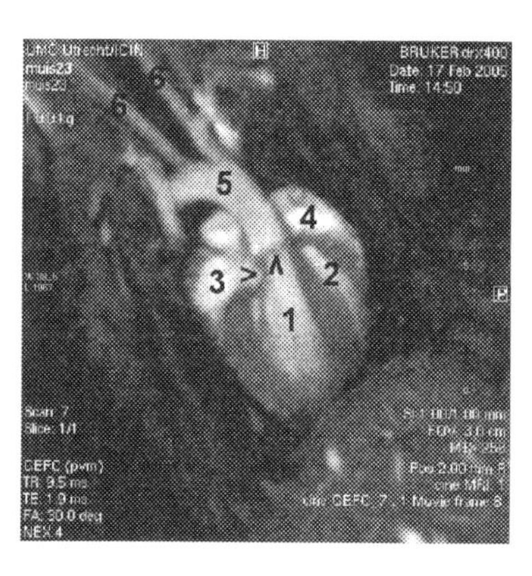

FIGURE B9.1.2 MRI scan of a healthy mouse heart. *Source: Cees van Echteld, University Medical Center, Utrecht, Netherlands/Stamcellen Veen Magazines.*

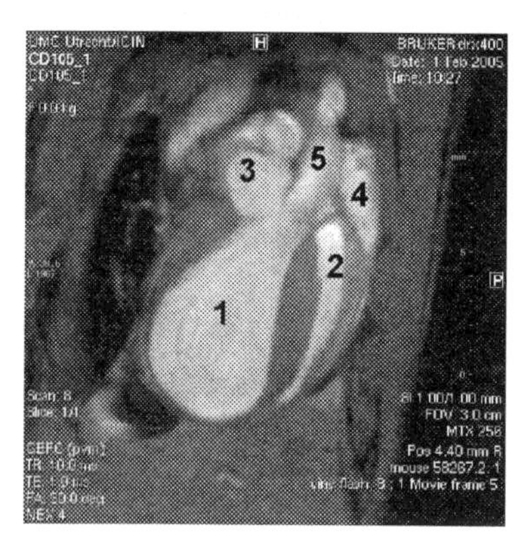

FIGURE B9.1.3 MRI scan of a mouse heart after a myocardial infarction. The dark area is the cavity of the heart chamber; it is much larger than in a healthy heart. In addition, the heart muscle wall is thinner than that of a mouse that has not had a myocardial infarction. *Source: Cees van Echteld, University Medical Center, Utrecht, Netherlands/Stamcellen Veen Magazines.*

heart of a patient, the transplanted cells do not act independently as a kind of artificial pacemaker, sending inappropriate signals to the heart muscle tissue to contract. Such a mechanism could lead to potentially fatal arrhythmias and an artificial pacemaker would be required to prevent them occurring. Fitting a pacemaker simultaneously with cardiomyocyte transplantation may in the future be a standard part of the procedure. These issues are aside from the risk of teratoma (tumor) formation due to the presence of undifferentiated stem cells in the transplant. A high priority is making sure that all undifferentiated stem cells have been carefully removed before transplantation.

Current thought is that stem cell derived heart cells may best be attached to special scaffolds to encourage the heart cells to connect and align before transplantation. This could function as a kind of "patch" for the heart, a little like a patch on a punctured tire.

Preclinical studies in nonhuman primates may lead to solutions to these problems. Only then will it be possible to make a better assessment of the true perspectives for therapy in humans to replenish cells lost after a myocardial infarction to prevent or treat heart failure.

9.1 THE HEART AND CARDIAC REPAIR

If the blood supply to the heart is blocked, for example, by a blood clot, cardiac muscle cells (cardiomyocytes) will die because of lack of oxygen. This is called a *myocardial infarction*. The number of cells that die depends on how long and by how much the blood supply is reduced, but has been estimated to be up to a billion cells. This type of ischemic damage caused by a lack of oxygen resulting from a blood clot can also take place in the brain. It is then referred to as *stroke*. Wherever the blood supply is interrupted in the body, ischemic damage can occur.

The small number of stem cells thought to be present in the heart cannot repair the amount of damage resulting from a myocardial infarction. The loss of muscle cells and their replacement by scar tissue reduces the heart's contractile muscle mass, so that less blood can be effectively pumped around the body (Figure 9.1). This is known as heart failure and is a serious, sometimes life-threatening, condition. The failing heart cannot provide organs and tissues with sufficient blood, so they receive too little oxygen to sustain normal function. Blood that cannot be adequately pumped into the aorta accumulates behind the left cardiac chamber and, as a consequence, the pressure in the small blood vessels in the lung increases. Fluid may, thus, leak out into the lung tissue. Transport of oxygen from air, inhaled by the lungs, to the

FIGURE 9.1 Myocardial infarction (more commonly referred to as a heart attack) is caused by lack of blood to the heart, usually the result of thrombosis or a blood clot, blocking blood vessels to the heart muscle. Without a blood supply, the cells lack oxygen and die. The wall of the heart thins locally at the site of the dead tissue and has to be replaced quickly by scar tissue (fibrosis) to prevent the heart rupturing under the pressure of the blood that it pumps around the body. This picture shows a cross section of a mouse that has had an experimentally induced myocardial infarction. The heart wall is thin and the blue color indicates the extensive areas of scarring. The scar tissue contains no living cardiomyocytes, so is, therefore, no longer able to contract with the rest of the heart. *Source: Linda van Laake, University Medical Center, Utrecht, Netherlands.*

blood becomes lower than it should be, and the amount of oxygen in the blood decreases. Shortness of breath, one of the most noticeable symptoms of heart failure, is just one of its consequences. Because the heart cannot pump as well, blood pressure to the kidneys is too low for proper filtration of the blood to occur and even more fluid accumulates. Lifelong use of drugs is required to alleviate the symptoms, and even then these tend to worsen progressively, as the disease cannot be cured.

Let us suppose, however, that the cardiac muscle cells that had been destroyed by the infarct could be replaced with new healthy cardiomyocytes, either immediately after they have been lost or later when scar tissue has formed. The ability of the heart to pump blood around the body might improve or, even better, development of heart failure might actually be prevented. Symptoms would improve but, more importantly, replacing lost cardiomyocytes might cure the disease and reduce or eliminate the need for drugs to increase the contractile force of the heart. Alternatively, it might be possible to enhances the heart's own ability to produce heart cells, either from resident stem cells or by encouraging mature cardiomyocytes to divide.

9.2

A FLUORESCENCE MICROSCOPE

Objects can only be studied well under the microscope when they are transparent. Cells that are cultured one layer thick in a culture dish are still transparent and can be easily studied under the microscope, but to study tissue made up of more layers of cells, very thin slices are cut from the object to allow sufficient light transmission so that the tissue can be seen. The tissue sample is first embedded in carrier material, such as paraffin or plastic, and then cut by a very sharp metal, glass, or even diamond knife into extremely thin slices. These thin tissue slices are then transferred to a glass slide for examination under the microscope.

The first microscope was made around 1595; but, unfortunately, its maker is not known. Over time, the quality of the simple light microscope greatly improved, but the principle remained unchanged; in essence the microscope is a very strong magnifying glass. Details of

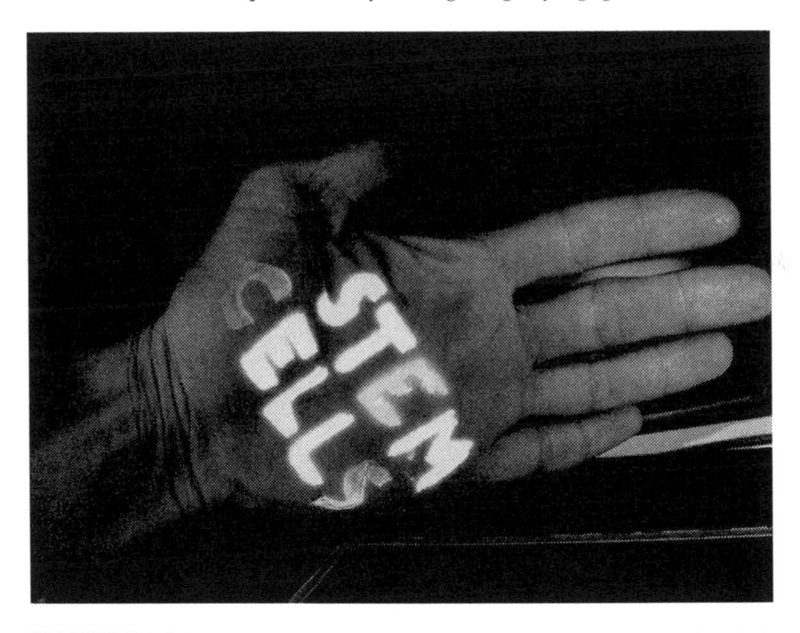

FIGURE B9.2.1 Fluorophores will emit light of a certain wavelength, which can be seen by the eye as having a specific color after illumination with light of another wavelength. Dyes in many commonly used highlighter pens contain fluorophores. These become visible under ultraviolet light and we see them as fluorescent.

9.2 (*cont'd*)

objects that are too small to be seen by the naked eye can be visualized under the microscope. However, scientists would often like to know whether a specific protein is present in a cell, because that teaches us something about the function of the cell. Unfortunately, proteins are so small that they cannot be seen with even the most powerful light microscopes. To be able to "see" a specific protein in a cell, we often use indirect methods: the protein is coupled to another protein, an antibody. The antibody recognizes and binds specifically to the protein of interest—a unique property of antibodies. A fluorescent molecule can be coupled to the antibody, which emits light when irradiated with monochromatic light of a certain wavelength (Figure B9.2.1). In the cells or tissue slices on the glass, fluorescence will now indicate if and where the protein is present. The disadvantage of this technique is that the cells or tissue slice are continuously irradiated with light during the analysis. Since fluorescence may only be short-lived under bright light, the fluorescent proteins can only be studied for short periods (Figure B9.2.2).

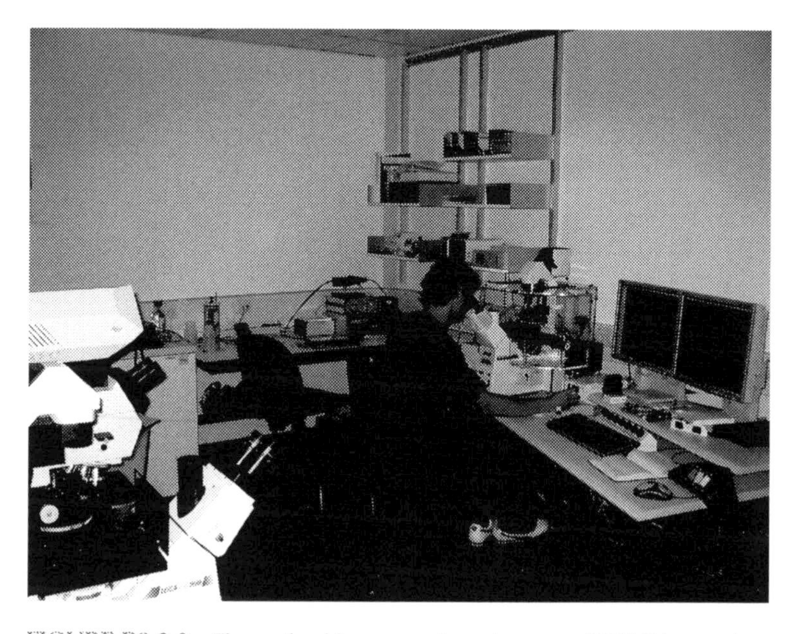

FIGURE B9.2.2 The confocal laser scanning microscope (CLSM) is a modern variant of the fluorescence microscope. The laser allows focus on very narrow optical planes, so a tissue that has been stained can be scanned by the microscope to build up a fluorescent 3D, rather than 2D, image. *Source: Hubrecht Institute, Netherlands/Stamcellen Veen Magazines.*

FIGURE B9.2.3 Different cell structures can be seen if marked using immunofluorescence. In the cell shown here, the nucleus can be seen in blue and the structural protein smooth muscle actin can be seen in green. *Source: Ewart Kuijk.*

Instead of a normal light source (made monochromatic by using a filter), one can also use a laser. The advantage of a laser is that light reaches the object only at a very small spot, while the other parts stay in the dark and can be examined at a later time. Using sensitive sensors and a computer, the whole sample preparation on the glass slide can be scanned, the computer integrates all the information and generates a complete picture, and this is stored and visible at the monitor. The microscope, laser, and computer that are used together are called a *confocal laser scanning microscope* (CLSM). Despite the instrument being very expensive, it is currently used routinely in modern cell biology research (Figure B9.2.3).

Research is ongoing in all of these areas: deriving cardiomyocytes from stem cells and transplanting into the heart, or trying to increase the rate at which new heart cells form in the heart by natural mechanisms. It is perhaps easier to imagine that repair using stem cells might be possible soon after a heart attack occurred than much later, when scar tissue and perhaps heart failure were already fairly extensive. Early treatments would require that new heart cells have to be present in the heart within the first week or so of the attack, which in itself limits the number of therapeutic options. Off-the-shelf cardiomyocytes, for example, obtained from banked human embryonic stem cells or induced pluripotent stem (iPS) cell lines, would in this case be an option, although there are many more hurdles to their use clinically than simply producing them (say, from pluripotent stem cells) in the billions that would be required for treatment of one patient. For example, even if the cell transplantation itself was successful, this may not provide a sustained cure for those patients in which the underlying disease has been present for many years and has caused a change in the shape of the heart from the normal ellipse (like a rugby ball) to a less functional spherical form (like a soccer ball). Restoring the original shape may require additional skillful surgery, and not just new cardiomyocytes. Nevertheless, preclinical studies in animals have been started to try and repair the heart with heart cells. In several laboratories in the world, mouse and human cardiomyocytes derived from pluripotent stem cells, as well as the heart's own progenitor cells, have been transplanted into mice and rats with experimental myocardial infarctions and, more recently, large animal models such as pigs and sheep have been used for this purpose. Major challenges that have been encountered include producing enough cardiomyocytes at the same time

FIGURE 9.2 Pigs are excellent model animals to study many human diseases and therapies.

(up to a billion) and ensuring they survive in the heart: up to 95% may disappear or die in the hostile inflamed environment of infarcted tissue or just simply be squeezed out as the heart contracts each time during the delivery procedure. In addition, the new cardiomyocytes should align in the correct direction to enable sufficient contraction to eject ~70% of the blood from the heart with each beat. The transplanted heart cells should connect correctly to host heart tissue to prevent dangerous disturbances in the normal electrical activity of the heart and the beat rate. These are called *arrhythmias*. They might not occur in small rodents, as their heart rates may be up to 500 beats per minute and this is barely disturbed by more slowly beating transplanted cells. However, arrhythmias could be revealed in the larger animals where the heart is much more like that in humans. This is the reason that many translational studies toward clinical application are carried out in farm animal species (Figure 9.2).

9.3

PATCH CLAMP ELECTROPHYSIOLOGY TO MEASURE ACTION POTENTIALS

Embryonic stem cells can be induced to differentiate to beating heart muscle cells (cardiomyocytes). One way to confirm that they really are cardiomyocytes is to visualize certain proteins that are typically produced by heart cells. An example is the proteins of the sarcomeres, the structures in the cell that are the motor for contraction. The presence of these proteins, however, does not prove that the cell is indeed a functional heart muscle cell. For a heart muscle cell to contract, there needs to be a voltage difference between the inside and the outside of the cell. This voltage difference is defined as the *action potential*. It is in part determined by activity of certain *ion channels* in the cell membrane. Ion channels consist of one or more proteins in the cell membrane. Multiple types of ion channels exist which can selectively open to let a specific ion, such as calcium, into the cell or actively transport ions against a concentration gradient in or out of the cell. The activity of these ion channels and the resulting membrane potential can be directly measured by a technique called *patch-clamp electrophysiology* (Figure B9.3.1). This technique was invented by the German scientists Erwin Neher and Bert Sakmann. Its importance is illustrated by their award of the Nobel Prize in Medicine in 1991.

9.3 (*cont'd*)

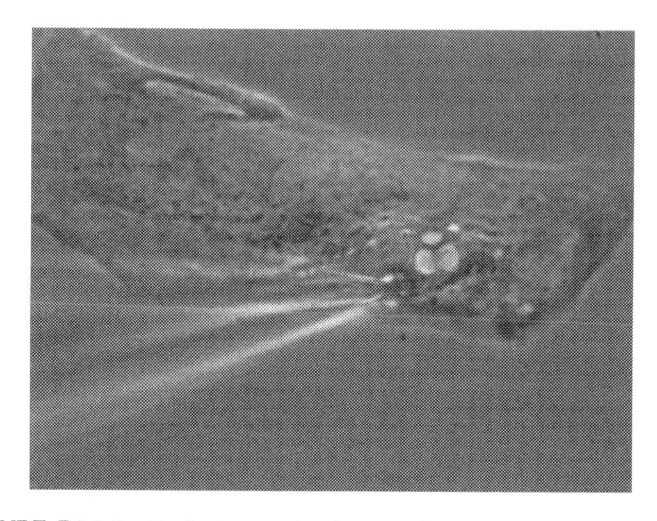

FIGURE B9.3.1 Patch-clamp technology can be used to analyze the transport of ions over a cell membrane and the associated electrical activity. Micromanipulators bring a glass electrode very close to the surface of a cell under the microscope so that it just touches the plasma membrane. A seal then forms and the electrode pinches off a very small piece from the membrane (the cell membrane is "patched" and the electrical potential of the membrane "clamped" or fixed) so that the electrode is then actually in direct contact with the inside of the cell. It is then possible to measure the electrical currents in and out of the cell and the corresponding changes in voltage. *Source: Stefan Braam, Hubrecht Institute, Netherlands/Stamcellen Veen Magazines.*

To perform patch-clamp electrophysiology, an electrode needs to make contact with the cell. This is done by putting a narrow glass pipette with a diameter of roughly one micron (one thousandth of a millimeter) on the cell membrane. By applying a slight "under-pressure," a small part of the membrane is pulled into the pipette, creating a tight contact. This is called a *giga-ohm* seal, because the resulting electrical resistance is in the order of giga-ohms. Using this technology, measurements can be performed on individual ion channels in the membrane. After successfully creating a giga-ohm seal, the membrane can also be ruptured in the pipette. This causes an open connection between the fluid in the pipette and the inner part (cytoplasm) of the cell; the voltage over the cell membrane can now be accurately determined. By comparing the results of these measurements between a normal adult heart muscle cell and a heart cell derived by

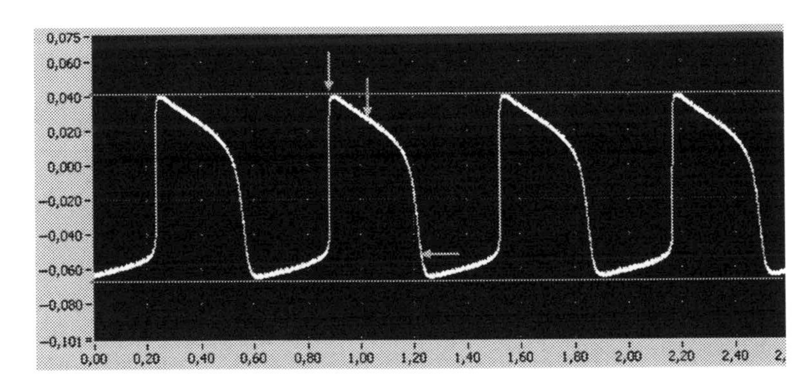

FIGURE B9.3.2 The result of a successful patch clamp experiment: the electrical impulses of an individual contracting heart muscle cell can be visualized. *Source: Stefan Braam, Hubrecht Institute, Netherlands/Stamcellen Veen Magazines.*

differentiation of a pluripotent stem cell, the scientist can compare functionally how similar or how different these two types of heart cells are (Figure B9.3.2).

This technique is of growing importance now that it is possible to derive induced pluripotent stem cells from patients with cardiac disease due to a mutation in one of the genes coding for the proteins that make up these ion channels. It will soon be possible to test drugs for their effects on diseased as well as normal human heart cells derived from stem cells.

Since the cardiomyocytes do not align well if injected as cell suspensions in the heart, research is being carried out to see whether they could be prealigned in some way before transplantation. Tissue engineers, in particular, are working on this: they use various natural and synthetic materials in which narrow lines have been patterned, about the width of a heart cell. When the stem cell derived heart cells are plated on to these sheets, the whole sheet can contract in synchrony. The idea is to then wrap these sheets around the heart or suture them on to it, to see whether they will squeeze the heart from the outside rather than pulling it by contraction from within. Only time and a great deal more research will tell whether these problems can be solved and the approach will develop into a safe and effective treatment for human patients.

Quite another type of research that has recently started is based on using induced pluripotent stem cells from patients. These are being used for characterization of normal and abnormal development of the human heart and better understanding of its function, as well to study heart genetic disease (see Chapter 4, "Of Mice and Men: The History of Embryonic Stem Cells" for more on induced pluripotent stem cells and what they can do). It is hoped that this will lead, among other things, to the identification of genes that play a role in inherited or congenital defects in cardiac structure and function as well as cardiac disease. Heart defects are among the most common congenital defects at birth and, in contrast to the past, patients are now much more likely to reach adulthood. Some of these patients have an increased risk that any children they have may suffer from the same cardiac abnormalities. If the genes causing the defects were known, it might be possible to offer prenatal genetic diagnosis long before the defects are evident by echocardiography of the fetus. Early diagnosis may help prevent late (and possibly traumatic) abortion. In some countries it may even be possible for couples with increased risk to carry out *in vitro* fertilization to establish pregnancy. For genetic diagnosis, one cell from the embryo can be removed in the laboratory at the eight-cell stage, and subsequently only embryos without the defect selected for transfer to the mother. Selection of embryos in this way is, however, not acceptable in many countries for disorders that are not necessarily fatal, or can be treated.

9.4

IMMUNOFLUORESCENCE AND FLOW CYTOMETRY

Immunofluorescence techniques are used to determine whether cells contain specific proteins. The presence of receptors to which hormones and growth factors can bind, transcription factors that indicate cell identity or function, or structural and signaling proteins, can all be shown by immunofluorescent staining. The underlying principle of this technique is that the molecule of interest (in general, a protein or sugar) in the cells is first bound by an antibody that specifically recognizes the target molecule. A second antibody, which can be coupled to a fluorescent molecule, is then added and binds to the first antibody already attached to the cells. In the dark, the fluorescent antibody is excited with light of a specific wavelength. The induced emission of light from the fluorescent molecule has another wavelength, usually red or green. This signal can be seen under the fluorescence microscope when tissue or cells on a glass slide or in a tissue culture dish are being studied. The signal is detected using a flow cytometer when cells are floating freely in the culture fluid ("in suspension"). Flow cytometry counts the number of fluorescent cells among the total cell population and can even sort them out from the rest of the cells (Figure B9.4.1).

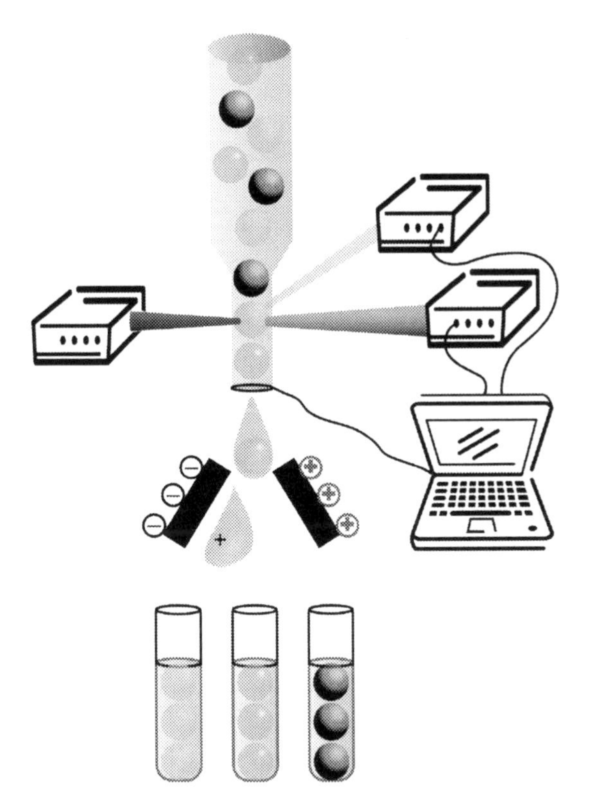

FIGURE B9.4.1 Schematic representation of a fluorescence activated cell sorter (FACS). Cells suspended in fluid are first fluorescently labeled either using an antibody to stain a protein in combination with a fluorescent dye or a gene construct expressing, for example, green fluorescent protein (GFP). In a population of cells that needs "sorting," some cells will express more fluorescence than others. The whole cell population is put into the cell sorting machine irradiated with a laser (red), while the emitted light (yellow or green) from each cell is detected. When a cell is fluorescently labeled at a preset cut-off level, the fluid drop with the cell in it will become positively charged. This drop is deflected in an electrical field and collected in a tube. In this way these cells are separated from the more negatively charged drops containing cells with less or no fluorescence. The cells are thus sorted into groups with different levels of fluorescence. This tells us what proportion of the original population contained a particular cell type (say, how many stem cells that had differentiated into nerve cells if a nerve cell marker was used). This makes them available as pure cell populations for subsequent experiments, for example, transplantation into a mouse or investigating which messenger RNAs and proteins are present. *Source: Stamcellen Veen Magazines.*

Pharmaceutical companies are also increasingly interested in using human cardiomyocytes from stem cells to test new drugs for possible toxic side effects on the heart. Stem cell derived cardiomyocytes represent an alternative to, for example, cardiomyocytes obtained from the hearts of dogs that are currently used. In addition, pharmaceutical companies are interested in having human models for cardiac diseases that could be used to develop new drugs to treat cardiac diseases. See Chapter 14, "Stem Cells for Discovery of Effective and Safe New Drugs" for more details on how these cells are being used to predict toxic effects of drugs on the heart and drugs that may be used to treat heart disease.

9.5

MASS SPECTROMETRY FOR PROTEIN ANALYSIS

Mass spectrometry is commonly used to analyze the proteins present, for example, in blood, or in specific cell types. By coupling this information to complementary information on the presence of messenger RNA molecules obtained, for example, by microarray analysis, an impression of the way in which the cell functions can be obtained. How does this technique work? Proteins are separated from each other using *gel-electrophoresis*, and then cut into fragments (peptides) with an enzyme (such as trypsin). With a mass spectrometer, the size (or *mass*) of the peptides can be measured very accurately. Each one of the peptides can subsequently be further fragmented into random pieces in the mass spectrometer, enabling determination of the exact amino acid

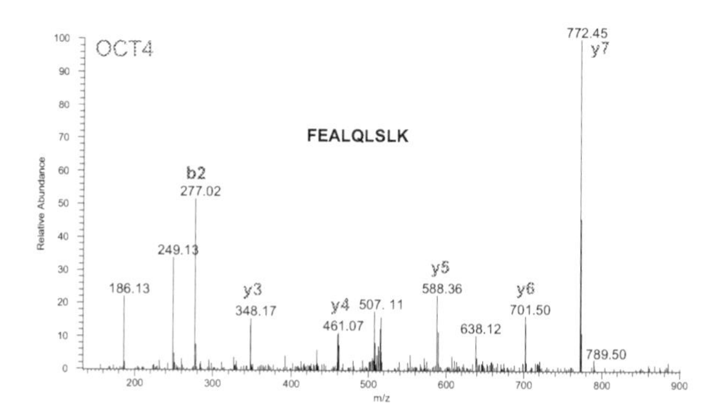

FIGURE B9.5.1 Fragmentation spectrum of a protein where the peptide fragments are separated by their relative mass or weight. *Source: Dennis van Hoof/Jeroen Krijgsveld, Hubrecht Institute, Netherlands/Stamcellen Veen Magazines.*

sequence of the peptide (Figure B9.5.1). By matching the amino acid sequences of the different peptides with amino acid sequences of whole proteins in a protein database, the proteins that were originally present in the sample can be identified.

9.2 FROM PLURIPOTENT STEM CELLS TO CARDIOMYOCYTES

Research during the first decade after James Thomson first isolated stem cells from human embryos showed that human embryonic stem cell lines may not all be the same. Later, this also seemed to be true for induced pluripotent stem cell lines. It seems that their properties differ somewhat depending on the way in which the cell lines had been derived or generated, possibly their genetic background, but most particularly the way they are maintained in culture and passaged. The greatest differences are found in the efficiencies with which they make specific cell types. Experience has shown that most human embryonic and induced pluripotent stem cell lines will differentiate fairly efficiently to ectoderm (and form neurons, for example); some, but not all, will form mesoderm derivatives such as cardiomyocytes and blood vessel endothelial cells; but relatively few will form endoderm derivatives such as insulin-producing β-islet cells of the pancreas and lung alveolar cells.

Taking cardiomyocyte differentiation as an example, variations in differentiation efficiency become evident when human embryonic stem cells cultured on top of a feeder layer in the presence of fetal calf serum and passaged "mechanically" are compared with cell lines grown in defined culture medium (all constituents known) on a synthetic protein matrix and passaged enzymatically instead of mechanically. In the first case, growing embryonic stem cells or induced pluripotent stem cells on top of a feeder layer consisting of endoderm-like cells (which mimics the conditions under which heart cell differentiation takes place in the embryo) leads to the formation of small "islands" of tissue containing beating heart cells—cardiomyocytes—within a week to 12 days (Figure 9.3). Under a microscope, the cells can be seen to contract in a coordinated manner, just like normal heart tissue. Using special electrodes (glass needles attached to conducting fluid and an oscilloscope), it is possible to measure the electrical activity (or *action potentials*) of the beating cells and see whether they resemble any of the three major cell types of the heart: the contracting cells located in the ventricular muscle wall, the cells in the relatively thin atrial wall, or the pacemaker cells that control the regular beating action of the heart. All these cells

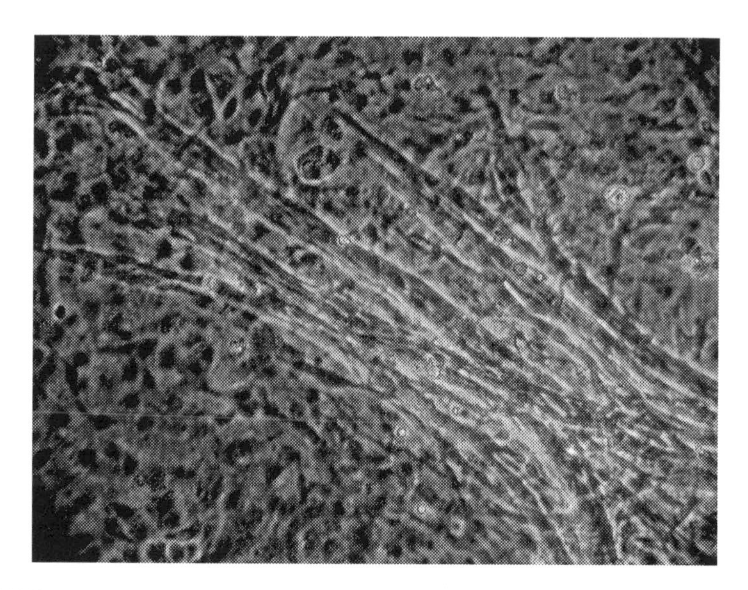

FIGURE 9.3 Mouse embryonic stem cells, cultured on top of endodermal cells, show elongated bundles of heart muscle cells beating synchronously. *Source: Stieneke van den Brink, Hubrecht Institute, Netherlands/Stamcellen Veen Magazines.*

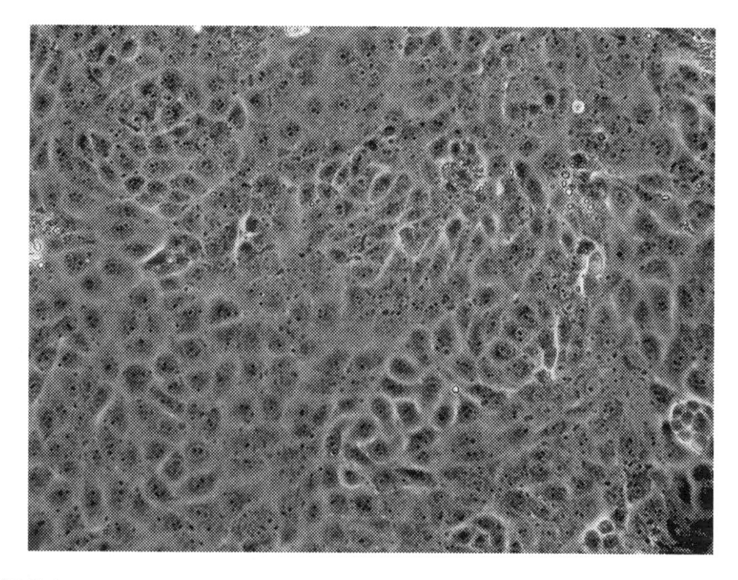

FIGURE 9.4 A cultured layer of endoderm cells on which human embryonic stem cells can be grown to induce differentiation to cardiomyocytes. *Source: Stieneke van den Brink, Hubrecht Institute, Netherlands/Stamcellen Veen Magazines.*

normally have their own specific function in the heart and can be classified according to their electrical activity. These cell types form after coculture with the endoderm-like feeder layer, provided the human embryonic stem cells had been mechanically passaged prior to initiating differentiation (Figure 9.4). By contrast, this method does not work well for enzymatically passaged human embryonic stem cells. For these, growth as small cell aggregates in suspension (as so-called *embryoid bodies*) in the presence of specific growth factors is more effective in producing cardiomyocytes. Even then, however, not all embryonic stem cell lines produce cardiomyocytes with equal efficiency, even though all three types of cardiomyocytes present in the heart are formed.

9.6

MICROARRAY CHIP TO INVESTIGATE WHICH MESSENGER RNAS ARE MADE BY A CELL

Microarrays are typically used to investigate which and how much messenger RNA molecules are made by a specific cell type. A series of different oligonucleotide probes, consisting of short DNA sequences that are complementary to a part of all known messenger RNAs, are coupled to microscopically small spots on a plastic plate. The DNA is spotted on the plate in arrayed series, hence the name and identity of each probe can be easily traced back by its position on the array. In general, RNA is extracted from a large number of cells, the molecules are converted or copied into DNA, and the DNA labeled with a fluorescent molecule. Subsequently, this is allowed to bind to the probes on the array (hybridization), and the fluorescent signal is measured.

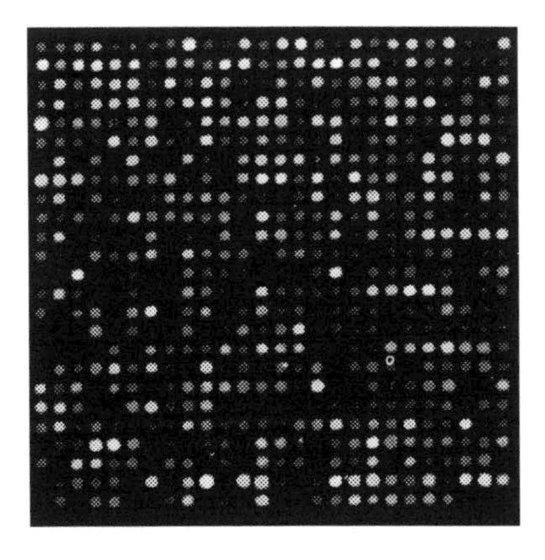

FIGURE B9.6.1 Part of a microarray after hybridization. Each spot represents expression of a specific messenger RNA. *Source: Adapted from "Germline Genomics" by Valerie Reinke, Department of Genetics, Yale University, New Haven, Conn., U.S.A.; NCBI/NLM, open domain.*

9.6 (*cont'd*)

The strength of the fluorescent signal at a specific spot on the array indicates how much of the messenger RNA was present in the cells. Alternatively, two RNA samples can be hybridized together to the array, where one sample is labeled with a red fluorescent molecule and the other with a green label. On each spot on the array, the red versus green signal indicates which sample contained most of the specific mRNA, relative to the other sample (Figure B9.6.1).

The reason for these differences in differentiation properties has yet to be explained, but as the growth and differentiation media supplied by commercial companies has improved and the protocols for passaging has become more defined, the differences between lines has become much smaller. There are now very robust protocols that work on many human pluripotent stem cells to produce cardiomyocytes with high efficiency. The short history of how this has evolved indicates how complex and what a great deal of work it has been to be able to produce human cardiomyocytes from stem cells on a regular basis in large numbers in many laboratories throughout the world (Figure 9.5).

Independently of how the cardiomyocytes are formed, they are always relatively immature and have the properties of very young heart cells, almost like those in the fetal heart. This is a distinct advantage for transplantation purposes, because the cells may still be capable of dividing a few times after transplantation, thus increasing the cell numbers available for contraction. Moreover, they survive much better than adult cardiomyocytes after transplantation into the heart, and usually further mature under the influence of the local cardiac environment. It is a disadvantage, however, when either trying to model heart disease of adults using stem cell derived cardiomyocytes or carrying out research on some types of drugs. Many researchers are currently investigating how to make the cardiomyocytes (and other stem cell derivatives) become more adult-like in culture. See Chapter 13, "Human Stem Cells for Organs-on-Chips: Clinical Trials without Patients?" for ways in which this is being tackled (Figure 9.6).

In cultures *in vitro*, the islands of beating cardiomyocytes are often surrounded by less differentiated cells, or cells that have differentiated to cell types other than heart cells. For many purposes, including cell

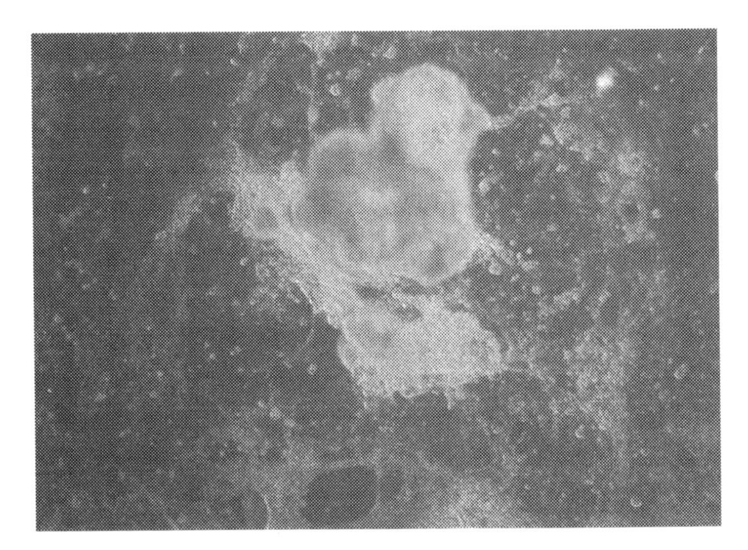

FIGURE 9.5 Picture of a clump of human embryonic stem cells that have differentiated to heart muscle. The clumps typically show rhythmic contraction (about 100 beats per minute) rather like a real heart. Addition of adrenaline, which makes a normal heart beat faster, also makes these cells beat faster. *Source: Dorien Ward, Hubrecht Institute, Netherlands/Stamcellen Veen magazines.*

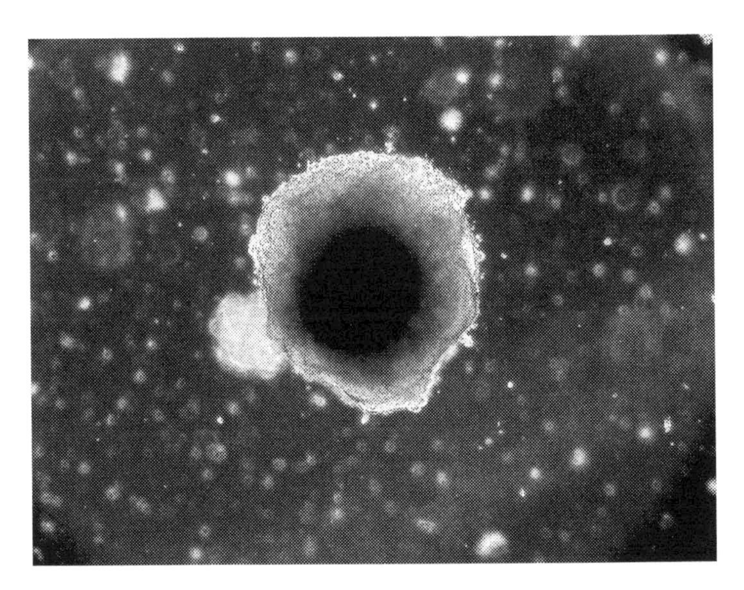

FIGURE 9.6 By allowing embryonic stem cells to aggregate together, a structure is formed that resembles the inner part of a blastocyst-stage embryo; this is called an embryoid body. The cells present in the embryoid body influence each other and start to differentiate to form many different tissue cell types. *Source: Stieneke van den Brink, Hubrecht Institute, Netherlands/Stamcellen Veen Magazines.*

transplantation, a pure population of heart cells is required and, in some cases (e.g., for testing drugs), the cells need to be actually beating. This is one of the major scientific challenges still to be addressed. One way to at least enrich for heart cells is manually dissecting the beating islands from the rest of the culture under the microscope, using very small glass knives. This is time consuming and limits scaling up the production of cardiomyocytes. Alternatively, antibodies coupled to a fluorescent or magnetic label could be used to specifically bind to proteins on the cardiomyocyte cell. The labeled cardiomyocytes could then be separated from the rest using a flow cytometer or a strong magnet. The problem here, however, is that it has been very difficult to identify protein–antibody combinations for the cardiomyocyte cell surface, although there are now cell surface proteins that allow early heart cells to be selected from mixed populations and these are in wide use in research.

Another approach is based on the higher metabolic rate of cardiomyocytes compared with other cells, which is reflected in a large number of mitochondria in the cell. Fluorescent labeling of the mitochondria enables separation of heart cells from other cell types simply by selecting the cells with the highest fluorescence (Figure 9.7).

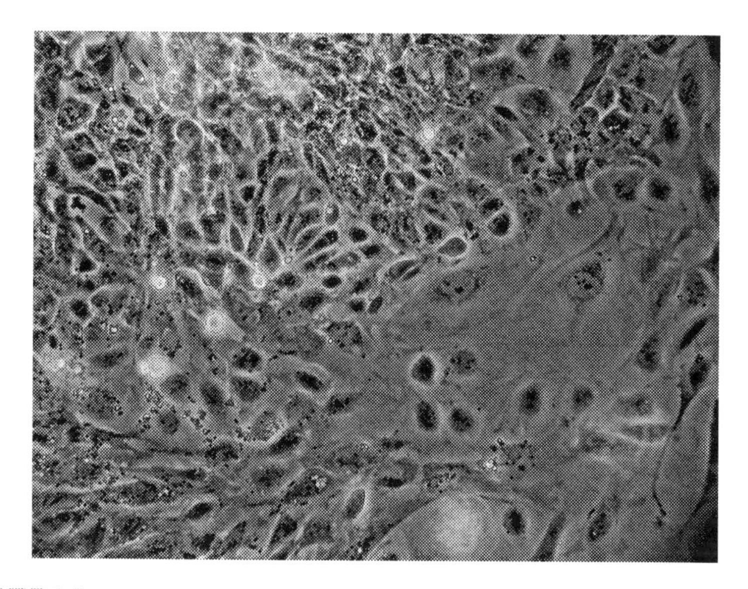

FIGURE 9.7 This microscope picture shows cells formed an embryoid body and were subsequently transferred to a culture dish. The cells clearly differ from each other in shape and size, a sign that they are differentiating. *Source: Stieneke van den Brink, Hubrecht Institute, Netherlands/Stamcellen Veen Magazines.*

9.7

TISSUE ENGINEERING TO REPAIR A DAMAGED HEART

To create a cardiac tissue patch to repair a damaged heart, heart cells have been obtained from rat heart muscle and cultured on special oval-shaped substrates that look rather like elastic bands. Five of these artificial pieces of heart muscle shown in Figure B9.7.1 were stitched onto the surface of a heart of a rat with an experimentally induced myocardial infarction. The cells survived, cardiac function improved, and closer inspection four weeks after the operation showed small blood vessels had grown into the engineered patches of tissue. The blood vessels appeared to contain rat erythrocytes (red blood cells), indicating that they were properly connected to the blood circulation of the rat.

One of the challenges now is to find the right source of heart cells, which would preferably be autologous (from the patients themselves), to advance this work toward future clinical applications. This research is largely being carried out in Germany. An alternative method being investigated in Japan is to culture cell mixtures (cardiomyocytes, cardiac fibroblasts, and endothelial cells in the same ratios as present in real heart tissue) from stem cells on synthetic sheets of biocompatible

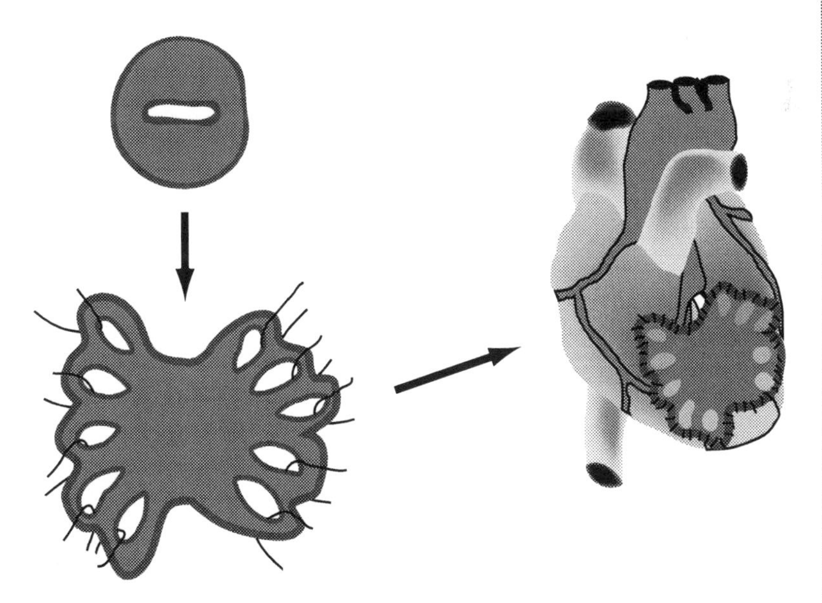

FIGURE B9.7.1 A tissue engineered patch to repair the heart.

9.7 *(cont'd)*

polymers or natural substances. These contracting sheets can be wrapped around the heart surgically and have been used successfully in rats to improve the force of contraction of the heart.

Other approaches include constructs made from a stack of different cell layers; cells are mixed with natural supportive materials; cells are seeded on preformed structures; or cells and supportive materials are "printed" in an organized manner. Although still preliminary, costly, and time consuming, these approaches may contribute to more effective application of stem cell therapy.

FIGURE 9.8 Embryonic stem cells can differentiate to cardiomyocytes (heart muscle cells). Immunofluorescence techniques, in which proteins typically present in real heart cells are stained with an antibody carrying a fluorescent marker, can be used to confirm that these cells are indeed cardiomyocytes. The striped pattern shows "sarcomeres," the contractile machinery of a heart cell. *Source: Anna de Lisio/Susana Chuva de Sousa Lopes, Hubrecht Institute, Netherlands/Stamcellen Veen Magazines.*

Differentiation in a culture dish is relatively inefficient, so selection only partly helps to solve the problem of obtaining large numbers of cardiomyocytes. Considerable effort is still going into discovering ways to improve the efficiency of stem cell differentiation to cardiomyocytes to increase the overall yield. Methods being developed include mimicking the signals that control normal cardiac development in the embryo, for example, using specific growth factors and hormones. Alternatively, small molecules that activate or inhibit signaling pathways are being tested. Analysis of messenger RNA expression during differentiation by microarray chips, and protein expression by mass spectrometry, are ways to identify these signals. This type of research should eventually lead to an overall picture of which proteins play a role in differentiation, how they influence one other, and how we can best use them to obtain high yields of the right cells. Bioreactors have also been developed for the production of cardiomyocytes in large numbers. These are rather like small versions of the vats used for brewing beer. They keep the cells in suspension and use specific reagents based on antibiotic resistance to kill all of the noncardiomyocytes present so that only the heart cells of interest are left over (Figure 9.8).

9.8

CLINICAL TRIALS WITH PATIENTS TO INVESTIGATE STEM CELL TRANSPLANTATION AS A TREATMENT FOR HEART FAILURE

Delivering stem cells as therapy to patients with a myocardial infarction or heart failure is still being investigated for its ability to improve heart function after myocardial infarction and during heart failure and angina pectoris (chest pain). The original idea for this came from experiments in mice early in 2000, which showed heart function improved if any of a variety of noncardiac stem cells (e.g., from bone marrow) were injected into blood vessels or directly into the heart muscle. The work quickly moved to studies in humans, where incidental reports based on treatment of individual patients with their own bone marrow cells after a myocardial infarction were encouraging. This led the way to more extensive clinical trials using the same approach. The results showed that while the treatment was safe, improvement in heart function was limited and rarely sustained, even though in some cases exercise capability did appear to increase. The results were, however, less consistent and not as large as in mice, where overall survival from myocardial infarction was clearly improved.

The present interpretation of these human studies is that any beneficial effect is probably indirect and caused by growth factors produced and secreted by the bone marrow cells which ended up in the heart, and not by differentiation of the bone marrow stem cells to contracting

9.8 (*cont'd*)

heart cells. This effect is probably not limited to the bone marrow stem cells; other stem cells (such as those from umbilical cord blood or adipose tissue) seem to act in a similar fashion when injected directly into the heart, or even when injected into the blood stream to find their way to the heart. However, the long-term outcome of these experimental treatments, in terms of extended lifespan or reduced disability, remains largely inconclusive.

Recently, mesenchymal stem (or stromal) cells, present as a subpopulation in the bone marrow of the patient, were injected into the hearts of patients with chronic heart failure or chest pain not caused by acute myocardial infarction. Again, cardiac function did not greatly improve, but chest pain, or *angina*, did seem to be significantly reduced. So far, these results suggest that it may be more relevant to supply the damaged heart muscle with the right factors, such as growth factors, to stimulate the growth of new blood vessels in the heart, than to actually try to replenish contractile heart cells. It is also possible that these non-cardiac stem cells activate cardiac progenitor cells already in the heart or stimulate the cardiomyocytes to divide.

Clinical trials, largely supported by commercial stem cell suppliers, are still ongoing and it is expected to be a few years before it is clear which patients, if any, benefit from the treatment and for how long.

Most research on the differentiation of human stem cells to beating heart cells has so far been carried out using human embryonic stem cells. To what extent these experimental results will apply to human-induced pluripotent stem cells remains to be fully explored, but the expectation is that much of the knowledge already gained will be directly transferable.

9.9

ENGINEERING HUMAN MYOCARDIUM

An intriguing new research field is cardiac tissue engineering, which combines knowledge from materials chemistry with cell biology and medicine. The idea here is that preformed structures can be generated from biologically compatible polymers or natural substances and combined with a cell mixture (cardiomyocytes, cardiac fibroblasts and endothelial cells in the same proportions as present in real heart tissue). This would result in an organized synthetic tissue construct that (partly) mimics that normally present in the heart and could possibly be attached to the heart as a kind of band-aid, or injected as smaller fragments.

Several attempts to generate artificial cardiac tissue have already been described; constructs are made from a stack of different cell layers, cells are mixed with natural supportive materials, cells are seeded on preformed structures, or cells and supportive materials are "printed" in an organized manner. Although still preliminary, costly, and time consuming, these approaches will hopefully advance the field towards more effective application of cell therapy. In the short term though, it is thought they will be very useful as surrogates for human heart tissue to test drugs and examine the effects of stress such as that resulting from oxygen shortage or pressure overload.

9.10

HOW AND WHEN TO ADMINISTER STEM CELLS TO THE HEART

An important point that needs to be addressed in stem cell therapy for the heart is the strategy for delivering cells into the damaged heart tissue. It is possible to deliver cells locally through infusion into arteries or veins, or directly via injection into the heart tissue. Each approach has its pros and cons in practice: infusion of cells is relatively safe and simple, but because the cells are not delivered into the tissue and are easily flushed out of the target area, delivery effectiveness is limited. In fact they often accumulate in the capillaries (tiny blood vessels) in the lung. Intracardial injection of cells will lead to improved homing of cells to the affected tissue, because cells are inserted into the cardiac tissue via a needle. Unfortunately, homogeneous spreading through the tissue is limited and there is an increased risk of developing arrhythmias, particularly if electrically active heart cells were to be delivered.

9.10 *(cont'd)*

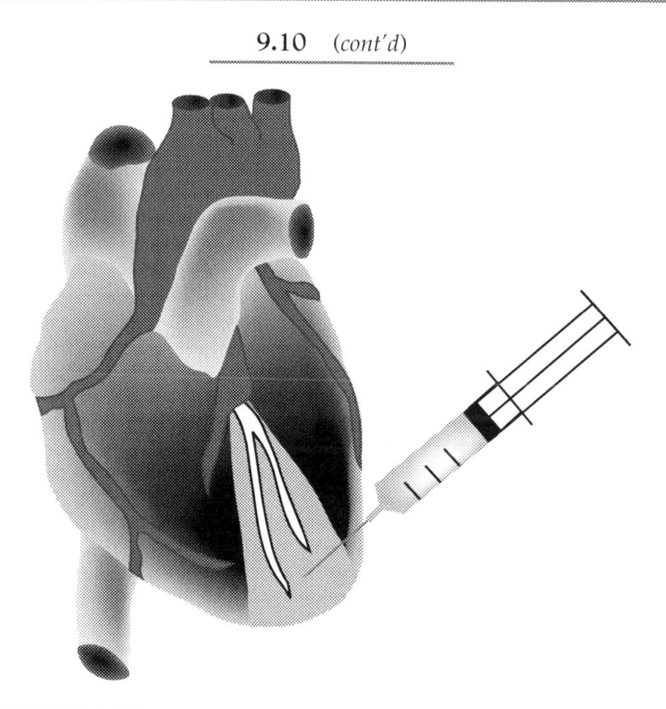

FIGURE B9.10.1 In this example, adult stem cells or cardiomyocytes derived from pluripotent stem cells can be injected into the heart after a myocardial infarction. The infarcted area is shown in gray. *Source: Stamcellen Veen Magazines.*

An optimal time frame for cell delivery has still not yet been determined precisely, although, of course, damaged tissue should be replaced by new cells as soon as possible to avoid any additional complications and further progression of the disease (Figure B9.10.1). However, the hostile environment after, for example, an acute myocardial infarction, with a lack of oxygen and nutrients, inflammatory reactions, and a large number of cells dying, may not be the best place for cells to be injected. In addition, some cell types for autologous transplantation that could potentially be used need additional time, maybe several weeks, to be collected and processed before they are ready for transplantation. This limits the time frame available for possible injections.

9.11

CARDIAC PROGENITOR CELLS FOR HEART REPAIR CONTRIBUTED BY DR.JOOST P.G. SLUIJT

Although the heart is among the least regenerative organs in our body, a progenitor cell population has been identified that naturally resides within it. These cardiac progenitor cells may provide new options for heart repair if they can be stimulated to grow within the heart, or can be expanded in the laboratory after isolation from the heart, and subsequently transplanted back to the patient as a cell therapy. Cardiac progenitor cells have been isolated from fetal and adult human heart tissue. Although they have successfully been isolated during heart surgery from the adult human heart auricle, which is an appendage of the atrium, the numbers that can be retrieved from this source decrease with age. The growth capacity of fetal and adult cardiac progenitor cells is very similar, but because the starting numbers of cells is lower in a small auricle biopsy from the adult heart, it takes much more time to obtain sufficient cell numbers for transplantation. Cardiac progenitor cells grown in the right culture medium differentiate into heart muscle cells and, when derived from young cells, beat spontaneously. They closely resemble the cardiac progenitor cells derived from human pluripotent stem cells. Interestingly, when grown in the presence of certain angiogenic growth factors, they can also form blood vessel and smooth muscle cells, the building blocks of the vascular wall.

To explore their potential for cell-based therapy, cardiac progenitor cells have been used in preclinical studies. Using a mouse model for myocardial infarction, injection of the cells into the injured heart has been shown to improve cardiac performance up to three months after injection, but the total numbers of human cells still present at that time is often less than 3% of those originally injected.

When noncardiac cells, for example from bone marrow or fat tissue, have been injected or infused into the heart, there is some evidence that the endogenous progenitor cells in the heart may be recruited to repair and differentiate into cardiomyocytes. Likewise, the noncardiac stem cells, which do not form cardiomyocytes, may also be able to induce resident cardiomyocytes, which normally only divide or "turn over" very slowly, and may actually encourage this process to speed up.

Contributed by Dr. Joost P.G. Sluijter and Professor Pieter A. Doevendans University Medical Center Utrecht and Professor. Marie José Goumans, Leiden University Medical Centre, The Netherlands

9.12

OPTOGENETICS AND MAKING A BIOLOGICAL PACEMAKER

Pacemakers for the heart are widely used to control the heart rhythm of patients in which the normal beat rate is disrupted. The heart can beat irregularly or too slowly for many reasons, particularly as people grow older, and, in many cases, a pacemaker will solve the problem. This involves inserting electrodes into the heart muscle and connecting it by wires to a battery that is usually inserted under the skin near the shoulder. The problem with pacemakers, however, is the batteries and sometimes the wires need replacing, and this is a fairly invasive process that creates some tissue damage.

The natural pacemaking center of the heart is located at the node and consists of a few hundred specialized cardiomyocytes, which produce currents that induce sequential contraction of the heart chambers as they pass through the conduction system. Replenishment of the pacemaker cells by "biological pacemakers" rather than electrical pacemakers has been considered as a therapeutic option for a decade or more, and new developments in a field called *optogenetics* has resulted in some realistic options.

Optogenetics is based on light-sensitive molecules, which, in this case, would open and close ion channels in the plasma membrane of the cardiomyocyte in response to light. One way to do this that already works in the laboratory is to genetically engineer one of these light-sensitive molecules into human pluripotent stem cells (Figure B9.12.1). When these cells are differentiated to cardiomyocytes in culture, shining an LED light onto the cells is sufficient to open the ion channels and make the cell contract. When the light goes off, the channel closes and the heart cell relaxes again. If the LED light is switched on-and-off at the rate of a normal heartbeat (around 60 times per minute

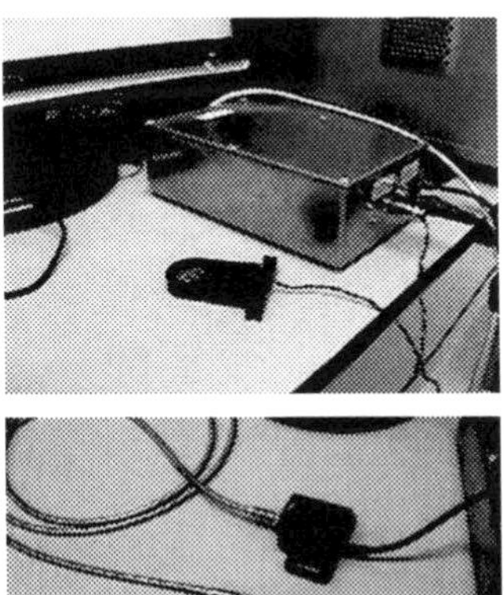

FIGURE B9.12.1 Optogenetics uses light of different wavelengths (different colors) to open ion channels in cells. *Source: Leon Tertoolen, Leiden University Medical Center, Netherlands.*

in humans), it can actually *pace* the heart at its normal rate. The battery for the LED light is far smaller than that for an electrical pacemaker because it uses much less current. This advantage of small size and longer life, together with glass fiber optics being sufficient to bring the light to the heart, means that transplanting just a small number of cells in the node could result in a biological pacemaker that can be paced by the LED light. While much more work will be needed to bring this to clinical practice, optogenetics combined with stem cell biology opens new perspectives for the future.

Contributed by Dr. Leon Tertoolen
Leiden University Medical Centre, Dr. Stefan Braam and Dr. Caterina Martens Grandela, Pluriomics bv, The Netherlands

CHAPTER

10

Adult Stem Cells: Generation of Self-Organizing Mini-Organs in a Dish

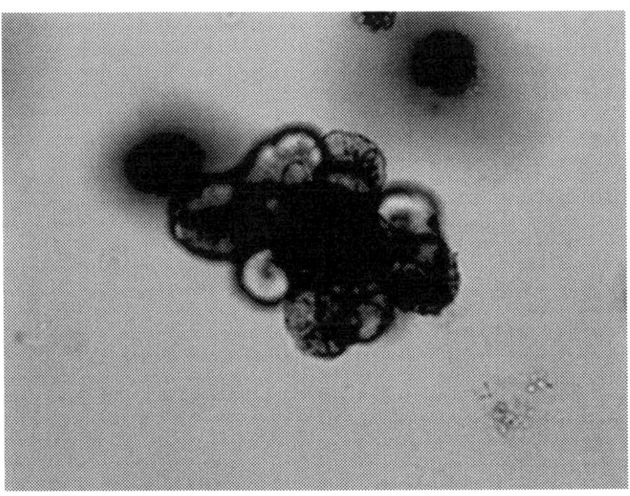

Anja van de Stolpe, Philips, Eindhoven, The Netherlands

OUTLINE

In Chapter 5, "Origins and Types of Stem Cells: What's in a Name?" we introduced the different types of human stem cells and, for simplicity, categorized them as *pluripotent* (embryonic and induced pluripotent stem cells) and *multipotent* (adult stem cells). When pluripotent stem cells were first identified, their ability to divide indefinitely and form all cell types of the body made them in some ways more interesting for both basic research and future therapeutic applications. As far as we knew, adult stem cells had only a limited ability to divide in culture, were difficult to find in the body, and could only form relatively few cell types.

This view changed dramatically during the first decade of the twenty-first century. Firstly new *markers* of the cells were discovered, which allowed them to be identified in many tissues of the body. Secondly, culture conditions and proteins were found that supported the growth of adult stem cells without compromising their ability to differentiate. Adult stem cells have thus returned to center stage in the search for therapies that can lead to tissue and organ regeneration. It is clear that their applications will develop far beyond the present use in hematopoietic stem cell therapies to reconstitute bone marrow after chemotherapy. For this reason, we devote this chapter to more detailed information about their biology and discovery, with a view to making it clear why they have become particularly interesting both for clinical use and perhaps drug discovery.

Like each stem cell source, however, adult stem cells do have shortcomings and, like pluripotent stem cells, their differentiated derivatives often remain rather immature and little like adult tissue. This can have advantages, such as better survival of the cells after transplantation, but also disadvantages. Approaches to solving this issue are similar to those of differentiated cells from pluripotent stem cells.

10.1 ADULT STEM CELLS IN INTERNAL ORGANS

The area of basic research that provided the greatest clues to what adult stem cells actually look like and what they can do is the study of the intestines. Interest arose not from any wish to apply adult stem cells in regenerative medicine, but from an interest in understanding how the intracellular Wnt signal transduction pathway (see Chapter 12,

"Cancer Stem Cells: Where Do They Come from and Where Are They Going?") was involved in causing colon cancer. Many years were devoted to very fundamental research trying to figure out why disruption of genes that play a role in controlling the activation of this Wnt pathway in the cell made epithelial cells in the intestine grow abnormally to form a tumor.

The first important consequences of this knowledge were that it allowed the identification of families with mutations in some of these genes who were at high risk of developing colon cancer. It is often difficult for the general public, public and private research funders, and politicians to understand the value of fundamental research, but this research success provides an example of how things can be discovered serendipitously and have huge long-term consequences for society, in this case health. To illustrate how this process works, we will tell the story of the intestinal stem cell.

10.1

ORGANOID CULTURE OF ADULT STEM CELLS

The term *organoid culture* was coined to describe a way of culturing adult stem cells in much the same way *embryoid body* was coined to describe a way of culturing pluripotent stem cells to make them differentiate. In mice, fluorescent Lgr5 or Lgr6 protein-containing adult stem cells can be isolated from several organs, such as the intestines, pancreas, stomach, kidney, liver, or skin. These green fluorescent stem cells can be collected in fairly large numbers using a fluorescence activated cell sorter (FACS). The isolated stem cells are carefully placed in a culture dish containing a gel (called Matrigel) that contains a mixture of extracellular matrix scaffold proteins, somewhat similar to what is present in the intestinal crypts. A protein called R-spondin, which activates the Wnt signal transduction pathway in the stem cells, is also added. This is the "magic bullet" for organoid culture. In addition, the gel surrounding the stem cells provides a soft, 3D environment for the cells, mimicking a real "niche."

Interestingly, during the isolation of the stem cells, most of them remain attached to their neighboring Paneth cell from the intestinal crypt. Thus, the stem cells hang on to their "mate," which is thought to be necessary for start of the self-renewal process. If stem cells do not have a Paneth cell, they grow very badly, and, in fact, often create their own Paneth cell by initiating differentiation in that direction first if necessary.

10.1 (*cont'd*)

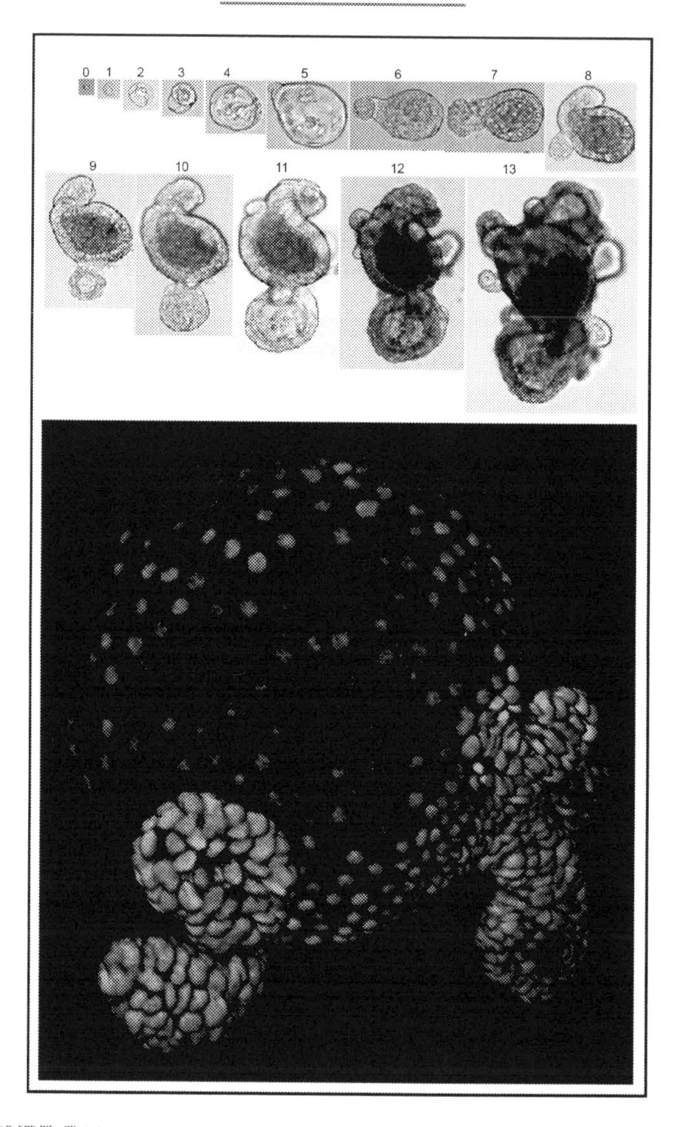

FIGURE B10.1.1 Intestinal organoids (also called mini-guts) can be grown from just one single cell, as shown in the images from 1 through 13. The bottom image shows a mini-gut in which the nuclei of the individual cells are stained in green. Different crypts grow out from this mini-gut (two on the left and three on the right), very much like the structure of normal intestines. *Source: Reproduced from Hans Clevers, The intestinal crypt, a prototype stem cell compartment, Cell, 2013, 154, pp. 274–284.*

Once in the dish, each stem cell—Paneth cell combination starts to form a new crypt—villus structure in three dimensions, which represents the smallest structural unit of the intestinal lining (Figure B10.1.1). This self-organizing process is much like that seen in normal intestines, and the 3D structures have come to be called *organoids*. They consist of circular, crypt-like cell aggregates that transition into villus-like epithelium, which forms a kind of a lumen (the hollow part of the tubular intestine). These organoid structures, as simple as they are, can behave as intestinal tissue, but also still contain a small reservoir of stem cells. In fact, they are exactly like a real piece of intestine, a *mini-gut*. Within the mini-gut, the stem cells differentiate and form the different cell types that make up the villus structures, and then they die through a suicide mechanism, called apoptosis. The whole process mimics the turnover of the intestinal epithelium that takes place in animals and humans alike.

Such organoids from mice have already been transplanted via the anus into mice in which the intestinal tissue was severely damaged. Here they attached, flattened out, and integrated to form a new functioning intestinal epithelial layer! This experiment provided crucial evidence for the full functionality of the intestinal organoids formed from adult stem cells, and elegantly illustrated the regenerative medicine potential of the organoid culture approach. This research in mice paves the way for similar studies in humans.

10.2 ADULT STEM CELLS IN THE INTESTINE

The discovery of stem cells in the intestine began with a project in the lab of Hans Clevers, at the Hubrecht Institute in the Netherlands, to identify genes in the Wnt signaling pathway. Cellular signaling initiated by Wnt proteins is disturbed in colon cancer, but which particular steps are affected was not clear. The project set out to introduce fluorescent proteins into genes thought to be part of the Wnt pathway through homologous recombination in mouse embryonic stem cells. These were used to make chimeric mice (see Chapter 3, "What Are Stem Cells?"). After almost a decade of systematic work using this approach, the research group struck gold. They found a gene, called Lgr5, of which the expression marked cells with all of the properties of a stem cell: the ability to divide and give rise to more than one other cell type. These cells were very small and located at the bottom

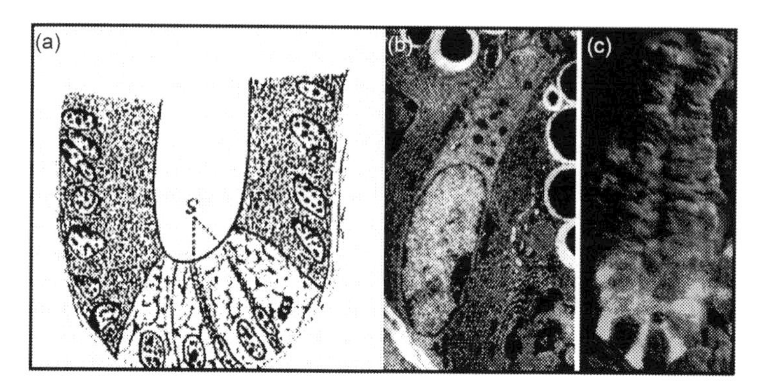

FIGURE 10.1 (a) The bottom of intestinal crypts, as shown in the hand-drawn picture from a publication by Paneth who had already identified these special stem cells in 1887, contains small cells (S) and Paneth cells (large white cells). In an electron microscopic picture (b) the stem cell appears light gray, flanked by two darker Paneth cells. The stem cell is the source of all other cells on the surface of the crypt and villus that form the lining of the intestine. (c) shows a whole intestinal crypt of a mouse with green (fluorescent) stem cells separated by larger, dark-gray, Paneth cells. *Source: Reproduced from Hans Clevers, The intestinal crypt, a prototype stem cell compartment, Cell, 2013, 154, pp. 274–284.*

of the crypts of the intestine, invariably next to another cell called a *Paneth cell,* named after the pathologist who first identified them (Figure 10.1). The Paneth cells form a kind of niche in which the stem cells can grow and survive. Later work showed that the stem cells can form all of the cells lining the intestine right up to the tips of the villi, the finger-like structures pointing to the interior of the intestine that form the large surface area needed to absorb food. The stem cells move up the crypts, differentiating along the way, and are finally shed off into the intestine itself.

It turned out that in mice, many tissues of the gastrointestinal tract (small intestine, stomach, pancreas, liver, etc.) and other tissues (lung, skin, ear, etc.) have Lgr5 (or Lgr6) expressing cells that show as green in transgenic mice, in which the presence of the protein in a cell is associated with a fluorescent signal. All of these adult stem cells can differentiate to the cells of the organ to which they belong, but, curiously, they cannot form other tissues if transferred to another site in the body. Lgr5-expressing stem cells from the large intestine, for example, cannot make cells of the small intestine if transferred to the small intestine.

Based on this ability to identify the stem cells through the green fluorescent staining, it became possible to isolate the cells, and even grow them in culture where they could organize spontaneously into "mini-guts" (see Box 10.1, "Organoid Culture of Adult Stem Cells").

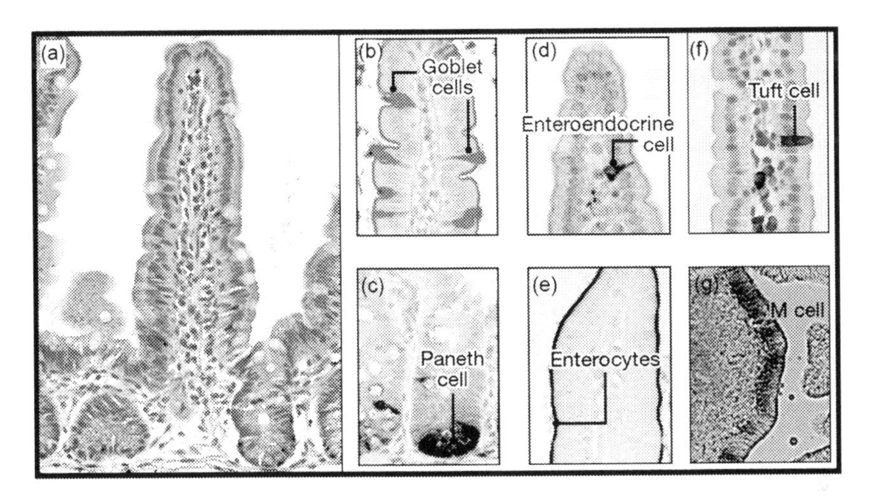

FIGURE 10.2 Different cell types can be distinguished in the small intestine. Panel A shows a complete villus, while various cell types making up the intestine are shown in the other panels, each marked using different staining techniques. *Source: Reproduced from Hans Clevers, The intestinal crypt, a prototype stem cell compartment,* Cell, *2013, 154, pp. 274–284.*

It was then possible to discover which signals made the cells grow and which made them differentiate. It became really exciting when similar stem cells were identified in the human intestine (as well as liver, pancreas, and the like). In addition, when they were derived from a biopsy from a patient with cystic fibrosis (a genetic disease caused by a mutation in an ion channel gene) the mini-guts that formed in culture were also abnormal and showed symptoms of the disease. That simple basic research could lead to findings with such far-reaching implications was completely unpredicted at the outset. It was a leap of faith among the researchers and grant funders alike.

For an idea of the numbers: Every crypt in the intestinal wall (and we have billions of crypts in our intestines) contains a few, around 15, adult stem cells in a stem cell niche (Figure 10.2). Within this niche, each intestinal stem cell is coupled to its neighboring Paneth cell (around 10 per crypt), creating a 3D spatial microenvironment which provides the necessary spatial, physical, and biochemical cues to the stem cells to enable it to maintain this delicate stem cell compartment and enable the continuous, and surprisingly fast, production of differentiating cells to replace dying cells in the intestinal epithelium. The intestine is exposed to one of the harshest environments in the body: it's no wonder the cells forming its lining do not last long and need replacing very rapidly.

10.3 ADULT STEM CELLS IN MUSCLE TISSUE

Besides the stem cells in organs of the body that contain epithelial cells, connective tissues (derived from the mesoderm layer in development) can also contain stem cells. One example is skeletal muscle, which contains a stem cell population called satellite cells (see also Chapter 5, "Origins and Types of Stem Cells: What's in a Name?"). These cells are again very small, are able to divide when the muscle is damaged, and can give rise to fully differentiated muscles of the limbs and back, for example. As people age, muscle tone usually decreases irreversibly and the ability of satellite cells to replace muscle tissue declines. What is interesting for regenerative medicine is the question of whether satellite cells can be grown outside the body and used to repair or rejuvenate skeletal muscle in decline as a result of aging, damage, or disease (e.g., muscular dystrophy).

Some researchers have tried to find cell surface markers for identifying different types of muscle cells. They have found, for example, markers that allow the precursor cells of both muscle and satellite cells to be identified. These studies in the lab of Amy Wagers at Harvard University in the United States revealed a fascinating aspect of aging in muscle: if the blood supplies of young and old mice were linked together, then the muscle was rejuvenated. Not only that, if Wagers restricted the calorie intake of the mice, skeletal muscle stem cell function was enhanced. Through linking young and old mice, she also observed a similar rejuvenation in the heart. While there is no evidence of this taking place in humans yet, it certainly opens an original line of research. Helen Blau at Stanford University, also in the United States, discovered that a stem cell niche was needed in muscle tissue to allow satellite cells to flourish well into old age, much like the niche in the bone marrow and intestines. She also discovered, however, that in addition to biochemical factors, such as hormones or growth factors that induce signal transduction pathways, satellite cells also needed the physical properties of the niche to be appropriately regulated. The stiffness, or softness for that matter, of the environment in which the cells lived was crucial for them to be able to grow, particularly in cell culture. When she isolated muscle stem cells and cultured them on standard tissue culture plastic surfaces, which are very hard, they failed to grow well and even died. However, when she tried different degrees of stiffness, she found that the stem cell would divide, or "self renew," on substrates with a very defined stiffness, neither too soft, nor too stiff—in fact, rather close to that found in real muscle (Figure 10.3). This demonstrated that the muscle stem cells could only be cultured when the biophysical aspect of the stem cell niche is well mimicked in culture.

The proof of the pudding came when she transplanted the muscle stem cells cultured on hard and soft surfaces into the muscles of mice. The survival of cells grown on the soft substrate was spectacular compared with those from the hard substrate.

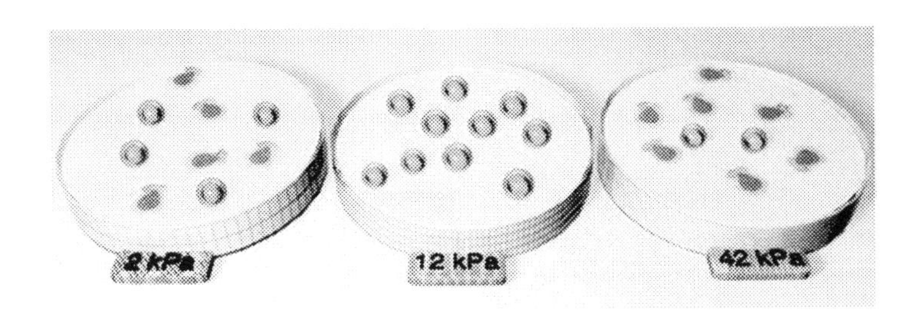

FIGURE 10.3 The stiffness of the surface to which a stem cell is attached can determine its survival and fate. When muscle stem cells (green) are cultured on surfaces that are soft (left, 2 kiloPascals or "kPa") or rigid (right, 42 kPa), their survival is less than when they are cultured on a surface with a medium stiffness (12 kPa), much like it would be in normal muscle tissue. *Source: Reproduced with permission from Gilbert, P.M. and Blau, H.M., Engineering a stem cell house into a home, Stem Cell Res. Ther., 2011, 2, p. 3.*

10.2

RIBBONS AND LINEAGE TRACING

How do you demonstrate that an adult stem cell is multipotent and can actually form all the cell types in the organ it comes from? The answer sounds simple: give the stem cell a color, such that all its daughter cells inherit the same color. Then see where these daughter cells end up in the organ, and what function they finally get in the organ. In practice it is not simple at all, but years of experience in carrying our fundamental research have made it possible for many labs to carry this out. For the studies of the intestine, however, the approach was fairly new when it was first carried out. By giving stem cells in the intestinal crypts of mouse intestine a "colored" gene (by inserting it in the DNA of the mouse), it could be shown that these stem cells produce cells that migrate upward in the crypt toward the villus, forming a beautiful colored "ribbon," and there they appear to have divided among them the different cellular functions that are needed for the gut (Figure B10.2.1). All of the cells with the same color shared the same parent cell in the crypt.

10.2 *(cont'd)*

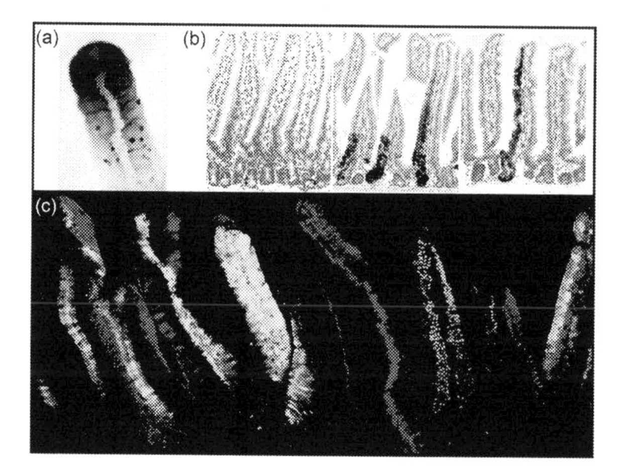

FIGURE B10.2.1 Entire intestinal crypts can be formed from single stem cells located at the bottom of intestinal crypts. (a) shows the top of a villus. (b) shows villi at different time points, with the descendants of stem cells appearing in blue. (c) shows a similar picture, but here the descendants of single stem cells each have a different color. It is clear that "ribbons" of one color run from bottom to top of a crypt and, therefore, all the cells within the ribbon must be the descendant of a single stem cell, forming all the different cell types of the crypt–villus structure. *Source: Reproduced from Hans Clevers, The intestinal crypt, a prototype stem cell compartment, Cell, 2013, 154, pp. 274–284.*

10.4 WHAT WE HAVE LEARNT ABOUT ADULT STEM CELLS

From these examples, even though they are not comprehensive, it is clear that several requirements need to be met to culture adult stem cells. Firstly, it is likely that each stem cell niche has unique characteristics, both biophysical and biochemical, that need to be known before the stem cells can be isolated and cultured successfully. The culture niche should provide an environment that mimics their own niche in the human body as closely as possible. Secondly, the spatial (3D) environment of the stem cell needs to be taken into account in designing the culture setting. Thirdly, interactions with other organ or tissue cell types that normally interact with the stem cells should be known and mimicked as closely as possible in culture. In this way, it may be possible to expand adult stem cells safely and in large numbers, something that had not been possible before, and this limitation was, for many years, the major bottleneck to using adult stem cells therapeutically, or for any other purpose.

10.5 THE FUTURE: ORGANOIDS TO REPAIR TISSUES AND ORGANS

We are witnessing just the beginning of new aspects of stem cell medicine and regenerative therapies based on adult stem cell culture. Organoid structures can now be grown from adult stem cells isolated from organs such as the stomach, prostate, lung, liver, pancreas, testis, skin, and mammary gland, and the kidney is likely just around the corner. Many challenges undoubtedly lie ahead in translating studies in mice into humans, and it will take years and years before a first patient may be cured by an "organoid-based" stem cell treatment. Until this has been successfully carried out, we cannot predict with certainty that it will succeed.

10.3

BRAIN ORGANOIDS

Can we mimic the structural and functional complexity of human organs or tissues *in vitro* to the extent that drug responses can be predicted? For safety pharmacology and drug discovery the saying "the simpler the better" is most appropriate for low-cost, high-throughput screening of compounds or drugs at an early-stage preclinical study. However, more complex human three-dimensional *in vitro* models encompassing multiple interacting cell types and extracellular substrates are expected to better mimic *in vivo* physiology and pathophysiology, and thus may be better for understanding how drugs affect healthy and diseased organs and tissues.

A case in point is the human brain. Human brain diseases such as Alzheimer's disease and Parkinson's disease have been very difficult to model in animals. During human development, neural progenitor cells (radial glial stem cells) are localized in a region of the brain called the *subventricular zone*. By contrast, these progenitor zones are present to only a limited extent, or not at all, in rodents. This largely explains the difference in brain size and function between human and rodent brains and is reflected in their limited value as brain disease models.

Human stem cells can, however, form many of the cell types in the human brain and may offer a solution. Recently, researchers succeeded in generating organized brain tissue they called *cerebral organoids* from human-induced pluripotent stem cells (Figure B10.3.1). In these *in vitro* cultured cerebral organoids, different regions of the brain, such as cerebral cortex, choroid plexus, and ventral forebrain could be identified.

10.3 *(cont'd)*

FIGURE B10.3.1 In this cross section of a brain organoid, neurons can be seen in green and neural progenitor cells in red. *Source: Reproduced with permission from Lancaster, M.A. et al., Cerebral organoids model human brain development and microcephaly, Nature, 2013, 501, pp. 373–379.*

Specific differentiation steps induced by neural growth factors, followed by three-dimensional culture in spinning bioreactors, were sufficient to produce parts of the human brain. The cerebral organoids displayed a remarkable three-dimensional self-organizing capacity (similar to that described for intestinal organoids) and were strikingly similar in morphological organization and molecular characterization to human brains. Moreover, identical differentiation and culture procedures using induced pluripotent stem cells from patients suffering from microcephaly, a disorder leading to a severe reduction in the size of the brain associated with mental retardation, resulted in cerebral organoids much smaller than their normal counterparts. Most likely this was a result of premature neural differentiation of radial glial stem cells, which caused a depletion of the progenitor cell population and impaired expansion of brain tissue. Under these culture conditions, cerebral organoids reached a maximal diameter of 4 mm. For more complex and bigger organoids, it would be crucial to incorporate other cell types or cellular structures, such as a vascular network for the delivery of oxygen and nutrients, to avoid necrosis of cells in the central regions of the organoids.

It is expected that these developments will facilitate the use of complex human stem cell derived brain models for studying brain development, for discovery and advancement of drugs to treat brain disease, and for evaluation of neuronal safety of drugs.

CHAPTER

11

Stem Cell Tourism

OUTLINE

11.1 DEFINITION OF STEM CELL TOURISM

Stem cell tourism is a term recently coined to describe a growing practice among patients to pay large sums of money to private clinics for often unproven stem cell therapies. Patients can be desperate because conventional medicine has failed to provide a solution for their particular condition. For diseases affecting children, emotions may run particularly high: the children themselves cannot make properly informed decisions, so parents face the additional conflict of wanting the best for them while, at the same time, having to protect them from undue risk. Advertisements for these clinics, often outside the patient's own country, claim that stem cell treatment can benefit or cure complaints ranging from diabetes, stroke, paralysis caused by spinal cord injury, cerebral palsy, and Lou Gehrig's disease (ALS), to wrinkles in skin and age-related hair loss in men (Figure 11.1).

The stem cells in these clinics may be autologous and derived from patients own bone marrow or fat, for example, or from another source

FIGURE 11.1 Lou Gehrig was the best and most famous baseball player of his generation, but he had to stop his professional career because of a devastating disease that was ultimately fatal. There is presently no effective treatment for this disease, which still carries his name: Lou Gehrig's disease. It is also now known more correctly as amyotrophic lateral sclerosis (ALS). Some clinicians hope that this disease will be curable in the future through stem cell treatment, but it is more likely that the generation of induced pluripotent stem cells from patients with the disease will teach us something about how it develops and ways in which disease progression can be slowed down. Induced pluripotent stem cells have already shown us why motor neurons die and cause the paralysis: it seems there is nothing wrong with the neurons as such, but they are killed by the cells that surround them. It is these surrounding cells that are now the focus of attention for therapy development, based on new drugs or specific other interventions. *Source: www.lougehrig.com.*

and be based on infusion or injection of umbilical cord blood (containing blood stem cells) collected from donors immediately after birth. Donors in this case are likely to have been paid for donation and their health history may be unknown; this conceivably entails a risk. The donated stem cells may not, for example, have been tested for hepatitis or HIV infection, which could give rise to AIDS if the cells are used for transplantation. Alternatively, the cord cells may have been collected at birth from the person to be treated or be from a state-run cell bank. In these cases, the cells may have a lower risk of being contaminated if properly collected by an expert, although there is no guarantee they will have any therapeutic benefit (discussed in detail in Box 7.14, "Stem Cell Therapy for Macular Degeneration"). The growing interest in these clinics among patients and their families derives, to a large extent, from the extensive media coverage of stem cell research and the sometimes sensation-seeking publicity that has surrounded their therapeutic potential (Figure 11.2). This has been particularly emotive in the United States and Italy. Polarization of public and political opinion of the field by the ethical issues associated with human embryonic stem cell research has contributed to interest in adult stem cell sources for therapy. Human embryonic stem cell research is often defended because of its potential

FIGURE 11.2 Despite very many articles in newspapers and magazines on stem cells and stem cell treatments, it is not always easy to distinguish facts from fiction in stem cell research. The fiction is often wishful thinking based on the hope of patients and families looking for cures.

use in future cell transplantation. If there were equivalent human stem cell alternatives, then it would be more difficult to justify the use of human embryos to derive stem cells. Induced pluripotent stem (iPS) cells may change this aversion to pluripotent stem cells on ethical grounds in the foreseeable future.

11.1

THE STORY OF JOKE KNIEST

In 2002, Joke Kniest, an active 52-year-old woman who enjoyed gardening and cycling and was a trained nurse working full-time, quite suddenly had an infarct (blockage) in a vessel in her spinal cord. This severely damaged the nerves. Within a few hours she was paralyzed from her chest down. Through intensive physiotherapy and rehabilitation in the months that followed, she was eventually able to stand and walk short distances, although for longer distances she needed an electric wheelchair. Her doctors were nevertheless pleased with her progress. Then Joke read about a stem cell clinic on the internet that claimed to be able to treat many chronic ailments. It was called the XCell-Center. She contacted the clinic and was told that partial spinal cord lesions like hers were particularly suitable for treatment. She was also told that she might be able to walk again properly if she had the treatment. The fee would be 15,000€. No mention of any risks or, in her own words, "that I might get worse." She would have to be at a clinic in Turkey within three days to be included in the next round of treatments and, as it was unclear when the next flight would leave, she would have to decide quickly. Despite feeling pressurized, she took the flight to Istanbul with her sister. Once there, she was asked by the director of the clinic to sign a consent document for the operation in Turkish. The operation then took place. "When I came out of the anesthetic, I felt terrible, so much pain, as if I had been run over by a train." She returned, sick and in great discomfort, to the Netherlands. There, she remained ill in hospital for the next three weeks and underwent several new operations to try and fix what had gone wrong. To little avail: when her own doctor examined her on her next visit he found that what little function she had in her legs had been lost. She could no longer stand, let alone walk a little. "It made life so much easier when I could stand to get dressed, but now I have even lost that. It is criminal what these clinics do to people." Did the clinic contact her ever again? Only while she was in hospital to request the remainder of their fee (Figure B11.1.1).

FIGURE B11.1.1 Joke Kniest. *Source: Photograph courtesy of Joke Kniest.*

Joke Kniest tried to claim damages in a civil case against the stem cell company but, unfortunately, she had signed a document, she says under pressure, that confirmed that she knew the risks of the treatment. A second patient with a spinal cord injury treated in the same clinic had a similar story: before the "treatment" he was able to lift and use both arms, but afterwards not only did he no longer have any strength in his right arm, he was also unable to get in and out of his wheelchair and into his car. He also began a court case for maltreatment and fraud.

According to the second patient, "I began the court case to protect potential new victims. The owner presented himself as a doctor, but he is not. He exploited our desperation as spinal cord lesion patients. He should be stopped. Looking back, I find it incredibly stupid that I believed it all."

In 2011, the XCell-Center in Dusseldorf was closed following an undercover investigation by the British newspaper *The Sunday Telegraph* into its activities. The center claimed on its website "Due to a new development in German law, stem cell therapy is currently not possible to perform at the XCell-Center. Regretfully for this reason, we must cancel your appointment until further notice. We will notify you for further updates about the matter."

Opponents to embryonic stem cells may be biased to interpret results of research on ethically acceptable (adult or umbilical cord blood) stem cells more positively than experiments show. Likewise, the proponents of human embryonic stem cell research may be less forthcoming on the potential risks of these cells in clinical use than might be necessary. In addition, there are many different types of adult stem cells: some that are rather undefined and simply grow out of many types of tissues such as bone marrow and fat and known as mesenchymal stromal (or stem) cells (MSCs), others that are specific to the organ from which they derive and have real potential for tissue regeneration in the future. The private stem cell transplantation clinics operate in this confusing space of an unmet medical need for millions of patients with untreatable ailments, limits to health research budgets, and widely differing ethical standpoints.

In this chapter, we will attempt to present a neutral perspective on this in an overview of the present clinical state of the art.

11.2

QUESTIONS A PATIENT COULD ASK WHEN CONSIDERING COMMERCIAL STEM CELL THERAPY

The ISSCR has produced a *Patient Handbook on Stem Cell Therapies* which contains a list of questions that could be addressed to a commercial clinic before making a decision on whether to proceed or not. This handbook is freely available on *www.isscr.org*. Among the questions are the following:

1. Have preclinical studies been published, reviewed, and repeated by other experts in the field?
2. Has the clinic received approval from an independent regulatory authority, such as an institutional review board, ethics review board, or a national or university hospital medical ethics committee, that has ensured that risks are as low as possible and worth any potential benefits? Are there independent scientists or clinicians who can provide independent advice?
3. Have there been earlier clinical trials using this treatment and what was the outcome? What benefits might be expected, how will this be measured, and how long will it take?
4. Are the patient's rights protected? What is the risk of the procedure itself and what will be done if there are bad side effects? Who will provide emergency medical care should it be necessary? Is the clinic

equipped to handle emergencies? Who is the doctor in charge of treatment and what specialized training does the doctor and support staff have?

5. What are the costs of the treatment and what other costs are there (e.g., travel and insurance)? What would be the cost of emergency treatment should something go wrong, and who pays for this?

The *Patient Handbook* also provides a list of warning signs to look for:

1. Claims based only on patient testimonials or claims that there is no risk. All effective medical interventions carry some risk.
2. Multiple diseases treated with the same cells. With the exception of blood diseases, which are often related, it would be expected that unrelated diseases, such as Parkinson's disease or heart disease, would need different treatments and different medical specialists.
3. The source of cells that will be used and how they will be delivered is not clearly documented.
4. High costs of treatment. It is not customary for someone to pay in a clinical trial.

FIGURE B11.2.1 Current and past presidents of the International Society of Stem Cell Research. Standing from left to right: Rudolph Jaenisch, Elaine Fuchs, George Daley, Janet Rossant, Fiona Watt, Fred Gage. Front row left to right: Gordon Keller, Irving Weissman, Leonard Zon, Shinya Yamanaka. *Source: Photograph courtesy of the ISSCR and Jim Ezell, photographer.*

11.2 *(cont'd)*

FIGURE B11.2.2 Nancy Witty, Chief Executive Officer of the International Society of Stem Cell Research. *Source: Photograph courtesy of Nancy Witty.*

11.2 WHAT'S THE DIFFERENCE BETWEEN TRIALS AND TREATMENT?

Deciphering the difference between commercial stem cell "treatments," early-phase medical intervention, and standard stem cell clinical practice puzzles many. It is important to understand the differences in any context where stem cells are offered as therapy. First, we provide a short explanation of how specific treatments for disease become standard clinical therapy for many patients in a regular hospital setting.

In many countries, private and public health insurance policies require that the costs of medical treatments they cover be evidence based. "Evidence-based medicine" means that the treatment prescribed by a physician has been demonstrated to be effective and safe in a broad patient group. After preclinical research in animals and preliminary tests for safety and feasibility in a small group of patients, most treatments will have then undergone one or more of what we call "double-blind, placebo-controlled" clinical trials. "Double-blind" means that neither the patient nor the doctor know whether the treatment being tested is actually delivered or not. "Placebo-controlled" means that the person in the control group undergoes a procedure very similar to those being treated but actually does not receive any of the substance (in this case, stem cells) being tested for therapeutic activity. This may, for example, be by simple physiological salt solution or cells that are not stem cells. Clinical trials are designed to test whether a new treatment is better than an existing therapy.[1] The patients taking part in the clinical trials are usually recruited as volunteers. They are divided randomly into two groups: one group receives conventional treatment, the other group (preferably matched, for example, with respect to age and gender) receives the conventional treatment plus the treatment being tested or, in some cases, the new treatment alone (Figure 11.3).

The first (phase I) step of a clinical trial is called "safety and feasibility": will it be safe to give the treatment to the patients and will it be feasible? For stem cell therapies, this usually means asking whether it will it be possible to obtain sufficient numbers of healthy stem cells at the right time using reagents and conditions that meet with regulatory requirements. Stem cells are, after all, a medicinal product for use in humans and, therefore, have to meet generally accepted safety specifications. They have to be proven virus- and bacteria-free, for example, and have no contaminating reagents toxic for humans in the cell suspension that is delivered to the patient. In addition, the regulatory authorities have to ensure that the discomfort and risk a patient will undergo is proportional to the severity of the disease. Permission will more likely be given for high risk experimental treatments or phase I trials if a disease is always

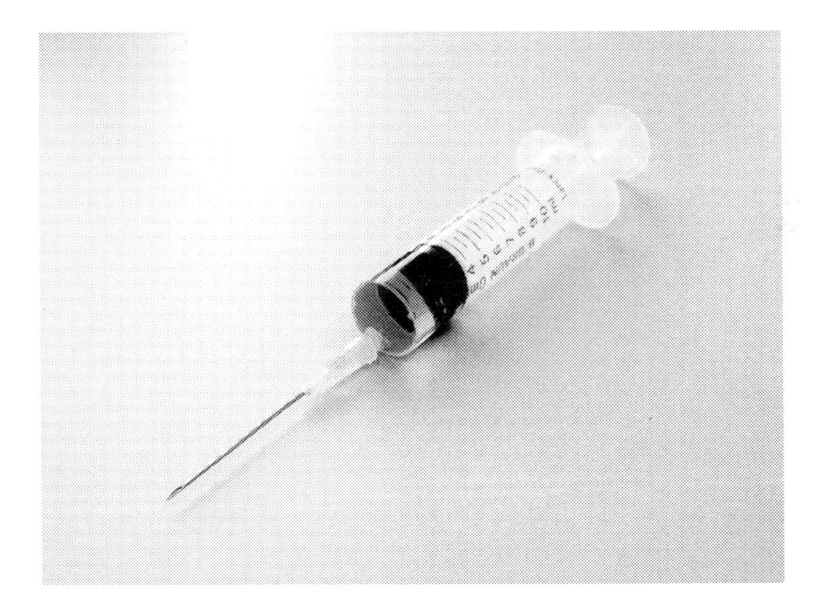

FIGURE 11.3 It will take many years and many clinical trials before stem cell therapy can be safely used to treat certain diseases.

[1]Further information explaining clinical trials, written by the National Institutes of Health in the United States, can be found at *www.clinicaltrials.gov/ct/info/resources*.

fatal and life expectation is short. For example, Batten disease, a rare and fatal metabolic disorder in children that affects the brain, is fatal before five years of age. Permission was granted by the U.S. Food and Drug Administration (FDA) to treat some of these children by transplanting well-characterized neural stem cells into the brain. The decision is supported by strong evidence in rats that the neural stem cells function properly after transplantation. The first part of a phase I trial on six children with advanced disease was completed in 2009, but the second part, to treat children with less advanced disease, was discontinued in 2011 because no children could be found that met the selection criteria for inclusion. Finding enough patients to include highlights a major problem in attempting a proper clinical trial for these rare diseases.

By contrast, for more common conditions, most regulatory authorities would not give permission to transplant bone marrow stem cells to the brain, for example, after stroke, because there is no evidence in animals that it would be effective, the transplantation would have high risk, and, even though stroke may be chronically disabling, after stabilization it is not acutely fatal.

11.3

THE INTERNATIONAL SOCIETY FOR STEM CELL RESEARCH

The International Society for Stem Cell Research (ISSCR; *www.isscr.org*) is an independent, nonprofit organization established to promote and foster the exchange and dissemination of information and ideas relating to stem cells, to encourage the general field of research involving stem cells, and to promote professional and public education in all areas of stem cell research and application. Members are admitted on the basis of their professional credentials as scientists or clinicians working in the field of stem cell research. The ISSCR holds an annual meeting open to members and nonmembers; any profits of the annual meeting support the day-to-day running of the society, including its website, elected board, and the costs of generating independent advisory documents. In 2013 the society was in its 12th year, had more than 4000 members worldwide, and was still growing.

The ISSCR is committed to ensuring the promise of stem cell research is delivered to patients in a safe, effective, and fair manner. In 2008, the ISSCR released *Guidelines for the Clinical Translation of Stem Cells* that call for rigorous standards in the development of stem cell therapies and outlining what needs to be accomplished to move stem cells from promising research to proven treatments. The *Guidelines* include a standalone appendix, a *Patient Handbook on Stem Cell Therapies*. In 2010, the ISSCR launched a website to further support patients, their families, and doctors as they consider a stem cell clinic or treatment (Figure B11.3.1).

FIGURE B11.3.1 Logo of the International Society of Stem Cell Research (*www.isscr.org*). The ISSCR was founded in 2002 and is now a large international organization for stem cell researchers with thousands of members throughout the world. It is an independent, nonprofit organization established to promote and foster the exchange and dissemination of information and ideas relating to stem cells, to encourage the general field of research involving stem cells, and to promote professional and public education in all areas of stem cell research and application.

Perspective of the ISSCR

The ISSCR website includes tools to assist patients as they consider a clinic or treatment. The website provides an alert where claims of efficacy are unsubstantiated and a list of diseases, injuries, or conditions for which, to the knowledge of leaders within the ISSCR, there is currently no proven benefit from stem cell therapies. There is also an extended list of questions a patient (if possible, working with their doctor) should ask as they consider and evaluate a clinic and treatment. The ISSCR will also investigate clinics and list those that do not provide evidence that the treatment they offer has credible scientific rationale and that appropriate regulatory oversight and other patient protections are in place.

One of the other considerations taken into account is how the stem cells have been treated prior to transplantation. The more they have been manipulated or expanded in culture in the clinical laboratory to produce sufficient cells for transplantation, the higher the chance that the cells have undergone chromosome changes (or mutations) that may make the cells malignant and able to cause tumors in a patient. Any tumors derived from a patient's own stem cells would be very difficult to treat because the body would recognize them as "self" and not reject them. Mesenchymal stem (or stromal) cells from bone marrow or fat (see Chapter 5, "Origins and Types of Stem Cells: What's in a Name?"), for example, have been cultured and expanded in the laboratory before use in patients. Regulatory authorities, such as the FDA, require that culture reagents should be free of animal products (such as fetal calf serum) or proven virus free, and the cells should have normal chromosomes (i.e., be karyotypically normal) after growth in culture (Figure 11.4). The cell preparations also need to be produced in special laboratories, rather like a clean room used for electronic chip production, by personnel wearing special clothing to protect the preparations from contamination. This is not a trivial requirement and may be very expensive, costing thousands of euros or dollars just for cells for one

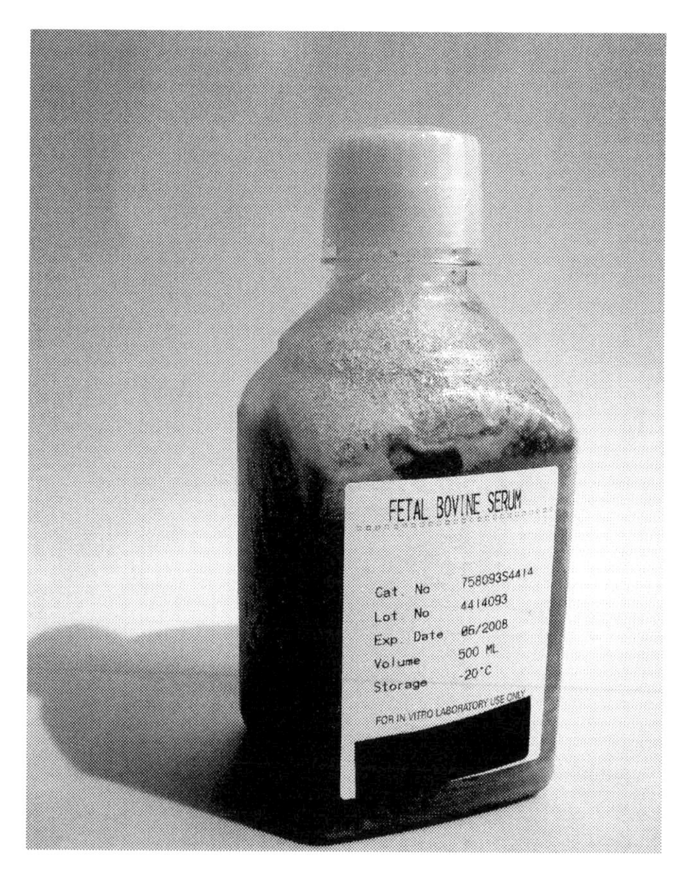

FIGURE 11.4 When stem cells are being prepared for transplantation in patients as a therapy, reagents used for culturing stem cells should be free of animal products ("xeno-free"). For example, they should not contain the widely used fetal calf serum, and any growth factors or insulin used as substitutes for the fetal calf serum should be synthetic (recombinant) or obtained from human sources. Reagents to induce differentiation should also not be of animal origin.

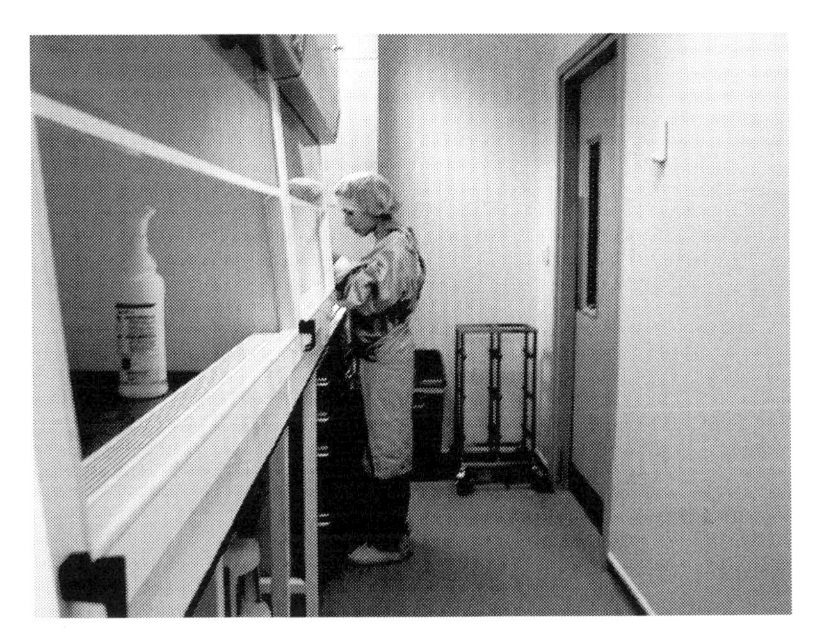

FIGURE 11.5 Cells for therapy need to be prepared by personnel wearing special clothing that is sterile and prevents any cells (e.g., from skin or hair) or other contaminants getting into the cells that will be used for transplantation. *Source: Susana Chuva de Sousa Lopes/Stamcellen Veen Magazines.*

patient (Figure 11.5). These are among the issues that private stem cell clinics often do not comply with: the procedures and conditions used for processing cells and their source are not made known to the patient; the regulatory authorities have often not approved or sometimes even seen the protocols for the methods of delivery to be used, have not checked whether the medical staff carrying out the procedure is properly qualified and experienced to do it, and have not verified that there is a proper procedure in place in case of an emergency as a result of something going wrong during the procedure.

11.4

COMMERCE BEFORE SCIENTIFIC PROOF: HISTORY OF A STEM CELL CLINIC

The XCell-Center was opened in Cologne, Germany in 2007 after legislation in the Netherlands blocked its operations (carried out under the name Cells4health) by banning private clinics offering unproven

11.4 *(cont'd)*

stem cell therapies on a commercial basis. Similar legislation was in place in the United Kingdom but not in Germany. The XCell-Center rapidly became the largest commercial stem cell clinic in Europe, in the first few years accepting more than 3000 patients with many types of ailment for "treatment" with stem cells derived from their own bone marrow. A second site opened later in Dusseldorf. From its (former) website, "At the XCell-Center patients with various degenerative conditions are treated at the highest medical level. The therapies range from the treatment of diabetes mellitus (type 1 and type 2, as well as their complications) to the consequences of stroke. Various neurological conditions also benefit from stem cell therapy, in particular, spinal cord injury, multiple sclerosis (MS), amytrophic lateral sclerosis (ALS), Parkinson's, Alzheimer's, and arthritis, heart, and vascular disease, macular degeneration (age-related blindness), neuropathy, and incontinence." Cerebral palsy was later added to the list (Figure B11.4.1).

FIGURE B11.4.1 The interior of the former XCell-Center. This is where patients were provided with information and had bone marrow collected for later transplantation at other sites in their body. The clinic closed in 2011, first claiming bankruptcy but in a posting on its website, the XCell-Center said: "Due to a new development in German law, stem cell therapy is currently not possible to perform at the XCell-Center. Regretfully, for this reason, we must cancel your appointment until further notice. We will notify you for further updates about the matter." It is believed that the center moved its operation to Lebanon.

Patients paid minimally 25,000€ (US$30,000) for the procedure from bone marrow collection to injection in their brain, spinal cord, or blood. Often a deposit was paid in advance. The center claimed that it had an official license and that its goals included clinical research. There is no evidence for this.

In 2010, an 18-month old boy from Romania died after receiving bone marrow injection into his brain. A doctor who worked at the clinic was placed under criminal investigation over the child's death and the serious injury caused to a second, 10-year-old, child from Azerbaijan with cerebral palsy in another allegedly botched operation three months earlier. The clinic was accused, not only of making serious errors in carrying out the operations, but also of failing to respond quickly enough when the surgery went wrong. Legislation in Germany was, by this time, such that the clinic could be closed and it ceased operation almost immediately. Patients and families who had paid deposits for planned treatments lost their money. The British newspaper *The Sunday Telegraph* investigated the clinic and found that the director had simply moved it to Beirut, Lebanon where there was no legislation governing its operation. *The Sunday Telegraph* headlined "The founder of the XCell-Center, a German stem cell clinic that closed after a Sunday Telegraph investigation, has been accused of embezzling funds to fund a lavish lifestyle," the claim being that expensive houses, cars, and yachts had been financed from the profit. Several patients attempted to file for damages but the costs of doing this are often prohibitive especially when a company no longer exists.

However, even a year later when the center appeared to have been entirely discredited, it was still possible to read the following in the press, in this case from the family of a boy with cerebral palsy who had received cells without negative side effects:

> The family are determined to continue the fight to improve his quality of life despite the closure of the center providing his stem cell treatment in Germany. "We have been told that the center must apply for a new license and naturally we are very disappointed," says the boy's dad, "and as yet we don't have any timescale on that."

Thus despite the lack of any evidence of beneficial effects, the high costs, as well as high risks, desperate patients and their families remain prey to unscrupulous practices of these private clinics. The history of one clinic illustrates that of many: when it becomes complicated, close down and start up again under another name elsewhere.

The second and third phases of an approved clinical trial assess "efficacy" of the treatment, firstly in a small group of patients, then in a larger group. This simply means "does it work or not?" Hundreds of patients with the same diagnosis may be included. For rare diseases, it may be difficult to find enough patients to include in the trial, and so several centers within one country, or even internationally, may work together. Both the placebo-control group and the group receiving the real treatment are closely followed by medical specialists during the trial and for long periods thereafter. These physicians should not know which patients are controls and which have been treated ("blind"). They will monitor the symptoms of the disease and how they change over time. They will also check for adverse events (side effects) and, should they occur, report these to an independent oversight committee. Hospitals taking part in these trials always have insurance covering adverse events, which includes any extra medical treatment the patient might need as a result. The hospital also makes sure that the patients know that there may be risks associated with trials and that they might not necessarily be in the group that will benefit, a procedure called informed consent. Again, private stem cell clinics often do not meet any of these standard regulatory requirements: usually they do not have insurance covering adverse events, they do not have specialist doctors on the premises to deal with emergencies, and they do not have long-term follow-up of the patients. Their doctors may not be specially trained for the procedures that are carried out on patients.

The final phase (phase IV) of a proper clinical trial determines whether the whole procedure using stem cells (or a new drug) is safe, feasible and effective, and covers all of the necessary legal paperwork for full approval. For a single drug to reach market, the complete process costs over US$1 billion. An example of the costs involved in bringing the first human embryonic stem cell therapy to a phase I clinical trial is the study planned by the company Geron to generate cells from human embryonic stem cells that might benefit recovery from acute spinal cord crush injuries. In total, preparatory costs for this first-in-man phase I clinical trial was US$200 million: US$45 million was required for 22,000 pages of documentation for the regulatory authorities and the permit, and US$13 million was invested in preclinical research. These costs became so prohibitive that in 2011, Geron discontinued the study after including only six patients. None had appeared to benefit from the treatment.

11.5

CALL FOR TIGHTER REGULATIONS TO CURB EXPLOITATIVE PRACTICES DOWN UNDER

Shocked by what's happening in their own backyard, in 2013 scientists from Stem Cells Australia[2] called for tighter regulations to prevent Australia becoming the new destination of choice for stem cell tourism.

For many years, Australian stem cell research benefitted from a high level of community support. The achievements of Australian scientists were celebrated in the media much as elsewhere in the world. There is still strong public support: federal and state governments provide dedicated funding and, following extensive public consultation and debate, laws were introduced to permit the use of human embryos in research provided specified safeguards were met.

Recognizing the hope invested by many in possible new treatments, the Australian stem cell research community worked closely with their international colleagues and with local patient support groups, community organizations, doctors, and other health professionals to make information about real progress in the field known. Like their colleagues internationally, Australian scientists have been outspoken critics of stem cell tourism where clinics overseas offer expensive stem cell "treatments" without substantial evidence to back up their claims.

The message to the Australian community was simple: proceed with caution, the treatments being offered abroad are highly experimental and potential unsafe. However, some Australians were not prepared to wait: many began to think about venturing overseas. However, in 2011 things became a bit more complicated down under.

Identifying the need to provide greater guidance and oversight on products made from cells and tissues, in 2011 the Australian Government introduced a new regulatory framework. Consistent with regulations in other jurisdictions, these new regulations set out acceptable standards to ensure that benefits to consumers outweighed any risks.

Potential risk was categorized on how closely the intended use of the product matched the original biological function, and on how far removed the cells or tissue were from their naturally occurring state. The greater the degree of manipulation in manufacturing, or the more extended the proposed use from what the cells or tissue usually do in the body, the more stringent the requirement placed by the regulators on those providing the proposed therapy.

For many possible stem cell treatments, it was envisaged that evidence on safety or efficacy would need to be collected in clinical trials prior to consideration for approval. However, due to a concern about restricting current medical practice, some products including autologous cell and tissue-based therapies (where the cells are taken from the patient for their own use) were excluded from having to comply with these stringent standards. Inadvertently, the Australian regulators created an opportunity for Australia to become the southern hemisphere's destination of choice for stem cell tourism.

Instead of traveling overseas, Australian patients could now visit one of the growing number of clinics in their own country. Many clinics also attempted to attract overseas patients. With implied legitimacy given Australia's high standards of health care, treatments were (and continue to be) offered for an astonishingly wide range of medical

11.5 *(cont'd)*

conditions at a considerable financial cost. The experimental nature of the treatments is rarely acknowledged and, given the lack of manufacturing oversight, it is also unclear whether the patients are even receiving "stem cells" from their fat or blood.

Shocked by this development, Stem Cells Australia called for the regulations to be tightened. Simple modifications to the current regulations, such as incorporating recognition of the inherent risks in nonhomologous use of autologous cells and making it a requirement that all cell manufacturing occurs in accredited laboratories, should curb exploitative practices but still allow medical innovation in Australia (Figure B11.5.1).

FIGURE B11.5.1 The logo of Stem Cells Australia. Stem Cells Australia has started a campaign in Australia against a loophole in the law that allows anyone with a medical degree to offer (unproven) stem cell treatments.

As recently highlighted by the International Society for Stem Cell Research, scientists at Stem Cells Australia believe that it should be considered unethical to market unproven cell-based interventions outside of clinical trials, even when the patient's own cells are used. With so much attention on the ethical issues associated with obtaining stem cells, Stem Cells Australia says we need to remain vigilant about the ethical issues associated with responsibly translating stem cell science to the clinic. Sometimes the simple message is that we are not there yet.

[2]*Stem Cells Australia is an Australian Government funded special research initiative in stem cell science. Researchers are using a multidisciplinary approach to explore stem cell regulation and differentiation. Stem Cells Australia was established by the University of Melbourne, University of Queensland, Monash University, University of NSW, Walter and Eliza Hall Institute for Medical Research, Victor Chang Cardiac Research Institute, Florey Neuroscience Institutes, and CSIRO. University of Melbourne is the Administrating Institute.* www.stemcellsaustralia.edu.au

For bringing an "off-the-shelf" stem cell product to market, the costs of production would likely be of the same order, depending on the controlling regulatory body. The least expensive stem cell products are, therefore, those that do not require culture or only minimal processing. Commercial clinics generally use the least expensive cell source. Cells selected from a patient's own bone marrow are therefore most commonly offered, although cord blood may also be included. These stem cell types are offered to patients for a wide variety of diseases, among them diabetes, spinal cord lesions, stroke, different types of neural disease causing paralysis, and the consequences of a heart attack. Outside treating diseases or conditions of the blood, none of these treatments have been proven effective in proper clinical trials. For example, injection in the brain or spinal cord can cause additional damage. It is also hard to imagine what the rationale is behind the assumption that "one stem cell type will cure all," when in conventional medicine we know a different drug is needed for each ailment. Penicillin is a fantastic drug for bacterial complaints but it will not help to get rid of a migraine!

Advertisements for commercial stem cell clinics often encourage patients to take their fate in their own hands, while portraying standard clinical medicine as reactionary, lacking imagination and innovation. This appeals strongly to the needs of chronically ill or disabled patients, for whom present medical treatment does not offer a solution. As a result, patients and their families spend many hours surfing the internet looking for stem cell treatments for their particular disease. Very often they will find not just one but tens or even hundreds of options, usually in countries where there is no legislation controlling the activity of commercial clinics. Clinics have also been known to change name and country fairly quickly if there is any attempt at "name and shame" or an action for libel may ensue. Detrimental effects on a patient's condition resulting from mala fide stem cell "therapy" are often difficult to prove, and patients are reluctant to come forward if their condition has become worse. Only rarely will they acknowledge not having benefitted from the treatment. Reasons for this can be embarrassment at having paid a large sum of money for something that has not worked, or a placebo effect that for some ailments can lead to significant, but temporary, improvements. It is important not only that patients but also physicians are correctly informed on the true state of the art. Unfortunately, our modern resource of internet is not always the best place to look. In addition, because of the lack of "on the record" evidence, many patients do not realize that things can actually get worse as a result of these "therapies."

11.6

THE STORY OF CARMINE VONA

On April 3, 2008, the Italian newspapers described the story of stroke victim Carmine Vona, a 54-year-old traveling salesman who was partially paralyzed on the left side of his body. This left him with a permanent mobility problem that was particularly damaging for his job. Like other people living in the Turin area suffering from neurological problems, Vona had heard of a "new method of treatment" discovered by Davide Vannoni based on "stem cells." The procedure consisted of intravenous and intrathecal infusions of mesenchymal stromal (or stem) cells (also known as MSCs) extracted from the hip bone (iliac crest) of the patient and prepared using a unique method of isolation in culture, and allegedly followed by *in vitro* differentiation into neurons. It was claimed that the MSCs, which normally generate bone, cartilage, and adipose tissue, converted into neurons after brief exposure to ethanol (alcohol) and retinoic acid (vitamin A). Carmine contacted Vannoni by phone. Vannoni "assured [him] that for that kind of problem he would be cured, immediately, and for 100%."

The steps that followed between this first phone contact and the first infusion followed a pattern that was identical for another 62 supposed victims, who had by July 2013 addressed patient abuse allegations against Vannoni and his Stamina Foundation. Carmine Vona was invited for a neurological examination prior to the procedure, which he got "in five minutes" in a modest, ambulatory care center in downtown Turin that was run by Leonardo Scarzella. Scarzella worked as a neurologist at the Valdese Hospital, also in Turin, in the mornings. Vona then had his first meeting with Vannoni at the Stamina Foundation located in the basement of a building, again in Turin. It later turned out that the other floors above ground hosted a private company called Cognition, which carried out market research and organized courses on advertising and "psychology of persuasive communication" (the latter being the title of a handbook written by Vannoni in 2001). Vannoni was president of this company and an expert in this field, holding an academic degree in humanities and serving as an Associate Professor of Psychology at a state university.

During the visit, Vannoni was called the "doctor" by his collaborators. This title can be used for both academic graduates with doctorates and physicians, but doctors of medicine (or MDs) can also use the term "medico" to distinguish the academic and clinical titles; the Italian language is, in this respect, ambiguous. The "doctor" showed the patient two video clips that claimed to show cures: an old man sitting in a wheelchair who was able to walk again and a virtually paralyzed young dancer once more able to dance (Figure B11.6.1). The cost for the stem cells treatment was 27,000€, but since Vona was hesitant, Vannoni offered him a discount, eventually asking for 21,600€.

18 | Cronache | LA STAMPA

"Così Stamina mi ha rovinato la vita"

Una marea di denunce contro il metodo inventato da Davide Vannoni, bocciature nel mondo scientifico
Ecco il racconto del paziente che ha convinto la Procura a indagare: "Con loro ho rischiato di morire"

Inchiesta

62 querele
Presentate da persone che contestano a Vannoni la validità del suo metodo di cura

27 mila euro
È la cifra che i pazienti di Stamina devono spendere per potersi sottoporre alla terapia

2009 l'inizio
La sperimentazione di Vannoni prosegue da quattro anni

Il primo centro
È stato aperto circa dieci anni fa in un ufficio di via Giolitti a Torino: qui ci lavoravano due medici ucraini

LA BEFFA
«Quando sono stato male volevano firmassi una liberatoria e ammettere che era colpa mia»

Tutte le stroncature

L'INCHIESTA DI GUARINIELLO
1 Nel 2009 apre un fascicolo sulla vicenda

LA PRIMA DENUNCIA
2 Arriva con la morte di un uomo torinese

LA RIVISTA NATURE
3 «Il metodo Stamina è errato»

L'APPELLO DEL MINISTRO
4 «Vannoni consegni il protocollo»

RACCOLTA FIRME DEI RICERCATORI
5 In duecento chiedono lo stop dei test

Ha detto
Il 1° agosto consegnerò all'Istituto Superiore di Sanità il protocollo per l'applicazione del metodo. A giorni vedremo il comitato scientifico per la sperimentazione
Davide Vannoni

«Curato in un centro estetico»
La denuncia di Carmine Vona è considerata uno dei cardini dell'indagine in corso alla Procura di Torino

FIGURE B11.6.1 "Stamina has ruined my life" reads this article in the Italian newspaper *La Stampa*, a statement from one patient who fell for the claims of the Stamina organization.

11.6 *(cont'd)*

After Vona agreed, the collection of stem cells from his bone marrow took place in a private clinic (Lisa di Carmagnola) and 10 days later he was called to arrange the infusion procedure. This was carried out in a beauty and aesthetic center located in the Repubblica di San Marino (a small state close to Rimini, on the border of the Italian regions of Emilia Romagna and Marche). Vona noticed that at the entrance of the clinic there were advertisements for slimming cures and that the cleaning man at certain point put down his broom, put on a white lab coat, and helped two nurses in preparing the liquid for the bone marrow injection. Desperate as he was about his condition he decided to carry on with the procedure despite these forebodings.

As part of the package, a room had been booked at the nearby Hotel Passepartout so his condition could be monitored in the 24 hours following the treatment. While watching a movie on the TV, for the first time in his life Vona had an epileptic attack, with tremendous convulsions and foaming at the mouth. He believes that his life was saved by a close friend who had accompanied him to the clinic and rushed him to the local hospital. What he had called his "journey of hope" had turned into a nightmare. Once awake and conscious, he looked for the "physicians" of the beauty center unsuccessfully. Realizing that they left, he then called them on the cell phone and found they were 85 miles away, on their way to Bologna. Deeply shocked, Vona was able to convince them to come back: "They were really embarrassed," Vona recalls, "and denied that they treated me with stem cells. In the following period, Vannoni himself tried in every way possible to convince me to sign a release form. He wanted me to assume the entire responsibility. He strongly insisted." Of the 62 "apparent victims" of the "Stamina case," nine decided to file a lawsuit against Vannoni. Carmine Vona, still limping, is number 52.

Contributed by Dr Andrea Grignolio,
Section of History of Medicine, University of Rome «La Sapienza»,
Senate Office Elena Cattaneo and Senator Dr Elena Cattaneo,
University of Milan, Italy.

11.3 PERSPECTIVE OF THE INTERNATIONAL SOCIETY FOR STEM CELL RESEARCH

In the meantime, bona fide stem cell biologists and medical specialists are carrying out serious research on how to develop safe and validated stem cell therapies and bring them to the clinic. The International Society for Stem Cell Research (ISSCR) produced a document in which guidelines were provided by member scientists to doctors, patients, and their families on how to deal with information provided by private stem cell clinics. This includes assessment of whether the claims of the clinic are based on real scientific facts, what the clinical evidence supports the therapeutic claims, and whether the therapy is likely to be harmless or prone to cause serious side effects. The risk of serious side effects is the greatest concern, and there are documented examples of patients who were significantly less well after receiving stem cell preparations than before. The ISSCR also provided suggestions and guidelines for proper "informed consent," and a separate handbook on the sort of questions patients should ask before considering enrolling in a clinic offering unregulated stem cell therapy. These *Guidelines for the Clinical Translation of Stem Cells*, in which the ISSCR openly condemns the use of unproven stem cell therapies, has an accompanying *Patient Handbook on Stem Cell Therapies*. Both can be found on the ISSCR website (*www.isscr.org*). The guidelines have been translated into 11 different languages.

11.7

EUROSTEMCELL

Europe's Stem Cell Hub: Information, Education, Conversation

EuroStemCell was established to help European citizens make sense of stem cells. It provides reliable, independent information and road-tested educational resources on stem cells and their impact on society in six European languages.

A partnership of scientists, clinicians, ethicists, social scientists, and science communicators, EuroStemCell also works closely with teachers and patient representatives, and aims to make research on stem cells and regenerative medicine accessible to all.

11.7 *(cont'd)*

The central focus of EuroStemCell is a dynamic multilingual website (eurostemcell.org), featuring stem cell fact sheets, FAQ, blogs, interviews, images, films, comics, teaching tools, and more. All factual content is written or reviewed by scientists from an ever-growing pool of specialist contributors—ensuring the site reflects the latest developments in the field—and is tailored for the non-specialist reader. The website also features a toolkit containing tried and tested stem cell resources for a variety of audiences and settings—from the school classroom to science centres, open days, festivals and other educational contexts.

Complementing online activities, EuroStemCell works collaboratively to facilitate public and schools engagement events and professional development for teachers. It also offers diverse opportunities for scientists to get involved in science communication, and engages a broad cross-section of European citizens through film, press and social media.

EuroStemCell has been funded by the European Commission since March 2010.

Contributed by Dr Jan Barfoot,
Public Engagement Manager, EuroStemCell, Scottish
Center for Regenerative Medicine, Edinburgh, U.K.

CHAPTER

12

Cancer Stem Cells: Where Do They Come From and Where Are They Going?

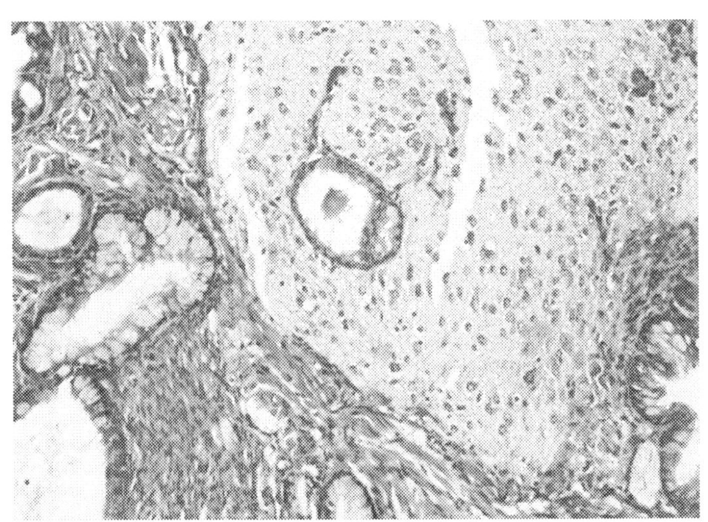

Rui Monteiro, Hubrecht Institute, Utrecht, The Netherlands

OUTLINE

12.1 CANCER: OBSERVATIONS AND QUESTIONS

Tumors arise from cells in the body that start to divide abnormally. They can be either benign (noncancerous) or malignant (cancerous). Cancer is generally taken to mean that the cells in the tumor grow uncontrollably and can metastasize, that is, detach from the main tumor and invade normal tissue elsewhere in the body. These additional satellite tumors (called metastases) often settle in preferred organ sites, such as the liver, brain, or bone. Sometimes a benign tumor can change over time into a tumor with malignant characteristics, including this ability to form metastasis in distant organs. A good example is a small polyp (called an adenoma) in the intestines (the gut), which can over decades transform into colon cancer. Although this is a rare event, once this transformation occurs, the tumor cells may metastasize to other organs if not treated appropriately and in time. In general, patients do not die from the initial, primary tumor, but from metastases that go out of control and do not respond to any treatment any more. Cancer is still a devastating disease which causes a lot of suffering, not only to patients but also to family and friends. Nearly everyone will know someone who has either undergone treatment for cancer and has been cured, or died from one of its many forms. If we are to change deaths into cures, we need to understand why tumors start to invade healthy tissue and how we can prevent this from happening.

Cancer is a very important topic in human health, but what does it have to do with stem cells? And if stem cells play a role in cancer, is it relevant to know what that role is? Will it lead to novel treatments or even cures for cancer? The short answer is: yes, (cancer) stem cells do probably play a key role in cancer growth as well as in the formation of distant metastases. Improving our understanding of what these stem cells look like and how they function is likely to be of great benefit in the future, although the road to the introduction of novel treatments based on this knowledge is expected to be long and arduous, full of roadblocks to be cleared. In this chapter we introduce the topic of cancer stem cells and discuss the clinical challenge of investigating their role in cancer.

12.2 INTRODUCTION TO STEM CELLS AND CANCER

Although the idea that cancer stem cells exist and are essential contributors to aggressive cancer growth dates far back into the last century, the question of the identity of these cells has proven extremely difficult to answer. Only recently has the concept gained wider adoption among the cancer research community. Confusion still arises about what they are because of unclear definitions of the actual terminology: the *cancer stem cell of origin*, for example, is not the *cancer stem cell*. The first is considered to be an adult stem cell in which a DNA mutation occurred and this, in turn, gave rise to growth of a tumor. The term *cancer stem cell*, however, is taken to mean cancer cells present in cancer tissue that have changed their appearance such that they resemble and behave like stem cells. Cancer stem cells are now thought to be the major cell type responsible for spreading cancer throughout the body. In recent years, the promise of more effective cancer therapies has been a powerful driver of cancer stem cell research, and several theories, supported by sound and sophisticated experimental evidence, now merge into a more consistent picture describing the mechanism by which they may promote metastasis.

12.2.1 A Brief History of Cancer Stem Cells

First indications that cancer stem cells may exist emerged at the end of the twentieth century at about the same time embryonic stem cells were discovered in mice. The idea evolved from research carried out to explain the cause of a relatively common form of blood cell cancer: acute myeloid leukemia (AML). The disease AML is characterized by large numbers of immature (not fully differentiated) white blood cells, *leukemic blast cells*, in the blood, caused by uncontrolled division of malignant precursor cells in the bone marrow. Leukemic cells all carry one specific DNA mutation in their genome, the leukemia-initiating mutation, but at the same time, they may carry other mutations added during progression of the disease. Investigating the properties of these leukemic cells surprisingly revealed that the majority of cells had lost the capacity to divide. Only a small fraction of the cell population was capable of dividing and initiating a new leukemia when transferred to a mouse. These *leukemia-initiating cells* closely resembled normal multipotent *blood stem cells*. Normal blood stem cells are rare cells in the bone marrow that can both self-renew and generate progenitor cells for the various types of blood cells, thus supplying them to the blood. The leukemia-initiating cells appeared to be resistant to chemotherapy, for reasons we still do not understand. This has since been blamed for the nearly unavoidable recurrence of AML after treatment that at first sight seems successful in eradicating the disease. Following chemotherapy the stem cell-like nature of these cancer stem cells enables them to rapidly replenish the leukemic cell population, leading to recurrence of the disease. In this perspective the cells are regarded as *leukemic stem cells*. This process is rather similar to what happens during repair of damaged tissue, for example, when stem cells in the skin are rapidly recruited to replenish lost skin cells in a wound and thus rebuild the damaged skin.

From these studies, it became clear that there is a certain hierarchy between all cells making up this form of leukemia. This hierarchy can be envisioned as a pyramid with relatively rare leukemia-initiating stem cells at the top, generating a larger population of rapidly dividing progenitor cells for all different blood cell types in the middle of the pyramid (Figure 12.1). The base of the pyramid is filled with the more differentiated and even more heterogeneic, nondividing cells that are the bulk of leukemic cells in the blood. The leukemic cell population in the blood contains immature forms of all blood cell types that could be placed in a pyramid-shaped hierarchical stem cell model. The way the leukemic cell population is built up actually resembles normal blood, only completely unregulated and uncontrolled, while the cells remain relatively immature. Having experimentally defined this small

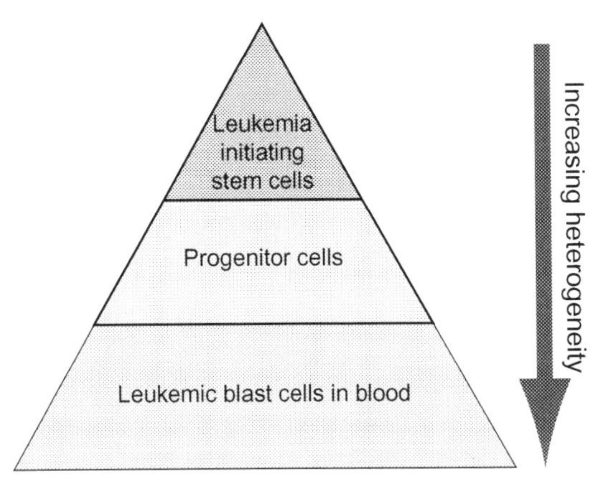

FIGURE 12.1 Cellular hierarchy in acute myeloid leukemia. Leukemia-initiating stem cells are relatively rare, generating a larger population of rapidly dividing progenitor cells, which further differentiate toward the nondividing bulk of leukemic blast cells in the blood.

population of leukemia-initiating cells, or leukemic stem cells, this does not yet address the question of the origin of these cells: where do they actually come from? Two options have been proposed, both of which have some truth. Either these cells belong to the blood stem cells in the bone marrow, the only difference between them and normal cells being the DNA mutation that caused the leukemia; or they are leukemic blood cells that have somehow reverted to a stem cell state, for example, by acquiring additional mutations. In the first case, these abnormal blood stem cells would represent the *stem cells of origin* of the leukemic disease. In the latter case, they are called *cancer stem cells*. The leukemia stem cell population could actually consist of a mixture of both (Figure 12.2).

Thus, the original concepts on the role of stem cells in cancer did not come from research on "solid" tumors, such as breast and colon cancer, but from research to discover the origin of leukemia. Based on the compelling evidence obtained from patients with AML, the search began for cancer stem cells in solid cancers, such as those in breast, colon, and prostate.

Early research on cancer stem cells in solid tumors was, however, slowed down by the lack of appropriate research tools. To prove that a cancer cell is a stem cell, it is necessary to show that a single cell can form a whole tumor. As cancer stem cells are so rare, it was almost an impossible task to sort through millions of cells from a solid tumor to see if they actually had a tumor-forming capacity. By the end of the twentieth century, however, fluorescent activated cell sorting (FACS) technology (see Box 9.4 "Immunofluorescence and Flow Cytometry") had sufficiently matured to enable isolation of individual cells from solid tumors for detailed study. Moreover, by that time, strains of mice had been developed in which the immune system was so defective that individual human tumor cells could be transplanted (a procedure known as xenotransplantation) without being rejected by the mouse immune system. If tumor growth occurred after transplantation of a single cell, this demonstrated that the xenotransplanted cell had stem cell properties such as self-renewal, and could form more differentiated, heterogeneic cells after dividing. In general, only a small number of cancer cells succeeded in forming a new malignant tumor with the same heterogeneity as the original tumor. These minority cells were tentatively called tumor- or cancer-initiating cells, analogous to the leukemia-initiating cells mentioned earlier. However, while in leukemia only a few cells had leukemia-initiating capabilities, in the solid cancers more cells appeared to have cancer-initiating capabilities, although they remained a minority. These cancer stem cells changed the way we think about tumors: they contain the DNA of the cancer cells, including the large number of DNA changes typical of cancer cells, but behave like

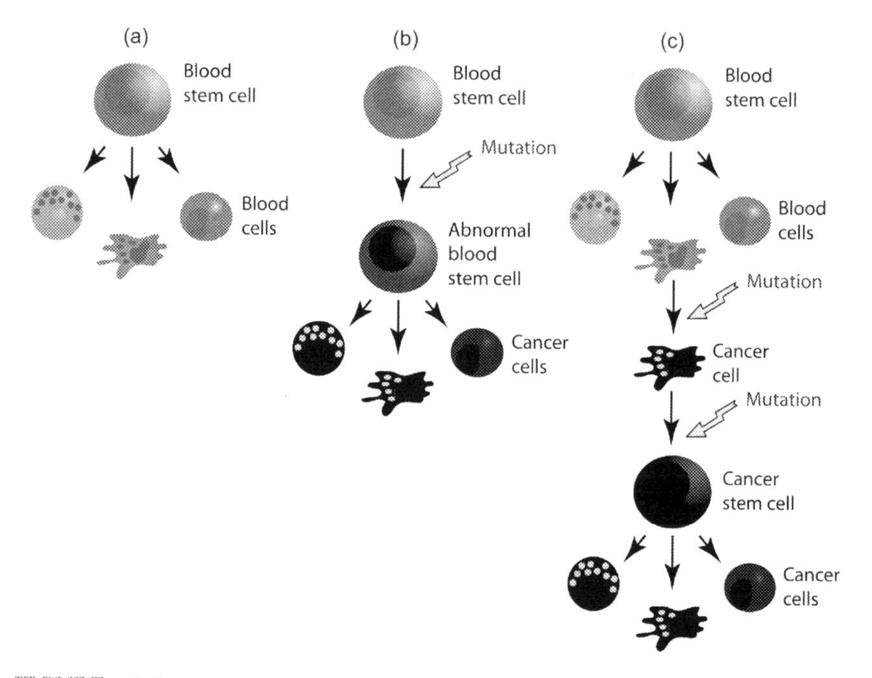

FIGURE 12.2 Healthy blood stem cells such as those present in bone marrow are responsible for a constant supply of healthy blood cells (a). Two scenarios have been proposed for the formation of leukemic cells which are not mutually exclusive. In (b) a mutation in the blood stem cell occurs so that it changes into an abnormal blood stem cell that can self-renew and give rise to cancer blood cells. In (c) a mutation occurs in already formed blood cells that changes them into cancer cells. One or more subsequent mutations may reprogram cancer cells into cancer stem cells, which have regained stem cell characteristics; they can self-renew and give rise to new cancer blood cells.

stem cells. Recently, much more has been discovered about their origin and the mechanism by which the cancer cell changes into this mysterious stem cell, which is not a real stem cell but a cancer cell.

12.3 THE BEHAVIOR OF CANCER CELLS: NOT ALL TUMORS AND NOT ALL CELLS WITHIN A TUMOR LOOK THE SAME

In a patient with cancer, for example of the breast, colon, prostate, or lung, tumor tissue usually appears to be heterogeneous when viewed under the microscope, which means that not all the cells that are part of the tumor look the same. And if they do not look the same, they might not be the same, and may also behave differently within the tumor (Figure 12.3). For example, the tumor may contain a dense network of blood vessels induced by the cancer cells to provide nutrition to the tumor as it grows in size; fibroblasts and immune cells can also be present in the tumor and the cancer cells themselves differ from each other with respect to morphology (the way they

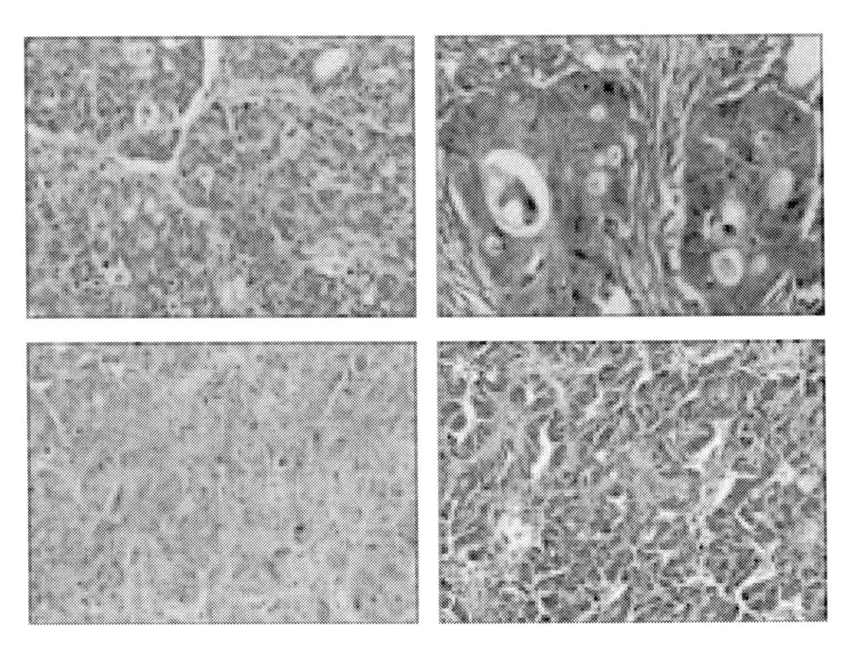

FIGURE 12.3 Four examples of colon cancer; the tissue slides are stained with hematoxylin. The heterogeneity (mixture of cell types) is clear even to the untrained eye. *Source: Reproduced with permission from Vermeulen L. et al., Wnt activity defines colon cancer stem cells and is regulated by the microenvironment, 2010, Nature Cell Biology 12, 468–476.*

look) and function (what they do). As a consequence, no two tumors look exactly the same under the microscope. And even a well-defined type of cancer (say, of the colon) can still behave in very different ways in different patients; for example, it can metastasize rapidly in one patient, but only slowly or not at all in another (we then say it is less "aggressive"), or respond very well to therapy in the one patient, but not in another. Understanding the cause of this heterogeneity is important for the development of new drugs for treating cancer and to prevent metastasis.

12.3.1 A Darwinian View: Evolution of a Tumor

One mechanism to explain tumor heterogeneity and differences in tumor behavior is the accumulation of many kinds of DNA changes (mutations) in the genome of cancer cells over time. When a cell divides to form two new cells, its DNA needs to be replicated (or doubled) to enable distribution over two daughter cells. This complex process carries a risk of incorporating DNA errors (mutations) into the genome of one of the two daughter cells. Research on the mechanism by which colon polyps arise clearly demonstrated that an accidental DNA mutation occurring in a multipotent adult stem cell could cause later tumor formation. If such a DNA change enables the stem cell to divide more frequently, then it may give rise to both a larger stem cell population carrying the same DNA mutation and a rapidly dividing daughter cell population: a small tumor forms that cannot yet metastasize and is still benign. In this tumor, the population of mutated stem cells among normal stem cells may function as a reservoir of *stem cells of origin of cancer*.

During tumor growth, additional DNA mutations may accumulate in the rapidly dividing tumor cells. Only if a change in the DNA of a tumor cell, or a combination of errors, enables it to divide even faster, do the progeny (daughter cells) of that cell have a chance of developing into the dominant cell population ("clone") in the tumor. If that happens, the same process of growth competition between the tumor cells continues within this population. In time, any one of the progeny cells that succeeds in dividing even faster will dominate the growth of the tumor. This process has been named *genetic tumor evolution*, in a sense quite analogous to Darwin's evolution theory. The concept of genetic tumor evolution was introduced by Professor Bert Vogelstein from Johns Hopkins University in the United States to explain the development of a polyp (called adenoma) in the intestines and its potential transition to colon cancer. At least in colon cancer, several decades may be needed for a tumor to acquire a sufficient number of mutations to become malignant and metastasize.

12.4 COLON ADENOMA: A CASE IN POINT FOR THE ROLE OF AN ADULT STEM CELL AS THE STEM CELL OF ORIGIN

Adenomas develop in humans in the colon region of the intestines and may sometimes progress to a malignant colon carcinoma. Colon adenoma represents a unique example illustrating the role of stem cells in development of a tumor. The lumen of the intestines is lined by an epithelial mucosal layer, which contains crypt and villus structures (Figure 12.4). Deep crypts create niches within the large intestinal lumen required for many kinds of functions, such as the absorption, transport, and secretion of molecules. The turnover rate of cells facing the lumen at the top of the villi is very high; these cells are continuously replenished, implicating rapid cell division for maintaining an intact cell layer.

12.4.1 Development of a Colon Adenoma

The colon part of the intestinal tract is the site at which a common carcinoma, colon cancer, develops. Colon cancer is one of the best-studied malignant tumors. Its development is preceded by a benign tumor, called a polyp or adenoma, which originates from cells in a crypt. Adenomas are frequently present in the colon of elderly people and may rarely, over a period of decades, develop into colon cancer. In the large majority of the adenomas and colon cancers, a characteristic change in the DNA of the cell is present right from the beginning. This DNA mutation results in the loss of an important protein, called APC. APC normally helps control cell growth and differentiation because it keeps the activity of an *intracellular signaling pathway*, called the Wnt pathway, in check (Figure 12.5). This is an *evolutionarily* well-conserved signaling pathway, which means that it is, in principle, the same in most species, from the fruit fly *Drosophila* to man. Pathways that are unchanged in so many species are usually those that are very important for certain cell functions. Among these are important roles in the development of embryos.

In a somewhat simplified form, this signaling pathway consists of a receptor in the cell membrane, which receives highly specific protein signals from the outside of the cell and transfers this signal into the cytoplasm, where it activates a transcription factor, called beta-catenin. Beta-catenin then travels to the nucleus of the cell where it binds to specific regions in the DNA. As a result, several genes start to produce messenger RNA from the regions of DNA that are activated, leading to the production of proteins. These proteins control different functions in the cell, including cell division. In real life, it is much more complicated than this and a number of proteins are involved in keeping this

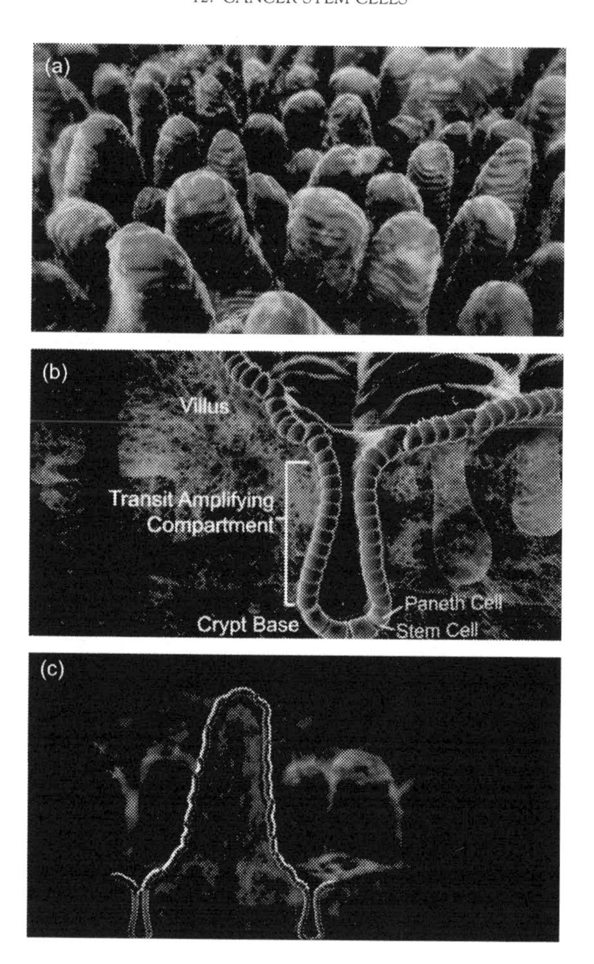

FIGURE 12.4 (a) The intestinal lumen is lined by an epithelial mucosal layer that contains deep crypt and elevated villus structures. (b) A crypt consists of an epithelial cell layer, with a few stem cells at the base (bottom) of the crypt. These intestinal stem cells divide to give rise to migrating cells in the "transit amplifying compartment," which differentiate toward all the cell types of the intestinal epithelium, for example the Paneth cells in the bottom of the crypt. (c) The epithelial layer continues from crypt to the top of the villus. *Source: Hans Clevers, Hubrecht Institute, Netherlands.*

pathway, and cell division, in check. APC is one of the proteins that prevent the beta-catenin protein from traveling to the DNA, as long as no external signal activates the receptor. When the APC gene in the cell does not function due to changes (mutations) in the DNA, the cell lacks the APC protein. As a result, beta-catenin becomes active, in much the same way that the Wnt pathway is activated by a normal external signal. The consequence is that the cell starts producing proteins that cause it to divide continuously.

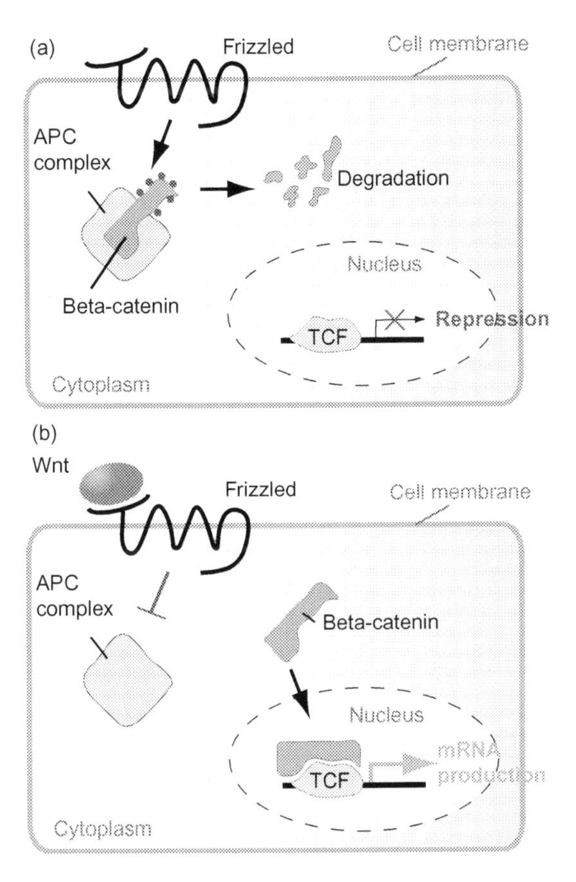

FIGURE 12.5 This intracellular signaling pathway plays a role in growth and differentiation in early embryonic development and in human embryonic stem cells, and is also active in specific adult stem cell populations and in cancer growth. During evolution, this pathway has been very well conserved, and even in the fruit fly *Drosophila*, it is quite similar. In *Drosophila*, the main component of this signaling pathway was first discovered because of some mutations in the gene that caused the flies to have a very odd appearance: they had no wings, hence the name "wingless." Only later were the human "homologs" (similar genes) discovered and named Wnt. The Wnt pathway is essential for formation of the intestinal wall, and mice that miss the transcription factor TCF4 belonging to this pathway do not form intestinal crypt stem cells and die before they are born. How does the Wnt pathway work? In the absence of the ligand (protein) that binds to and activates the Wnt receptor, beta-catenin is coupled to a protein complex containing APC (a). Within this complex, the beta-catenin protein becomes phosphorylated (red dots: phosphorylation sites on the protein), enabling its degradation. However, when Wnt ligand outside the cell binds to its receptor (called "frizzled") in the cell membrane, beta-catenin is no longer phosphorylated and degraded and accumulates in the cytoplasm, from which it can move to the nucleus (b). There it recognizes a transcription factor called TCF4 that is bound to specific sequences in the DNA. After binding, this transcription factor is activated and mRNA is made from the associated gene.

Professor Hans Clevers and his research group at the Hubrecht Institute in the Netherlands discovered that a few adult stem cells reside deep within each intestinal crypt. Because of an active Wnt pathway, these stem cells divide rapidly and give rise to a variety of cell types with different functions that together make up the crypts and villi of the intestinal epithelial layer. At each stem cell division, one new stem cell is created, while the other daughter cell starts to differentiate, a process that is associated with rapid inactivation of the Wnt pathway. When the APC protein is lost in one of these intestinal stem cells, the Wnt pathway cannot shut down and both cells created during cell division retain stem cell characteristics. Subsequent cell division leads to an accumulation of undifferentiated cells along the crypt: an adenoma has formed. Here the stem cell of origin for the adenoma is clearly defined: it is an intestinal stem cell with DNA changes resulting in a lack of APC protein. This tumor-forming process has been elegantly demonstrated in a mouse model and is thought to occur in a similar manner in human patients. However, in mice the adenomas do not progress toward cancer, in contrast to those in human patients, probably because mice simply do not live long enough. More research is needed to find the human equivalent of these "adenoma stem cells," to investigate whether they also play a role in the transition from adenoma to colon cancer and to find out how they might relate to the putative cancer stem cell in colon cancer.

12.4.2 When Does a Tumor Become Malignant and Metastasize to Different Organs?

Mutations (or the introduction of errors) are not the only things that can go wrong with DNA. A section of the DNA, or even whole chromosomes, can be lost completely or chromosomes can get mixed up and become abnormal. If this happens, the mistake is sometimes corrected by a special DNA repair mechanism or, if that is not possible, the cell may choose instead to die by "suicide," a process called *apoptosis*. However, when such mistakes occur in a tumor cell that had already acquired one or more DNA mutations that interfere with DNA repair and apoptosis, a real problem arises. In this case, extremely abnormal tumor cells may survive and continue to divide, and this is invariably associated with vary rapid accumulation of more severe DNA defects randomly in cells at different sites in the tumor. It will not be hard to understand that these mutations and changes in the genome can lead to rapid alterations in the behavior of tumor cells. This situation is referred to as *chromosomal instability* and marks the switch to a *genetically* highly heterogeneous tumor.

Chromosomal instability rapidly provides cells with the characteristics needed for invasion into normal tissue and migration via the blood stream to establish a metastatic tumor growth elsewhere: the benign tumor becomes a malignant cancer. In advanced cancer, a large number of genetically different cells, or "clones" of cancer cells may thus be present, while a dominant cancer cell with the most aggressive and invasive properties may generate most of the cancer cells in the tumor. The evolutionary mechanism underlying this genetic tumor heterogeneity and the associated differences in tumor behavior is well accepted by cancer researchers and, even if a difficult concept to understand, it will be clear that an important part of cancer heterogeneity has its origin at the DNA level. However, this mechanism cannot explain all of the observed cancer cell heterogeneity.

12.4.3 Interaction between Cancer Cells and Their Environment Leads to Phenotypic Heterogeneity

Aside from genetic (DNA) abnormalities that may directly influence the morphology (appearance) and behavior of cancer cells, all kinds of signals originating from the surroundings of a cancer cell can also modify its morphology and behavior, without changing the DNA. These contextual signals may vary depending on the location of a cancer cell in the tumor. We often refer to this local environment of the cancer cell as the *microenvironment* or *niche*. One microenvironmental factor that can affect the way cancer cells behave is the supply of oxygen, which is vital for the survival of all cells and may vary in the tumor depending on availability of local blood vessels. In addition, especially at the invasive border between cancer tissue and surrounding normal tissue, cancer cells interact with other, nontumor, cell types, such as fibroblasts, and inflammatory and immune cells, which are attracted to the tumor. At the same time, a variety of extracellular proteins, deposited in the cancer tissue by cells such as fibroblasts, also influence, for example, the ability of the cells to migrate (travel) through the tissue. DNA mutations and other DNA abnormalities in the cancer cell may together determine the outcome of these interactions with other cells and proteins in their microenvironment, leading to profound changes in the appearance and properties of cancer cells. It will be clear that these induced changes in cell characteristics can also contribute to cancer heterogeneity and may lead to morphological and functional differences between cancer cells containing the same DNA changes: they are phenotypically (the way they appear and function) different but are genotypically (at the DNA level) the same.

It is good to keep in mind that, in contrast to heterogeneity caused by differences in DNA mutations, these phenotypic changes are, in principle, reversible as they may depend on the continuing presence of specific signals in the microenvironment. Where does this cancer heterogeneity story bring us? One of the most important phenotypic changes occurring in cancer tissue is the switch from a cancer cell to a cancer stem cell.

12.1

THE DISCOVERY OF ADULT INTESTINAL STEM CELLS AND THE CELL OF ORIGIN OF COLON ADENOMA

In the laboratory of Hans Clevers (Hubrecht Institute in Utrecht, Netherlands) multipotent adult stem cells were identified and characterized for the first time at the bottom of the crypts of the intestinal wall. For this purpose a highly sophisticated molecular "toolbox" was developed, consisting of a series of complex artificial DNA sequences ("constructs"). These DNA constructs were introduced into the genome of mice to enable breeding of a series of genetically modified *transgenic* mouse strains (Figure B12.1.1).

One strain of mice was bred with a new gene, coding for green fluorescent protein (GFP), inserted in the genome of all of its cells. This GFP gene had replaced another gene in the genome that coded for a special protein, Lgr5. The Lgr5 protein had earlier been found to be a so-called "target gene" of the Wnt signaling pathway, meaning that activation of the Wnt pathway would lead to Lgr5 protein production by the cell. Lgr5 appeared to be produced by only a few cells in each of the crypts of the mouse intestine; in these cells the Wnt pathway was apparently "turned on."

Researchers wanted to prove that these cells at the bottom of the crypts were adult stem cells that could give rise to all cell types for the intestinal epithelial layer.

In the newly created transgenic mouse, production of GFP protein was regulated in exactly the same manner as the original Lgr5 protein: by the Wnt pathway. Thus, in this mouse, all of the cells that would normally produce Lgr5 protein now produced green fluorescent protein. This enabled Lgr5-expressing cells to be detected in the intestinal tissue of the mouse simply by looking for GFP using a fluorescent microscope. Indeed, in each intestinal crypt, around six well-defined cells showed green fluorescence (Figure B12.1.2).

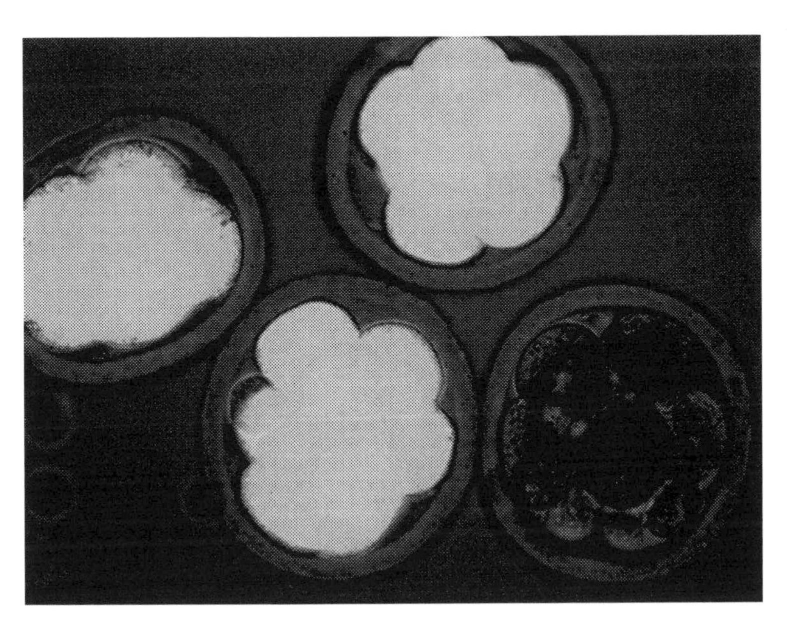

FIGURE B12.1.1 Four mouse embryos at the morula stage of development, about three days after fertilization of the egg. Three of the embryos express the gene for green fluorescent protein. These embryos are recognizable by their green color. *Source: Susana Chuva de Sousa Lopes.*

FIGURE B12.1.2 Green fluorescent protein (GFP) in intestinal stem cells in the crypts of mouse intestine. A few rare cells at the bottom of the crypts are green due to the expression of GFP; these represent the intestinal stem cells. *Source: Reprinted with permission from Barker N., et al., Identification of stem cells in small intestine and colon by marker gene Lgr5, 2007, Nature 449, 1003–1007.*

However, just showing GFP (instead of the Lgr5 protein) in crypt cells was not evidence that these were indeed stem cells. What was needed was additional proof that these cells acted like real stem cells

12.1 (*cont'd*)

and, therefore, could self-renew, as well as produce all the different cell types making up the intestinal epithelium from crypt to villus.

For this purpose another, genetically even more complex, mouse was created, this time with another gene sequence coding for an enzyme (recombinase) in its genome next to the GFP gene. The recombinase enzyme, like GFP, is only produced in the few crypt cells that make the Lgr5 protein. It is a very special enzyme in that it can recognize and subsequently remove another DNA sequence out of the genome of the cell, on condition that this DNA sequence is marked ("flagged") on both sides by a special DNA sequence, called loxP. However, to do this, the recombinase enzyme first needs to be activated in the cell and this only happens when the mouse is given a hormone, called tamoxifen. To actually see an effect of activated recombinase, another (third) DNA construct was introduced in the genome of this mouse that codes for the protein beta-galactosidase (lacZ). This enables cells in which the protein is present to be stained blue, and this can be done in the lab on tissue slices of the mouse intestines. However, in the absence of tamoxifen, production of this beta-galactosidase protein is prevented by the presence of a blocking DNA sequence ("STOP"), again with a loxP sequence on either side. The enzyme is activated if tamoxifen is now given to the mouse and taken up by the crypt cells in the intestine that have made recombinase. Activated recombinase removes the blocking DNA sequence STOP from the lacZ gene and the cell starts producing beta-galactosidase, which allows it to be stained blue. Beta-galactosidase is from then on continuously produced by the cell and by all of its later descendants. If the mice are examined at different time points and the intestinal mucosa stained so that the beta-galactosidase protein can be seen, then quite amazingly "ribbons" of blue-stained cells originating from a few green fluorescent crypt cells are seen, working their way upward along the crypt to the top of the villus. The blue-stained cells represented the descendants of the crypt cells. The latter cells, in addition to Lgr5, now produce both the green fluorescent protein and the beta-galactosidase protein. The fluorescent protein remained confined to the few crypt cells, while their descendant cells clearly migrated to the top of the villus. In those cells, the Wnt pathway is rapidly turned off. It takes a few days for a cell to reach the top of the villus, at which point it would normally commit suicide (apoptosis) to be released into the intestinal lumen.

The blue cells were shown to become all of the different cell types normally present in the intestinal mucosa, among which are the slime-producing goblet cells and the Paneth cells, which play a role in the resistance to specific pathogens. The persistent production of GFP in the few crypt cells, together with the blue ribbon of differentiating progeny cells, provided definitive proof that the green fluorescent crypt cell is indeed a true multipotent intestinal stem cell, which can both self-renew and give rise to the various cell types required to maintain an intact intestinal mucosal layer (Figure B12.1.3).

From an Intestinal Stem Cell to Adenoma

Intestinal stem cells normally divide asymmetrically, say, once per day: one daughter cell again becoming a stem cell while the other starts to differentiate. The Wnt pathway is active in the stem cell but is rapidly turned off in the differentiating daughter cell. When the APC protein is missing in one of the stem cells in an intestinal crypt, the Wnt pathway cannot be turned off any more in that cell and

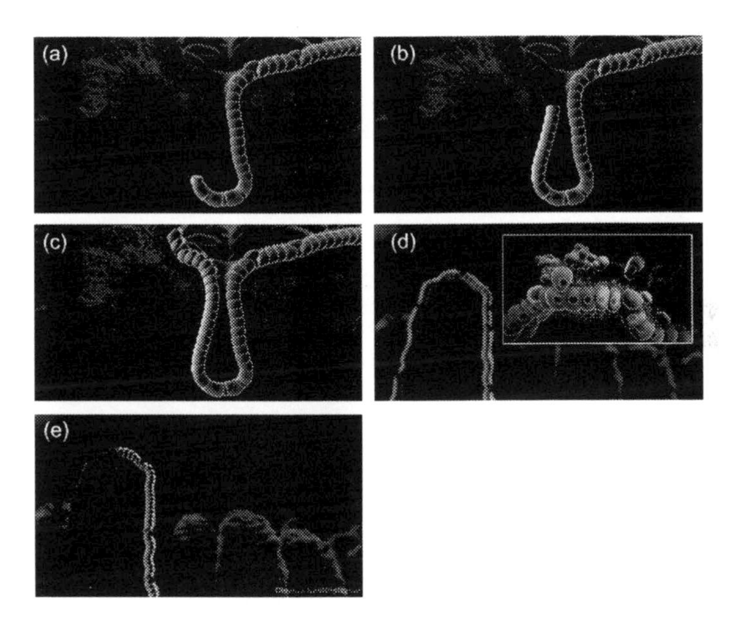

FIGURE B12.1.3 Computer simulation showing how intestinal stem cells form the epithelial cell layer of the crypt and villus. Several intestinal stem cells (blue) divide (a) and give rise to (blue) daughter cells ("transit amplifying compartment," a small group of cells that rapidly divide) (b). The daughter cells migrate upward to the top of the villus (c,d). The turnover of cells is rapid and at the top of the villus, the cells undergo apoptosis (commit suicide) and are shed from the villus into the lumen (d). *Source: Hans Clevers, Hubrecht Institute, Netherlands.*

12.1 *(cont'd)*

normal differentiation does not take place. This causes the epithelial cell layer from crypt to villus to grow and bulge out, forming an adenoma.

The challenge to answer the question of whether an intestinal stem cell could give rise to an adenoma—and potentially cancer—was investigated by the Clevers laboratory using a complex transgenic mouse strain. In this mouse, the production of the protein APC could be turned off in a living mouse at any chosen time in the stem cells in the crypts of the intestines. To do this, a mouse strain was first created in which both normal gene copies for the APC protein were replaced by DNA constructs containing the same APC gene but with a loxP DNA sequence on either side. This mouse strain was cross-bred with the strain previously described. What resulted were mice in which, after administration of the drug tamoxifen, two things occurred in the intestinal stem cells: they started to make the beta-galactosidase protein and the APC gene was removed from their genome. The latter left the intestinal stem cell with none of this important protein from the Wnt pathway. What happened? (Figure B12.1.4.)

Within two weeks, a number of adenomas developed in the crypts in the intestines of the mice. All of the adenoma cells produced the

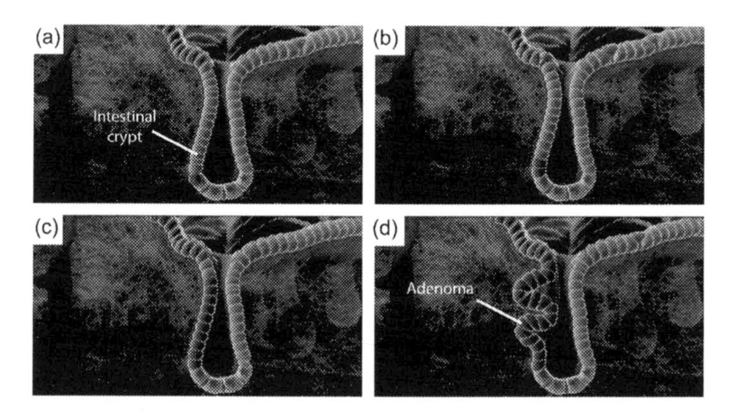

FIGURE B12.1.4 A computer simulation of adenoma tumor (a "polyp") formation in the colon. When the APC protein is missing in one of the stem cells in an intestinal crypt, the Wnt pathway cannot be turned off in that cell and the normal differentiation process does not take place. This causes the epithelial cell layer from crypt to villus to grow and bulge out, forming an adenoma (brown cells). Over time, an adenoma can turn into a colon carcinoma. *Source: Hans Clevers, Hubrecht Institute, Netherlands.*

beta-galactosidase protein and had an abnormally active Wnt pathway. This demonstrated that the adenoma had originated from an intestinal stem cell in which the production of APC protein had been eliminated. As a consequence, the cells kept on dividing, leading to the accumulation of a lump of cells along the intestinal crypt–villus structure. These exciting results proved that continuous activation of the Wnt pathway in an intestinal stem cell leads to an adenoma, and that an intestinal stem cell can act as *cell of origin* for an adenoma, at least in mice. Interestingly, the adenomas also contained sparse cells that expressed the Lgr5 and the green fluorescent protein, which were normally only found in intestinal stem cells in the bottom of the crypt. These stem cell-like cells were obviously in the wrong place. Thus the adenoma itself may contain what one might call "adenoma stem cells." Whether these stem cells are intrinsically the same as real intestinal stem cells, or whether they have somehow changed back from a more differentiated to a stem cell type, remains to be established.

What can we learn from this, when looking at colon adenomas and cancer in humans? Adenoma growth as observed in mice seems to resemble tumor development and growth as seen in humans quite closely. In human patients, lack of APC protein in intestinal cells is similarly associated with an abnormally active Wnt pathway, a key feature in the development of colon adenoma and cancer. However, the presence of actual stem cells in biopsies from intestinal crypts and from adenomas in humans has not yet been demonstrated. Mice do not develop colon cancer, perhaps because they do not live long enough to acquire the number of DNA mutations necessary for progression to the cancer stage. Therefore, even more sophisticated disease models are needed to study progression from colon adenoma to cancer and to the formation of metastatic tumors. Animal models to do this are being developed in the Clevers laboratory by artificially creating a combination of DNA mutations in a mouse, similar to those found by analyzing human colon cancer.

An exciting alternative, which may make some of the animal experiments redundant, could be a colon cancer model in a cell culture dish. The first pioneering steps in this direction have already been taken by isolating and culturing crypt stem cells from a mouse. Quite surprisingly, individual stem cells in culture fully recapitulate development of the crypt structure in which all variations of differentiated cells normally present in the intestinal epithelium can be recognized. Such innovative models will be a big help in further unraveling the role of the stem cell of origin and cancer stem cell, as

12.1 (*cont'd*)

well as the relationship between them, in colon cancer, and hopefully enable discovery of novel drugs to treat this deadly disease (Figure B12.1.5).

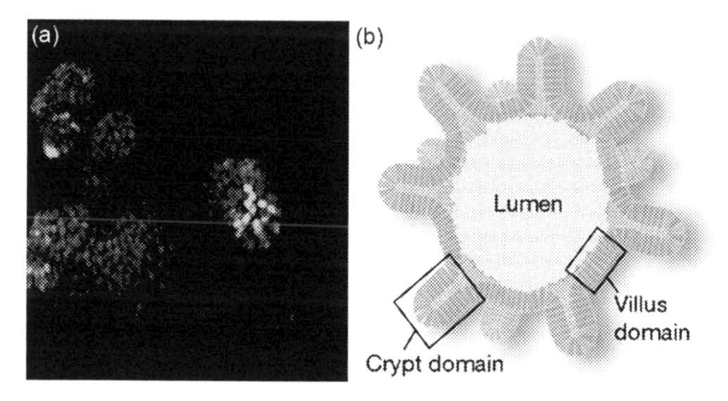

FIGURE B12.1.5 Intestinal stem cells isolated from mouse crypt structures can be made to grow in 3-dimensional (3D) culture where they differentiate into tubular, crypt-like structures, with all of the different cell types belonging to the crypt. (a) 3D structure of the crypts and villi in such a culture, with green fluorescent intestinal stem cells visible. (b) Schematic representation of cultured cells. *Source: Reprinted with permission from Sato T., et al., Single Lgr5 stem cells build crypt-villus structures in vitro without a mesenchymal niche, 2009, Nature 459, 262–265.*

12.5 HOW TO BECOME A CANCER STEM CELL: EPITHELIAL MESENCHYMAL TRANSITION

Professor Robert Weinberg, a founding member of the famous Whitehead Institute for Biomedical Research in Cambridge, United States is undoubtedly one of the most prominent opinion leaders in the cancer stem cell research area. Solid cancers, such as breast, colon, and prostate cancer, are called epithelial cell cancers, meaning that they arise from epithelial cells. Epithelial cells are cells in a tissue that look like closely aligned cobblestones, and they surround and line most solid organs, such as the stomach and intestines, the skin, but also the ducts in the breast. Epithelial cells are normally highly specialized and transport specific nutrients from one side of the intact epithelial layer to the other side. Cancer stem cells are thought to represent cancer cells that reverted from epithelial to a more stem cell-like *mesenchymal* cell type. Mesenchymal cells look quite different from epithelial cells: they have an elongated and flattened shape and are designed to be able to migrate (move) through connective tissue of the body. Mesenchymal-like cancer cells do exhibit some stem cell characteristics, such as self-renewal and generation of various differentiated daughter cells. In this way, they can contribute to tumor heterogeneity: some cells are more epithelial, others are more mesenchymal. However, they are not thought to be the stem cell of origin of the cancer, meaning the cell from which the cancer originally arose. Why such a transition to a cancer stem cell occurs within a malignant tumor is not yet fully understood. We will now take a look into what is known about the possible mechanism.

The epithelial–mesenchymal transition (also called EMT) that results in the formation of cancer stem cells shows some striking and fascinating similarities to processes that take place during embryonic development when tissues and organ structures are formed. During organ development, repeated switching between epithelial and mesenchymal cell types is actually common. EMTs enable cell migration (traveling) of cells to the appropriate location in the developing embryo, while the reverse process of mesenchymal–epithelial transition (MET) is necessary for differentiation to specialized epithelial cell types that line or surround organs as they form in the embryo. During EMT, the proteins necessary for epithelial cell functions disappear from the cells and are replaced by other proteins, which enable typical stem cell functions such as migration. This change in protein production is mediated by activation of specific *signal transduction pathways* in the cell, for example, the Wnt signaling pathway. Such signaling pathways instruct the cell to change its behavior and function, and are extremely important in the development of the embryo. For this reason, they are also called developmental signal transduction pathways.

The exact molecular mechanism behind the EMT that takes place in cancer tissue is not yet fully understood, but may provide an essential clue to understanding why cancer cells become migratory and can invade many different tissues and organs. Present evidence suggests intriguing similarities with EMT during embryo development, and use of the same signals and signal transduction mechanisms, such as the Wnt pathway. Activation of these developmental pathways in cancer stem cells is likely to be very important in the metastatic behavior of cancer, as we will discuss later in this chapter.

12.6 HOW DEVELOPMENTAL SIGNAL TRANSDUCTION PATHWAYS BECOME ACTIVE IN CANCER CELLS

As we have seen, both DNA changes as well as contextual cues from the niche of the cancer cell may induce changes in behavior of cancer cells. Changes in cell behavior are effected by signal transduction pathways, and some DNA mutations can directly alter activation of these pathways in cells. A very good example of this is loss of the APC protein in colon cancer: DNA mutations inactivate both gene copies of the APC gene so that cells lack the APC protein, which leads to constant activation of the Wnt signal transduction pathway and alterations in the stem cell behavior in the colon. On the other hand, the cancer cell microenvironment itself may also be the factor responsible for inducing abnormal signal transduction pathway activation. Both macrophages and other immune cells, as well as special fibroblasts in the tumor tissue, produce a variety of signal-proteins, which can reach the cancer cell and activate signal transduction pathways. As a consequence, the cancer cell may change into a mesenchymal cancer stem cell, which can break loose from other cancer cells and migrate away, or can self-renew like a real stem cell when the conditions are favorable. The most likely location in the tumor for initiation of EMT is thought to be the invasive front at the border between tumor and normal tissue, where signals that induce EMT are mostly found (Figure 12.6).

What happens if a cancer cell changes into a cancer stem cell and starts migrating through tissue?

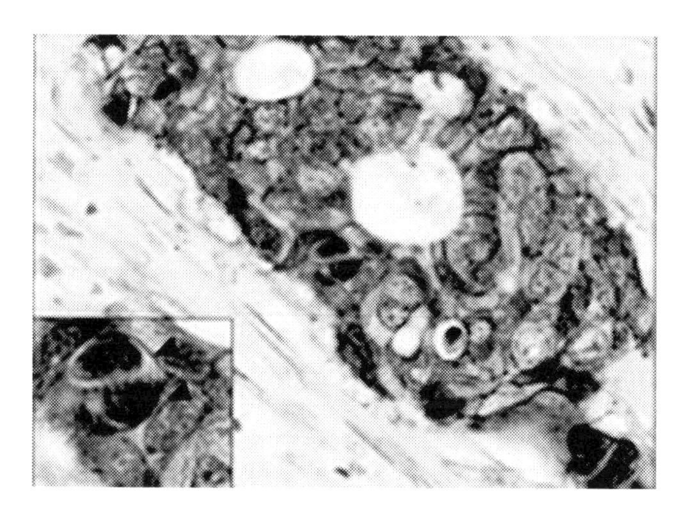

FIGURE 12.6 Histological slide of colon cancer tissue showing the heterogeneous pathology of colon cancer. Using this sort of slide, a pathologist can make the diagnosis of colon cancer. The brown staining indicates activity of the Wnt pathway in tumor cells. *Source: Reproduced with permission from Vermeulen L. et al., Wnt activity defines colon cancer stem cells and is regulated by the microenvironment, 2010, Nature Cell Biology 12, 468–476.*

12.7 CANCER STEM CELLS AS CIRCULATING TUMOR CELLS

To be able to create another tumor by metastasizing elsewhere in the body, cancer cells have to migrate through the endothelial lining of the blood vessels into the blood stream. In principle, the blood will then carry them to all organs, but to start a new tumor, one or more cells have to "seed" in a location where the circumstances are such that they can survive, and at a certain moment start dividing again. Cancer cells that are found in the blood circulation are often called circulating tumor cells (CTCs). They are very rare compared with all the blood cells that are present. While some of the CTCs look like epithelial cancer cells, it is now thought that many of the circulating tumor cells have the characteristics of cancer stem cells. As they become cancer stem cells, connections with other cells are lost and the cells also become more elastic, probably allowing them to squeeze one by one between the cells that make up the blood vessel wall and enter the blood stream. Their stem cell properties may provide protection against attack by the large number of immune cells in the blood. Indeed, the circulating tumor cells can survive passage even through the very narrow blood vessels (capillaries) in organs. At their smallest, the blood vessels are actually narrower than the tumor cell itself and the cell has to either squeeze itself through the small lumen or otherwise move out of the capillaries into the nearby tissue of an organ. Once in the tissue, in an environment where it can survive, the cell can remain in a "dormant" state until it is triggered to start dividing again and form a new tumor.

12.8 THE FINAL STEP: INITIATION OF METASTATIC GROWTH

When does metastatic growth start? Once we know this, we may learn how to block the process before it begins. Initiation of metastatic tumor growth depends on appropriate growth signals from the microenvironment in which the circulating tumor cell ends up—actually quite comparable to a normal stem cell niche in an organ. Normal stem cells rarely divide but are induced to do so when regeneration or "wound-healing" is required to replace cells lost during disease or following trauma. Many of these regeneration-activating signals probably come from inflammatory cells in the damaged tissue. Quite analogously, inflammatory cells are thought to deliver signals in the cancer stem cell niche, which induce growth of a new tumor. Such signals from the microenvironment can be received and translated by a signal transduction pathway in the cancer stem cell, resulting in the start of cell division and growth of a metastatic tumor. Thus, timing of metastatic tumor growth may be related to the local arrival of active inflammatory cells.

12.9 A CANCER STEM CELL: CAN IT DIFFERENTIATE TO ANOTHER CELL TYPE?

Pluripotent stem cells can give rise to all cell types in the body, and research on induced pluripotent stem cells (iPSC) (see Chapter 3, "What Are Stem Cells?" and Chapter 4, "Of Mice and Men: The History of Embryonic Stem Cells") has demonstrated that nuclei of differentiated cells can be reprogrammed to a pluripotent stem cell state. Would it be possible for a cancer stem cell to give rise to cell types other than the tissue that it originated from, depending on its microenvironment? Some experimental evidence suggests that this may be possible. Fibroblasts that are present between the cancer cells in tumor tissue are quite different from fibroblasts outside a tumor environment. Surprisingly, they sometimes appear to contain mutations identical to those found in the cancer cells, suggesting that the fibroblasts might originate from the cancer cells. In melanoma, a very aggressive skin cancer, the blood vessel endothelial cells within the cancer tissue appear to be cancer cells that mimic vascular cells. Thus, it is possible that some cancer stem cells, given the right microenvironment, can indeed give rise to other cell types. If true, this implies that the microenvironment of the cancer cell may to some extent be self-generated, and this may have major implications for the development of novel drugs. For example, targeting the fibroblasts for chemotherapy may be easier than the tumor cells and destroying the tumor environment may be sufficient to kill the cancer cells.

12.2

A THERAPY TO CURE CANCER BY DIFFERENTIATING THE CANCER STEM CELL

When stem cells, or cells with stem cell characteristics, form the basis of cancer, as with some forms of leukemia, a therapy that specifically induces differentiation of the abnormal stem cells so they stop or slow down dividing, may be a possible approach to treating the cancer. An interesting example is a special form of leukemia, promyelocytic leukemia. This leukemia is characterized by a DNA abnormality that causes fusion of two unrelated genes, such that the cell makes a novel "fusion" protein. This interferes with normal differentiation. Part of this novel protein is a receptor for vitamin A, which is an important molecule regulating differentiation in the embryo. By treating these patients with a derivative of vitamin A, the differentiation process is restored and the leukemic cells rapidly differentiate to the state where they can no longer divide; the disease seems cured. Unfortunately, when given as a single therapy, the leukemia inevitably returns, thus nowadays this differentiating drug is combined with chemotherapy to produce an effective cure.

12.10 CANCER STEM CELLS: DEVELOPMENT OF NEW DRUGS TO TREAT CANCER

Through either professional or personal experience, many of us are aware that advanced and metastasized cancers tend to recur after seemingly effective treatment with surgery, radiation, and chemotherapy, alone or in combination. This suggests that not all tumor cells are effectively killed by conventional treatments, or they have alternate routes they can use to continue to grow. The cancer stem cell has been partly blamed for being resistant to therapy and responsible for recurrence of the tumor, presumably because it divides only rarely, while treatments usually attack rapidly dividing cells. When we add to this our current knowledge on the role that cancer stem cells may play in causing metastatic tumors, the cancer stem cell and the epithelial–mesenchymal transition (EMT) mechanism by which it is formed become extremely interesting targets for development of new cancer drugs. Drugs that would prevent EMT or induce a switch back from the cancer stem cell type to the "normal" epithelial cancer cell could potentially inhibit metastasis. Obviously this is an exciting idea, on the premise that the EMT process is reversible. If we adopt the concept that a switch to a mesenchymal cell type is caused by interaction between the cancer cell and its microenvironment, maintenance of the cancer stem cell phenotype may be dependent on continuous availability of the proper signals. And this implies that interference with these signals could be one therapeutic approach.

At present, tumors are categorized by the site at which they occur (lung, intestine, brain) and by what they look like in histological sections (graded 1–5, invasive, benign, malignant), but in the future we may no longer do this, simply because of the heterogeneity we just discussed. Tumors are beginning to be categorized on the basis of their gene expression profile and the mutations they contain; in the future, drugs

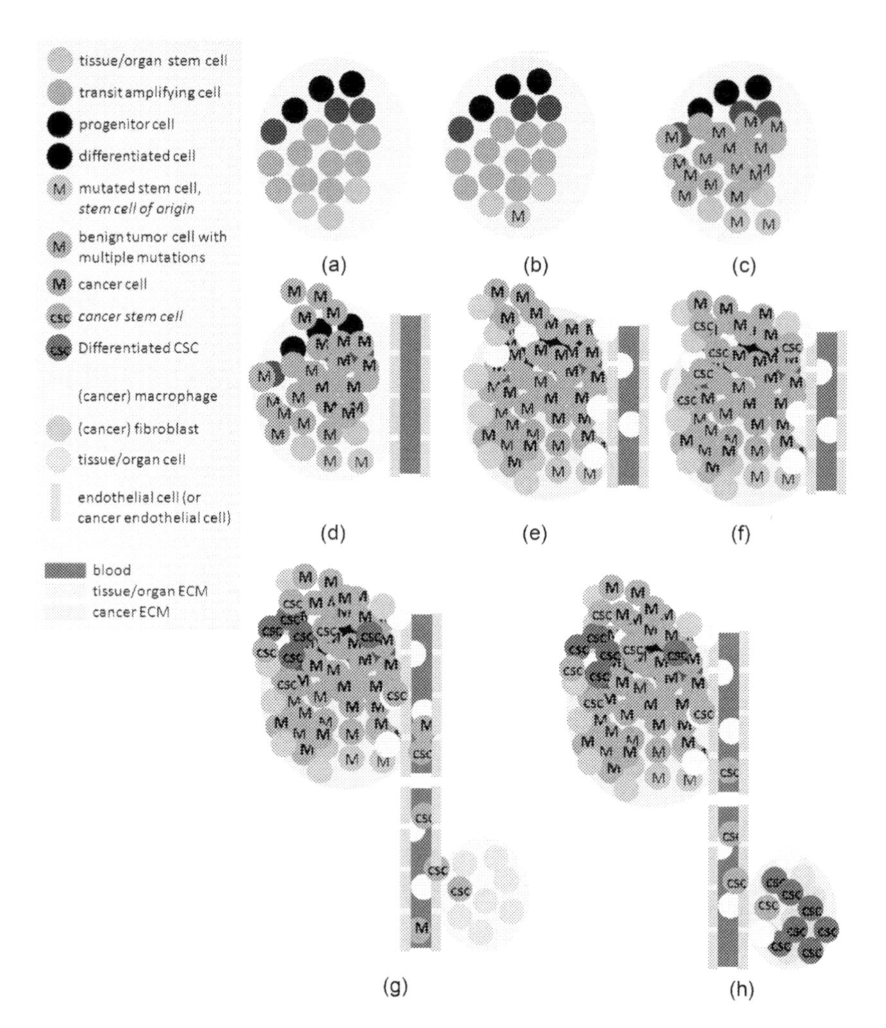

FIGURE 12.7 Simplified diagram of the concepts described in the main text. (a) Cells within the extracellular matrix (ECM) together form a tissue or organ. Stem cells in a "niche" both self-renew (divide to make new copies of themselves) and deliver progenitor cells, which form the differentiated cells that make the functional tissue or organ. (b) A mutation in a signaling pathway gene may occur and, as shown in (c), lead to increased cell division. This is often called a benign tumor. (d) Mutations accumulate, causing genotypic heterogeneity (each cell has different gene expression) and the most well-adapted cells initiate the transition to invasive cancer. (e) Immune cells, like macrophages, infiltrate the tumor. (f) Interaction between tumor cells, macrophages, abnormal fibroblasts, in combination with physical factors, such as a stiffer embedding (extracellular matrix) and increased hydrostatic pressure, induce the cells to change from an epithelial cell type with a "cobblestone" appearance into mesenchymal or stromal-like cells. These cells can migrate easier through tissue, resulting in mobile cancer stem cells (CSC). (g) Cancer stem cells give rise to a mixture of different cancer cells. Among these, some migrate into the blood, where they find their way to other tissue and organ sites. (h) If the growth conditions here are favorable, the cancer stem cells start to self-renew and produce more heterogeneous cancer cell mixtures through partial differentiation and switching back to epithelial-like cancer cells. These are then said to be metastatic tumors. *Source: Reprinted with permission from Anja van de Stolpe: On the origin and destination of cancer stem cells: a conceptual evaluation, 2013, Am. J. Cancer Res. 3, 107–116.*

or other treatments may be selected on this basis. We are already seeing dramatic effects from using this approach in treating colon cancer: two drugs that in themselves are not effective on the tumor are turning out to be very effective at inhibiting tumor growth when used together. The drugs appear to simultaneously shut down two pathways that the cancer cells need to grow so that there are no alternative growth routes to follow. Outsmarting the tumor cells and cancer stem cells in their attempt to take over the body seems to be the new way to go.

12.11 CONCLUSIONS AND RESEARCH CHALLENGES

In the concept described in this chapter, *cancer stem cells* do not represent the *stem cell of origin* of the cancer, but originate from a cancer cell that lost its epithelial properties and instead acquired certain stem cell characteristics. This enables it to contribute to invasion of normal tissue and to metastasize (Figure 12.7). The concept, though compelling, has not been completely proven by experimental evidence and many uncertainties still exist. These offer many exciting opportunities for creative scientists to contribute to better understanding, and ultimately curing, the complex and multifaceted disease of cancer. "The war against cancer" has yet to be won, but many battles on the way have been successful so that for many patients and their families, cancer has become a chronic disease or has even been cured.

CHAPTER

13

Human Stem Cells for Organs-on-Chips: Clinical Trials Without Patients?[1]

Ronald Dekker, Philips, Eindhoven, The Netherlands

[1]Part of this chapter was adapted from an Organs-on-Chips meeting report, Lab on a Chip 2013, DOI: 10.1039/c3lc50248a.

OUTLINE

13.1 INTRODUCTION

An exciting multidisciplinary area of scientific research has recently emerged around a new concept now referred to as "organ-on-a-chip." "Organ" is, of course, clear and "chip" is familiar to many as the miniaturized electrical circuitry in their mobile phone or laptop, but organ-on-a-chip? The term was coined to describe a cell culture-based model system in which cells of different kinds are placed on small structures (chips), usually made of synthetic polymers, that had been "patterned" into grooves, channels, or spirals using the same kind of technology used for making electrical chips. These chip and cell combinations can mimic (or model) the smallest functional subunit of an organ or tissue. Using organs-on-chips, it is becoming possible to mimic many organs and tissues in the human body but in a miniaturized format outside the body. Depending on the cell types used and the format of the patterning on the chip, the alveoli of a lung, small numbers of synchronously contracting heart cells, or even mini kidney and liver-like structures can be created. By adding bacteria, immune cells, drugs, or even cells from diseased tissue to the chips, it may even be possible to model human disease states, not only reducing the use of experimental animals for research, but also providing better ways of looking for cures for human ailments. In this chapter we will explain what making organs-on-chips entails, what role human stem cells could play in making organs-on-chips a success, what the applications could be, and what challenges will be faced in getting there.

13.2 ORGANS-ON-CHIPS

As we have seen elsewhere in this book, many cell types are conventionally grown in a culture flask or dish in an incubator at body temperature (see Chapter 1, "The Biology of the Cell"). In the organ-on-a-chip concept, cells are cultured inside a so-called "chip." This chip is not a microprocessor or integrated circuit such as those present in many electronic devices, but it is often made using the same type of microfabrication process in a clean (completely dust-free) room (Figure 13.1), hence the term.

The chip provides the basic housing for the cells that will form the tissue or organ model. As such, it replaces the conventional culture dish.

The chip is about the size of a microscope slide ($\sim$5 cm $\times$ 2 cm) and it contains one or more open or closed small chambers (of a variable format) to which cells can be added. In the chambers, the cells can grow, much as in normal cell culture, differentiate, or even mature (or age). The material that the chip is made of is transparent, so that the cells can still be seen with a microscope, just as in a normal culture dish (Figure 13.2). The surface on which the cells grow in the microchamber

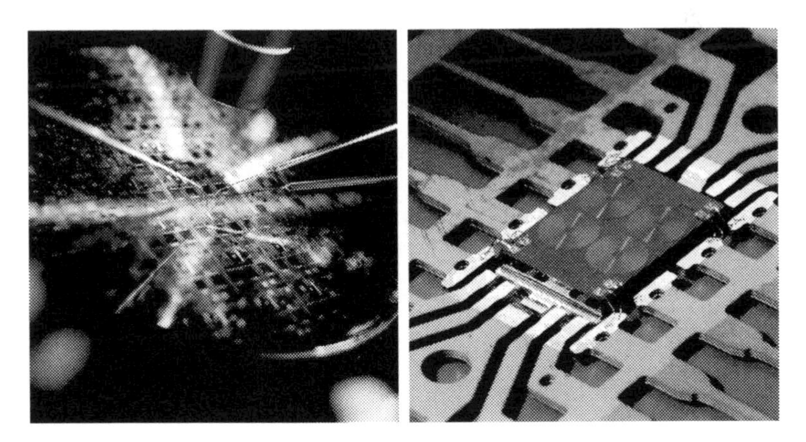

FIGURE 13.1 Many of the electronic devices we use today contain one or more "chips" or tiny printed circuits. The microfabrication process necessary to make these electronic chips takes place in a specialized clean room facility. Clean rooms have extremely low dust and particle levels in the air because of specialized filter systems. *Source: Photographs courtesy of Ronald Dekker, Philips, Eindhoven, Netherlands.*

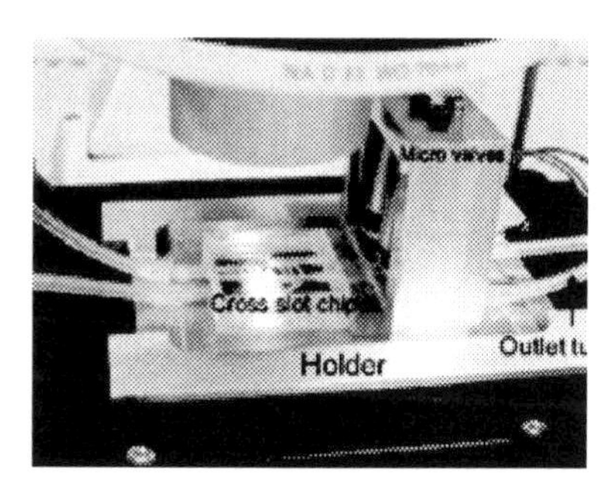

FIGURE 13.2 Example of a microfluidic chip manufactured from polymer material. Cells can be cultured in microincubation chambers on the chip, which are connected to each other by tiny channels and kept at body temperature. Liquids can be pumped through the cell culture chambers, and processes or changes taking place in the cells can be measured over time. This can be done in several different ways, but here is an example in which the chip containing cells is monitored under a microscope. The microscope can be seen at the top of the picture. *Source: Reproduced with permission from van de Stolpe A. and den Toonder J, Workshop meeting report Organs-on-Chips: human disease models. 2013, Lab Chip 2013, 13, 3449–3470.*

does not have to be standard culture plastic, but can be made of different kinds of materials (e.g., natural and synthetic polymers, silicone based rubber, alginate derived from seaweed and present in chewing gum, etc.) which could have different levels of stiffness, or roughness, mimicking real tissue. Bone, for example, we know is very hard and stiff, while muscle and skin are softer and more pliable. An advantage of these surface materials is that they can be very thin, flexible, and stretchable, more like a membrane (thickness in the range of microns, for example), they are not toxic for the cells, and they are permeable to oxygen or nutrients. In addition, the surface can be patterned or, for example, designed to contain pores or grooves of a defined diameter—all from the nanometer to micrometer size. The surface can also be coated with extracellular matrix proteins, proteins that normally reside in the space between cells and form structures that help create the tissue architecture. These proteins also influence the behavior of cells attached to them. The surface patterns can be used to guide the cultured cells in a certain spatial direction so that they all line up, much like the cells in skeletal muscle or the heart, so that they all "pull" in the same direction to create force (Figure 13.3).

It can sometimes be useful if the chamber membranes contain small diameter pores. This can, for example, enable white blood cells to travel through the pores to the other side, such as they would when leaving a blood vessel to attack bacteria during infection. One or more small diameter channels (microfluidic channels) connect the chamber with an external pumping system that controls the flow of, for example, culture

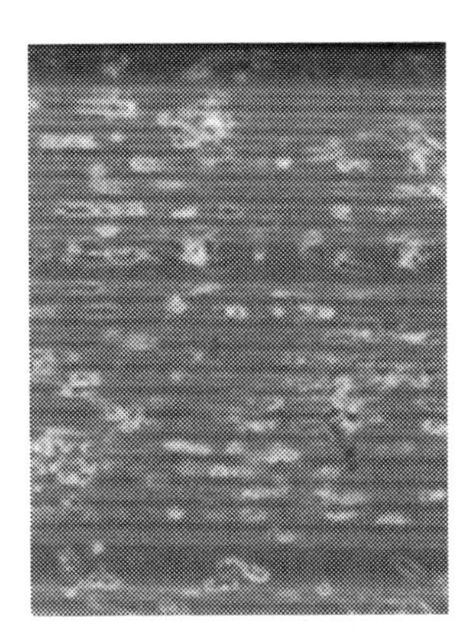

FIGURE 13.3 Cardiomyocytes from human embryonic stem cells attached to a thin patterned membrane (Cytostretch) align along the linear grooves. This mimics the way cardiomyocytes are aligned in the human heart so that they can all pull together in the same direction to make the heart contract. *Source: Photograph courtesy of Robert Passier and Ronald Dekker.*

medium through the channels and chamber, or either side of the chamber membrane. In a simple version, the culture medium can be continuously refreshed to keep the cells well supplied with nutrients and oxygen. Alternatively, the chip fluid system and culture chambers can be closed so that the chip behaves as an independent unit. It is even possible to link multiple culture chambers together, and thus mimic blood flow from the heart to multiple organs in the body such as the lung and liver. This can be important when studying the effects of drugs, for example, because the liver may change some drugs through metabolism and it is these metabolites that can adversely affect various organs and cause side effects of the drug.

There are, of course, endless variations on this basic principle, and multiple functionalities can be built into the chip devices.

13.1

QUESTION: WHY NOT JUST USE SMALL PIECES OF HEALTHY OR DISEASED HUMAN TISSUE FROM A BIOPSY TO STUDY?

A question people often ask is whether one could not just take a small biopsy (tissue sample) from the healthy or diseased organ or tissue of choice, which already contains all the different cell types, and put it in an organ-on-a-chip device. Indeed one could do so, and it depends on the scientific question asked as to what would be the best cell source to use. An intact piece of tissue has the advantage that the

13.1 (*cont'd*)

"architecture" of the tissue is more or less intact, as it was in the body. However, there are several limitations to this "primary tissue" approach. One important limitation is that the different cell types present in the biopsied tissue piece cannot be cultured or treated separately from each other. As an example, it is not possible to introduce into one specific tissue cell type of interest specific (fluorescent) markers necessary to track the position and behavior of individual cells, as well as the "on" versus "off" status of particular cellular processes, during their time in the culture. Other potential limitations are that the cell numbers making up the small piece of tissue cannot be controlled, and that the tissue slice needs to be very thin to allow, for example, proper oxygenation, because the original blood vessels do not function any more. Finally, if the diseased tissue of interest was the brain or some other internal organ, it would be extremely difficult to access this tissue without creating a great deal of damage. Which healthy individual would be willing to undergo a brain or heart biopsy for basic research, even if that were ethically acceptable?

13.3 WHY WE NEED HUMAN ORGAN AND DISEASE-ON-CHIP MODELS

The biological systems currently most widely used to develop drugs are based either on immortal cell lines (e.g., from tumors) or various animal models. No matter how they are modified and adapted, neither of these types of model truly recapitulate the human disease condition. In the case of animal models, this may be because of genetic background, size, or physiology, or in the case of cell lines, their lack of the intriguing complexity of real cells from the body. Even when primary cells can be isolated from the body in sufficient numbers for examining the effects of drugs, they are "mortal," which means they eventually stop growing and run out.

Partly for these reasons, the major pharmaceutical companies in the world have, over the past decade, started to change their strategy in developing new drugs to treat diseases. Put in a somewhat simplified way, one could say that in the past their goal was to discover drugs that would work against a range of diseases or symptoms, and would preferably be effective in most patients with one specific disease or condition. This type of drug was called a *blockbuster*. Aspirin and corticosteroids are good examples of blockbusters; they can be used as pain killers or anti-inflammatory drugs to treat the symptoms of a variety of conditions with different causes. In most cases, the way that these blockbuster drugs work may not even have been known at the time they started being prescribed to patients. However, most patients going to their doctor with a specific symptom, such as a headache in the case of aspirin, or an autoimmune disease in the case of corticosteroids, benefitted from taking the drug.

Over time, however, it has become more and more difficult to discover new blockbuster drugs, and pharmaceutical companies have needed to change their drug development strategy. They decided to switch toward the development of drugs that were directed specifically toward the defect in the cell that caused the disease or symptoms in the patient. Clearly, for a drug to be effective in treating a specific disease, it needs to correct the defect that caused the disease of the patient. To develop these more specific drugs, it is necessary to really understand the biology of the diseased tissue and which cellular mechanisms cause the disease-associated symptoms. Unfortunately, although symptoms associated with a specific disease may be more or less identical, the underlying disease biology can vary a lot between individual patients. As a result, the biological target of the drug may also be different in different patients, despite the fact that they seem to suffer from the same disease. This makes the whole issue extremely challenging and confusing!

There are many reasons why the same disease is more severe in some patients than in others, or one type of symptom is more prominent than another. It is thought that of prime importance is the genetic background of the patient, because individuals show many variants in their DNA sequence. These small variations have, in many cases, already been associated with the severity of the disease (although this has rarely been proven), but equally, these same variants may affect drug responses, so that a drug may work well in one patient against a specific disease but be ineffective in another. To minimize these variations in efficacy, drugs would need to be tested in patients from a diversity of genetic backgrounds, of different ages, and even of different gender. The first tests of drugs in humans are nearly always on young adult males; not children (for ethical reasons) and not females (for reasons of the menstrual cycle). It is not difficult to understand how testing drugs on large numbers of real human patients in these different categories is not ethically acceptable or financially sustainable for the drug companies.

Animal models of disease (particularly mice) have been suitable in the past for the blockbuster era but now increasingly fail in their role in the drug development process: animal genomes (except, perhaps, nonhuman primates or monkeys) are very different from the human genome. When a drug was found to be effective in treating a disease in an animal, the next step was often testing in a clinical trial. And while in the past this has been successful, in many recent cases the drugs have not been (sufficiently) effective in the majority of human patients for them to be worth marketing, or to obtain official approval from the regulatory authority (the Food and Drug Administration in the United States) for commercial use in patients. By the time a clinical trial revealed that the drug did not work well enough in humans, huge amounts of money had often been spent on its development. On the other hand, potentially valuable drugs may have actually been discarded because their effectiveness was considered unsatisfactory in animals with the disease, and the stage of testing in human patients was never reached.

The need for human disease model systems became increasingly evident. Ideally, model systems that accurately mimic the disease process should also have incorporated the genetic determinants that play a role in the disease, in that they either cause the disease or influence its manifestations (the disease symptoms), progression, and sensitivity to drug treatment. To date, no such human disease model system is available.

This is where the "marriage" between stem cells and organ-on-a-chip technology could provide a very promising and attractive solution. As we discuss, stem cells can give rise to many cell types, and when derived from human-induced pluripotent stem cell lines from a patient with a specific disease, the cells will contain the correct disease genetics. In combination with organ-on-a-chip technology, different cell types can be cultured in two or three dimensions and exposed to a variety of disease conditions to create human organ and disease models for drug development. Within an organ-on-a-chip model it is possible to imitate the physical environment of the cell that is present in a tissue. This is now recognized to be far more important for normal (or diseased) cell function than previously acknowledged. Examples include the soft embedding of the cell within the tissue, or stretching of heart muscle cells. The "chip" part of the technology also provides a means to enable long-term controlled culture of multiple cell types together that are normally present in the organ or tissue. The appropriate (micro) environment (both physical and biochemical) for the cells can, thus, be created, including, for example, interactions with immune cells that can flow through small channels through the whole system allowing migration into the cultured tissue, all under closely controlled conditions. Abnormal behavior of diseased cells or responses to drugs can be visualized under a microscope, for example, or changes in metabolism, growth, migration, and so on, can be monitored by a variety of biochemical techniques (Figure 13.4).

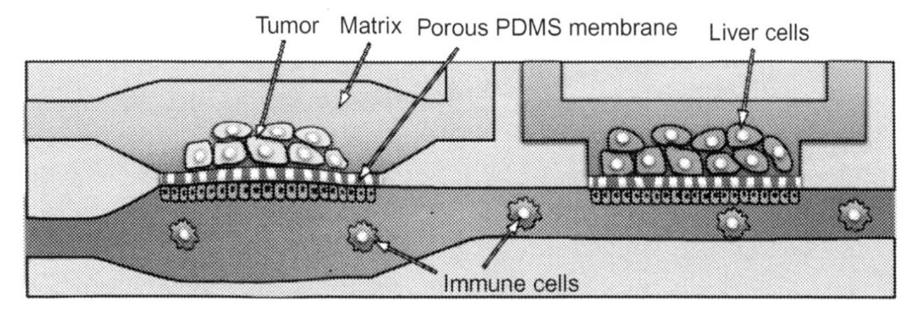

FIGURE 13.4 Vision of the future: cancer metastasis on a chip. Invasion of tumor cells from the primary tumor into surrounding matrix and subsequently into channels that represent the blood circulation. Once the cells have escaped into the circulatory system, they can "seed" into a new tissue or organ site and are said to become metastatic. This process can be mimicked on these chips, enabling development of drugs to prevent growth of metastatic tumors. A cross-section of the chip shows a metastatic seeding site compartment and the connection with the primary tumor site through microfluidic channels simulating the blood circulation. Additional channels can be designed for flowing through certain immune cells that are thought to play complex and yet not fully explained roles in both prevention and promotion of the metastatic process. *Source: Reproduced with permission from van de Stolpe A. and den Toonder J, Workshop meeting report Organs-on-Chips: human disease models. Lab Chip 2013, 13, 3449–3470.*

13.4 HUMAN ORGAN-ON-A-CHIP MODELS FOR CERTAIN DISEASES

The organ-on-a-chip approach is particularly relevant for diseases specific to humans for which no animal models are known. Organs such as the brain and the immune system are very typically human. For this reason, representative cell lines or animal model systems do not exist for many human neurological and psychiatric diseases, as well as for diseases in which the immune system plays a crucial role. These range from autoimmune diseases all the way to cancer. Just a few examples are amyotrophic lateral sclerosis, multiple sclerosis, depression, schizophrenia and psychosis, Alzheimer's, diabetes, rheumatic diseases and diseases such as systemic lupus erythematosus (SLE), skin diseases, fibrotic diseases, and all types of cancer. However, in view of the very human nature of our immune system, which is involved in the evolution and progression of basically every disease, development of drugs for nearly all diseases may benefit from the organ-on-a-chip approach when based on stem cells from patients with the disease of interest, as will be discussed next.

Will we be able to create models of depression or schizophrenia in this type of model? Time will tell, but both basic scientists and even funding agencies believe that it may be feasible, and the pharmaceutical industry is keen to see the outcome of this type of fundamental biomedical and bioengineering research.

13.5 HUMAN DISEASE MODELS AS ORGANS-ON-CHIPS: CHALLENGES

One of the major challenges in applying organ-on-a-chip technology to develop human disease models is to produce the different cell types that are needed from one single stem cell source, so that ideally all cell types have the same genetic background. These cells would be cultured together to build complex tissue-like structures, enabling interactions between different cell types, including immune cells, to be studied much as they take place in the body, but now within the defined environment of a chip. This needs to be sustained and the cells remain alive over long periods to enable the cells to organize and form the organ or tissue module, and allow characteristics of the disease to develop. In addition, the cells on the chip should be visible under a microscope, even when they are located deeper in a three-dimensional tissue structure. The functions of the different cells need to be examined, preferably also in the living cells, and functional changes occurring in response to drugs, for example, need to be examined. Tackling these challenges requires a multidisciplinary team of scientists (biologists, physicians, engineers, geneticists, pharmacologists, medicinal chemists, and physicists), a translational approach, and, ultimately, close collaboration with pharmaceutical companies to implement use of the model systems for drug development.

13.6 WHERE WE ARE NOW WITH ORGAN-ON-A-CHIP TECHNOLOGY

Organs-on-chips technologies have developed immensely over the past decade, influenced by initial pioneering studies with cell culture inside tiny microfluidic channels and on soft surfaces with micro- and nanopatterns designed on the surface to guide cells in a given direction. Three-dimensional "organoid" stem cell culture technology developed independently and is described in Chapter 10, "Adult Stem Cells: Generation of Self-Organizing Mini-Organs in a Dish." All kinds of scaffolds and carrier materials for cells to hold onto in a 3D environment, and natural or artificial gels in which cells can grow by themselves and even organize properly, have become available (see Chapter 7, "Regenerative Medicine: Clinical Applications of Stem Cells"). New microscopic technologies are being developed that make it possible to see live cells at very high resolution in great detail, and to follow changes in the whole tissue structure in a culture over time. A few examples illustrative of organ-on-a-chip models are "lung-on-a-chip" and "intestine-on-a-chip," developed by Don Ingber and colleagues at the Wyss Institute for Biologically Inspired Engineering at Harvard University in the United States (see Box 13.2, "Chips for Organ-on-a-Chip Applications"). These models are already being used to study organ and tissue physiology and disease, and to develop new drugs and best ways to administer them to patients. However, they are still two-dimensional tissue structures, which are easiest to create, and do not make use yet of human stem cells. We refer the reader to Chapter 3, "What Are Stem Cells?" and Chapter 4, "Of Mice and Men: The History of Embryonic Stem Cells" for an overview of the types of human stem cells and their derivative tissue cells that are becoming available for organ-on-a-chip applications.

13.2

CHIPS FOR ORGAN-ON-A-CHIP APPLICATIONS

The chip in which the cells are cultured for organ-on-chip models allows a wide variety of designs. Important aspects to consider to make best use of the chips are: microscope compatibility so that the tissue can be easily visualized and monitored, one or more microfluidic connections to transport liquid culture fluid or other solutions (e.g., containing immune cells or drugs) over or through the cell culture, sterility of the culture chambers and microfluidics connections so that no bacteria are present, ease of placing cells at specific locations in the chip, the cost and ease of chip production, and whether it is possible to transform the chips to higher-throughput devices. The challenges for the microfluidics integrated into the chip include control of flow through the microfluidic channels, and cellular connections able to exclude bacteria and ensure sterility during the long-term culture.

Example: Lung-on-a-Chip

The synthetic polymer (PDMS)-based organ-on-a-chip platform developed at the Wyss Institute for Biologically Inspired Engineering at Harvard University by Professor Don Ingber consists of two sterile fluidic compartments separated by a porous flexible membrane that

13.2 (*cont'd*)

can be mechanically stretched at various strains and frequencies by applying cyclic suction to the hollow side of the chambers in this flexible device. Cells can be grown on both sides of the membrane to recreate a cell–cell interface simulating the situation in the real lung in the body. This platform has been used to mimic the smallest functional unit of the lung, the alveolus, in the form of an interface between an alveolar epithelial cell layer and a vascular endothelial cell layer. By allowing air to flow over the epithelial layer and cell culture medium, with or without primary human white blood cells, over the vascular endothelium, the Wyss Institute team was able to mimic complex normal-organ-level function, including the vascular barrier and gas transport, as well as inflammatory responses to bacterial infection, simulating pneumonia (Figure B13.2.1).

Example: Multiple Organs-on-a-Chip

The group of Uwe Marx (with Reyk Horland) at the Technical University in Berlin, Germany was among the first to explore the possibilities of multiorgan models. The model for a "human lymph node" on a chip enables assessment of the toxic immune reaction (called a cytokine storm) to specific drugs. A multiorgan-chip platform was specially designed for long-term culture and maintenance of multiple different mini-organs or organoids on one chip. In these chips, organ

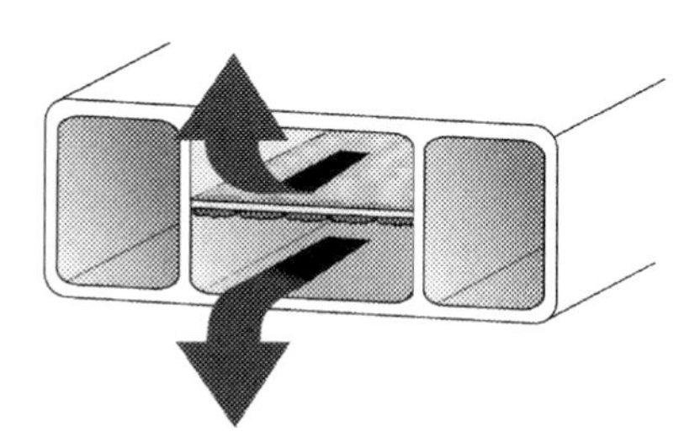

FIGURE B13.2.1 A schematic representation of a "lung-on-a-chip" as developed by the Wyss Institute. A flexible membrane (middle) is covered on the top with lung epithelial cells (green) and on the bottom with endothelial cells that are present in blood vessels. By allowing air (blue arrow) and (synthetic) blood (red arrow) to flow through the chambers, lung function can be examined. The chamber on either side of the device can be inflated and deflated to mimic the stretching of lungs. Bacteria and immune cells can also be added to mimic infection and certain proteins to mimic asthma. *Drawing adapted from Kim H.J., Human gut-on-a-chip inhabited by microbial flora that experiences intestinal peristalsis-like motions and flow. Lab Chip 2012, 12, 2165–2174.*

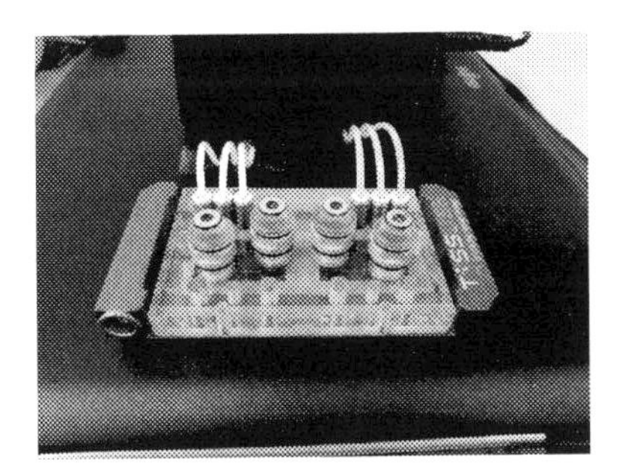

FIGURE B13.2.2 Multiple cell culture chambers in a chip, connected by microfluidic channels in which fluids can flow from one compartment to another. *Source: Uwe Marx.*

tissues are perfused using an artificial microfluidic circulation, which connects the different tissue chambers. Seeding microfluidic channels with vascular endothelial cells is being explored as a means to introduce vascularization in the organ tissue. This type of multiorgan platform is especially suited for drug toxicity screening (Figure B13.2.2).

13.6.1 Pluripotent or Adult Stem Cells

When would one use pluripotent (embryonic or induced) stem cells and when adult stem cells in an organ-on-a-chip model? Some diseases are caused by a highly predictable and well-known gene mutation in the DNA. In those cases it is possible to introduce the specific disease-causing DNA mutation into a stem cell line by the technique of *homologous recombination.* This can result in two human stem cell lines that only differ in the presence of the disease-causing DNA mutation. For this purpose, both pluripotent (embryonic or induced) stem cells can, in principle, be used. On the other hand, for diseases that can be caused by a whole spectrum of mutations in any part of the disease-causing gene, or diseases associated with a more complex genetic background (meaning a variable combination of DNA mutations or variations at multiple locations in the genome), this approach is presently difficult, although the introduction of 10 to 20 different mutations may soon be possible. In many cases, however, particularly when the causative gene is unknown, an induced pluripotent stem (iPS) cell line or adult stem cells derived from a patient with the disease could be used to recreate "the patient" on a chip.

So far, iPS cells are first choice, since, in contrast to adult stem cells, iPS cell lines can nowadays be relatively easily generated from, for example, a simple skin biopsy, or cells in blood or in a urine sample. Large libraries ("banks") of iPS cell lines from individuals with many kinds of genetic or familial diseases, as well as from "normal" individuals with different genetic backgrounds and from different ethnic groups, are being created, both in Europe (e.g., by the Wellcome Trust in the United Kingdom) and in the United States (e.g., National Institutes of Health and the California Institute of Regenerative Medicine). In the near future, hundreds of pluripotent stem cell lines are expected to become available to study many diseases, in principle providing the opportunity to recreate the disease *in vitro*, outside the patient. With respect to adult stem cells, one bank is already being created in the Netherlands (Hubrecht Organoid Technology), and this bank contains tissue samples from patients that can be used to isolate and culture adult stem cells using *organoid culture technology* (discussed in Chapter 10, "Adult Stem Cells: Generation of Self-Organizing Mini-Organs in a Dish").

13.6.2 Organ-on-a-Chip Technology to Mediate Formation of Functional Tissues and Organs

Differentiation of stem cells can be directed toward a required cell type by adding one or more biochemical factors or small molecules, but maturation and aging of the tissue once formed may require other types of signals. These may be mechanical, such as stretching or flow of fluid, electrical, such as receiving a signal from a pacemaker cell in the heart or a synapse in the brain, or be related to acidity, oxygen tension, or the presence of hormones and stress. In the case of cardiomyocytes of the heart, for example, cyclic stretch and electric stimulation are thought to be important for full maturation. Similarly, the stiffness and hardness of the surface on which the cells grow may play an important role in the proper functioning of stem cell niches in the body and in the further specialization of cells in specific directions.

In a conventional cell culture incubator, culture conditions are the same for all cultures in the same incubator, but in real life the O_2 concentration and humidity may actually differ between tissues and organs in the human body, depending on the normal physiology of the tissue or the changes associated with disease. Skin is a good example, where keratinocytes that form the major skin component should be exposed to low humidity air instead of a liquid culture medium. Also, in a standard culture dish, the stiffness of the plastic surface on which the cells are cultured obviously cannot be varied, and stretching the cells or applying electrical stimulation is not possible. One of the key features that make the concept of organs-on-chips so attractive is that by providing a microincubator setting for one or more cell or tissue types on the chip, such specific conditions can be mimicked very well. This also enables culture over longer times than is possible during conventional culture.

13.3

CHALLENGES IN DEVELOPING ORGAN-ON-A-CHIP MODELS

Standardization of stem cell sources, reagents, and differentiation protocols is highly relevant. Microincubators need to be incorporated into the microfluidics system to enable mimicking the *in vivo* microenvironment of the culture. Conditions for long-term culture maintenance need to be established, as well as microvasculature for 3D cultures. Natural extracellular matrix coatings should preferably replace conventional culture plastic. Technologies to monitor biochemical and cellular processes in the model systems, preferably in living cells, and in 3D engineered tissue are needed. For this purpose, both 3D microscopic imaging of cell morphology and extracellular matrix architecture are important, as well as fluorescent imaging, at high resolution (at least around 1 micron, one-thousandth of a millimeter) to distinguish nuclear or cytoplasmic localization of marker proteins in a cell. More technologies are still required to assess cellular effects, for example, of a drug compound, cell migration through the tissue, intracellular processes such as metabolism, gene expression, and DNA modification, and tissue architecture; in principle, all tissue and cell analysis methods. With respect to use by pharmaceutical companies, translation to a medium/high-throughput format is important; current organ-on-a-chip model systems are, in view of complexity, in principle, more appropriate for low-throughput use. However, dedicated design and device fabrication approaches may be well compatible with future high-throughput applications. Overall, it is expected that the field will create extensive knowledge of organ function and regeneration, which can be used as a step-up to later regenerative applications.

Quality Control

To be sure that the tissue or disease that is created on the chip resembles the *in vivo* condition, a quality control assay is needed. *In vivo* organ or disease histopathology analysis is the gold standard to compare with cell morphology and tissue architecture in the engineered organ or disease tissue on the chip. This can be analyzed by conventional histology

13.3 (*cont'd*)

(used by a pathologist in diagnosing diseases) and, as necessary, stained using antibodies against marker proteins or analyzing DNA or RNA. In addition, specific microincubator conditions should be monitored, for example, oxygen and CO_2 concentration, pH, and fluid flow rate.

13.7 APPLICATIONS OF ORGANS-ON-CHIPS

The main applications of organ-on-a-chip models are expected to be drug toxicity screening, particularly for the heart and liver as the most susceptible organs, and human disease models for development of novel drugs to treat the illness. In addition, it may be possible to use the models to investigate routes for optimal drug uptake in the body, for example, to measure the rate of transport through the intestinal wall after oral intake or via the lungs after inhalation. By capturing the genetic variation in the normal and sick human population, these models are expected to increase the speed and reduce costs of drug development. Moreover, safer and more effective drugs can be developed, in principle even tuned to the specific genetic profile of the patient. The future worldwide is rapidly moving toward personalized medicine (Figure 13.5).

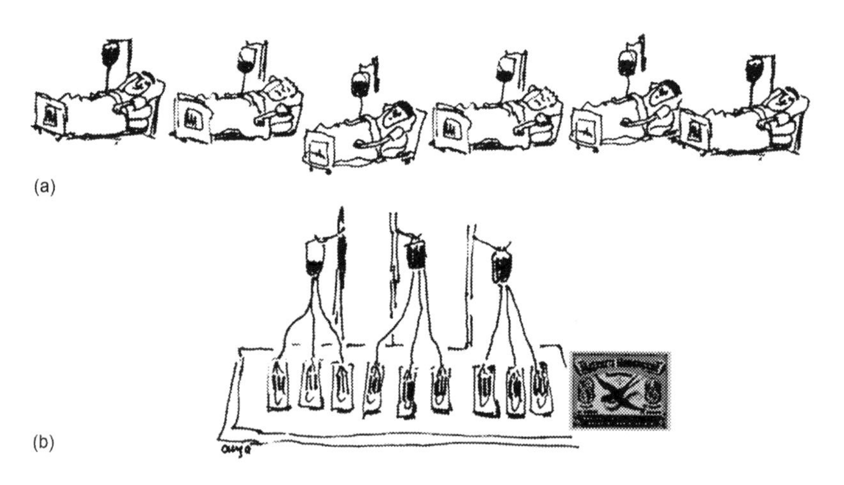

FIGURE 13.5 Vision toward the future of clinical trials: from patients (a) to a clinical trial "on a chip" (b). The matchbox indicates the relative size of the chips. *Source: Reproduced with permission from van de Stolpe A. and den Toonder J, Workshop meeting report Organs-on-Chips: human disease models. Lab Chip 2013, 13, 3449–3470.*

In addition, organ-on-a-chip models could provide the means to obtain a better understanding of organ physiology and the pathophysiology of human, but perhaps also animal, disease. Veterinary medicine is rapidly catching up with that in humans. Aside from providing superior organ and disease models, human organ-on-a-chip and disease-on-a-chip models are expected to promote "the 3 Rs" (reduce, refine, replace) in animal experiments, a goal broadly accepted by regulatory agencies worldwide. Last, but not least, organ-on-a-chip model systems are expected to represent an important step on the way to future regenerative medicine applications of stem cells, expected to be based on stem cell knowledge rather than stem cell transplantation as the means to therapy.

13.7.1 Model Systems for Drug Toxicity Screening

The U.S. Food and Drug Administration (FDA), the regulatory authority that controls approval for market introduction of drugs, has a stated interest in seeing organ-on-a-chip models developed for regulated pharmaceutical use to answer questions on drug toxicity. The highest priority is envisioned for organ-on-a-chip and, if possible, multiple organs on one chip (e.g., heart, liver, kidney, lung, intestine, lymph node), human blood–brain barrier models (to investigate whether drugs can enter the brain tissue from the blood circulation), and placenta-on-a-chip for teratogenicity testing (to investigate toxic effects on the unborn fetus). To this end, the National Institutes of Health (NIH) has made a multimillion dollar grant program available for top labs in the United States to advance these goals.

13.7.2 Human Disease Models for Drug Target Discovery and Drug Development: A Role for Organs-on-Chips

Libraries of iPS cell lines derived from patients with well-characterized diseases will make it possible to capture the genetics behind a disease in an organ-on-a-chip disease model. These models can be used to discover "drug targets," meaning the biological target on which the drug should act to treat the disease or its symptoms.

As the iPS cell line libraries expand, large numbers of iPS cell lines from patients with the same disease but with different severities or different responses to conventional drugs will become available, in part possibly caused by differences in individual genetic backgrounds. A series of such cell lines from patients with the same disease will eventually cover most DNA mutations and variations that affect progress of the disease and effectiveness of specific drugs.

Organ-on-a-chip model systems for a disease based on such a series of patient-derived iPS cell lines will provide new opportunities to test many drug compounds and discover which are likely to be most effective. Even more importantly, it may be possible to determine in which patient the disease responds to the drug and in which not. In this way, the association between genetic background, disease development, manifestation, and progression, and response to a drug could be established, which might not be possible using just a single cell type in a regular culture system, even if it is an appropriate cell type. In cases where this information can be translated into a simple blood test, it is referred to as a *companion diagnostic test*. This is perhaps the future of drug therapy.

For the purpose of efficient drug testing at the scale of thousands of drug compounds, so-called *high-throughput screening*, organ-on-a-chip disease models should be developed that can be handled and scanned by robots. This means a large number of identical model-chips, which can simultaneously be used to analyze the effect of the compounds alone or in combination under exactly the same controlled conditions. This brings us to the question of how complex an organ-on-a-chip model should be: the more complex, the more difficult to create high-throughput model systems. Organ-on-a-chip models can range from very simple to highly complex, as we discussed earlier. The degree of complexity chosen for any purpose only needs to be as complex as is necessary to answer the scientific question under study: the simpler the model, the easier to create a more high-throughput version of the model system.

Once all the hurdles discussed here have been cleared, it will become possible to think about replacing early clinical trials in human patients by "clinical-trials-on-chips"!

13.8 CONCLUSION

Organs are complex functional biological entities, consisting of specialized cell types in a 3D architecture undergoing highly controlled and organized interactions. The 3D environment of every single cell is an important determinant of its function, and needs to be carefully recaptured *in vitro* when developing models of organ tissue, or disease-on-a-chip. Because of the complexity and diversity of components to make organs-on-chips, the field is highly multidisciplinary: in addition to sophisticated cell culture and stem cell expertise, it requires integration between molecular and cell biology, organ physiology, microfluidics, microfabrication and materials science, mechanobiology, and medicine.

In the future, the combination of normal and diseased human stem cells on chips will facilitate replacing some animal experiments, and reduce the risks of first-in-man studies and early clinical trials that are necessary for development of new drugs by supplying a wide range of cells from different genetic backgrounds in the format of a disease-on-a-chip on which the drug can be tested ("clinical-trial-on-a-chip"). The development of simple blood-based companion diagnostic tests to identify patients who will benefit from a specific drug or, for example, are at high risk of specific drug toxicity, opens the path toward more personalized and safer use of drugs. This also enables "rescue" of a drug that may have been shelved at early developmental stages because of its failure to be effective in animal experiments or it may have been toxic.

CHAPTER

14

Stem Cells for Discovery of Effective and Safe New Drugs

Bernard Roelen

OUTLINE

14.1 DRUG DISCOVERY: A SHORT HISTORICAL PERSPECTIVE

What are drugs? Drugs are substances that can be absorbed by the body to alter cellular behavior and thereby correct abnormal body function, to relieve symptoms of the disease, or cure the disease when possible. Centuries of medicine as a "profession" in ancient, primitive, and contemporary cultures have led to the discovery of thousands of natural and synthetic drugs to treat a wide variety of acute and chronic diseases. In all cases, however, drugs interact at the molecular level with either the cell type or tissue in the body affected by and/or causing the disease, and, in the case of an infectious disease, also with the causative infectious microorganism, to exert a therapeutic effect. This interaction should be as specific as possible to avoid side effects.

The first drugs used historically were often based on plant extracts and minerals. However, in these cases patients usually receive complex mixtures of mostly undefined chemicals with low specific activity and, even though they may be effective treatments, they may only work slowly and have side effects. The simplest way to overcome this is to purify the active components and use these to treat the patient. This became possible at the beginning of the nineteenth century, as methods were developed to extract individual chemicals from complex mixtures. Distillation, centrifugation, and the identification of minerals that bind organic compounds were among the new methods used. Many active chemical compounds have been isolated in this way from plants and minerals, but only a few have proved to have a real therapeutic effect when put through objective evaluation in randomized clinical trials. Remarkably, less than 40 plant and mineral extracts are currently in mainstream clinical use. Equally remarkably, perhaps, plant-derived compounds tend to be among the most toxic drugs in clinical use. Examples are digitalis, from the Common Foxglove (*Digitalis purpurea*; Figure 14.1), which is used to treat heart failure and certain forms of abnormal heartbeat, and taxol, which is isolated from the Pacific Yew (*Taxus brevifolia*) and is in clinical use as a chemotherapeutic drug to treat ovarian and breast cancer (Figure 14.2).

FIGURE 14.1 Foxglove (*Digitalis purpurea*), the source of a widely used drug for treating heart failure and certain types of arrhythmias.

FIGURE 14.2 Yew (*Taxus baccata*), the source of drug used for treating a number of cancer types including breast and ovarian cancer.

The ancient Greek and Egyptian cultures studied the healing powers of plants but the world changed in 1856, when Professor August Wilhelm von Hofmann challenged a talented 18-year-old student, William Henry Perkin, to synthesize the antimalarial drug quinine, rather than extract it from the bark of South American Cinchona trees. Perkin never synthesized quinine, but serendipitously discovered mauveine, a purple compound that is considered to be the first synthetic organic dye. Synthetic organic chemistry became a new science. Rapid developments in the synthetic dye industry followed and created a wealth of knowledge on how to make organic molecules. For the first time in history it was no longer necessary to rely on nature as the source of new organic compounds. By the end of the nineteenth century the leading pharmaceutical companies were developing drugs in their own laboratories.

The twentieth century started with another landmark in biochemistry: the discovery of the first hormone, adrenaline (epinephrine). Four years later, two independent groups had already synthesized the hormone. Adrenaline was one of the first compounds modified by medicinal chemists to reduce its unwanted side effects, while leaving its mechanism of action intact. Since then, optimizing the biological activity of drug compounds through systematic chemical modification rapidly became the standard approach to nearly all modern drug development.

In the second half of the twentieth century, technologies were developed to change the chemical structure of a compound so that its transport through the body improved and, on arrival at the disease location, it could "dock" properly to its target molecule or cell. These approaches have led to the availability of hundreds of therapeutic agents to treat all kinds of diseases.

14.1

PREDICTING MAJOR CARDIOTOXIC SIDE EFFECTS: QT PROLONGATION OR HEART FAILURE

Drugs that block the *human ether-a-go-go related gene* (hERG) ion channel cause a condition called QT prolongation. This is defined as the period between the beginning of the QRS complex (depolarization of the ventricles) and the end of the T-wave (repolarization of the ventricles) on an electrocardiogram (ECG). QT elongation is an important risk factor for development of arrhythmias; an arrhythmia means that the normal heartbeat becomes extremely irregular and eventually goes out of control. It can be fatal and, in the past, was one of the main reasons that drugs were withdrawn from sale or their preclinical development was stopped. In 1998, the U.S. Food and Drug Administration (FDA) defined prolongation of the QT interval on an ECG as a major drug safety issue. Currently, assessing risk for delayed ventricular repolarization and QT interval prolongation is part of the standard preclinical evaluation of new chemical entities (or chemical compounds) as defined by the International Conference of Harmonization (ICH) Expert Working Group for all drugs in development. At present, both for cardiac and noncardiac drugs, primary dog or rabbit Purkinje fibers, or cell lines expressing specific ion channels, are the principal preclinical *in vitro* test systems.

Primary dog cells as test systems carry ethical issues associated with animal use, while, in addition, results obtained in animal cells still cannot be extrapolated directly to their human counterparts. Transgenic cell lines, expressing just one specific ion channel gene (e.g., the hERG channel), have the disadvantage that the ion channel then functions in isolation and is not part of the complex ion channel interactions that normally take place in heart cells. These transgenic cells may, therefore, be less predictive of real-life cardiotoxicity. Human pluripotent stem cell based test systems that give rise to real human cardiomyocytes are expected to be a valuable contribution to the field. In 2013, the FDA announced guidelines that would require the pharmaceutical industry to provide data on the effects of compounds on human induced pluripotent stem cell derived cardiomyocytes within two years. As ways of detecting drugs that could cause fatal arrhythmias improve, another side effect on the heart could replace this as the major problem in drug safety. And that is the problem of heart failure. Some very valuable drugs may cause heart failure, either as a result of acute heart damage or as a side effect of long-term drug use. Heart failure is also difficult to detect in animals or cell models and, again, the regulatory authorities and pharmaceutical industry are looking toward human stem cells for a solution. Heart failure, for example, has been observed as a long-term effect of chemotherapy in breast cancer patients. Up to 5% develop heart failure after being cured of the tumor. While this would not be a reason to stop using chemotherapy in these patients, a good model system might be helpful in finding new drugs that protect the patients against this side effect. Or help identify milder combinations of drugs that are equally effective on the tumor but do not damage the heart.

14.2 MODERN DRUG DISCOVERY

Drug discovery today is driven by ever-increasing knowledge of the biological processes that cause disease. In general, the biologically important molecule on which the drug should act (the *drug target*) is discovered first and a search then begins to identify an appropriate new drug. This can be done by actually designing and synthesizing a specific molecule or testing many different compounds that already exist for their ability to have a specific biological effect. These compounds, numbering perhaps many millions, have often been made all at once with just small differences between them that could alter the way they work. We call these collections of compounds "libraries." Each pharmaceutical company may have its own compound library. These libraries of thousand to millions of chemical compounds are "screened" to select a few candidate drug compounds that have the desired biological effect; these are called *leads* or *lead compounds*. The leads are then further improved by medicinal chemistry to achieve the best compromise between therapeutic effectiveness and drug safety. The most promising leads are selected for preclinical testing, which is the final phase before human clinical trials.

14.2

THE LIVER: THE GOOD AND THE BAD OF SIDE EFFECTS

The liver has a wide range of functions, among them, metabolizing and detoxifying drugs. It is made up of several different types of cells but the main one is the hepatocyte. Metabolism of drugs occurs in hepatocytes via cytochrome P450 (CYP) enzymes. These enzymes catalyze the oxidation (addition of oxygen or removal of hydrogen groups) of a wide variety of compounds. By metabolizing a drug in this manner, the liver can change the properties of the drug so that its effects on the body are altered. Medicinal chemists sometimes take advantage of this property by designing an inactive pro-drug that has no therapeutic effect, until it is converted into an active compound by the liver. There can be various reasons to develop a pro-drug, for example, to reduce toxicity in the stomach or intestines after oral administration. The pro-drug would only become active after it had passed and been taken up by these organs.

Many drugs, however, have an unwanted effect on the activity of CYP enzymes. When developing new drugs this is a major concern, as differences in CYP activity (either induced or inherited) may influence the rate at which some drugs are removed from the body by the liver. This could result in excessively high blood concentration of another drug that the patient is using simultaneously, for example, which then may cause toxic side effects. Examples include patients who uses both the anti-hay fever drug terfenadine and the antibiotic erythromycin. Erythromycin reduces activity of a CYP enzyme, which is required to metabolize and inactivate terfenadine. As a consequence, the blood concentration of terfenadine increases, predisposing the patient to potentially lethal arrhythmias. Pharmaceutical companies would like to choose or modify their drugs in such a way that these types of side effects on the liver do not occur.

Animal liver cells can be used to screen drugs and other chemical compounds for such side effects, although human liver cells (e.g., obtained from aborted fetuses) are also used. Both have the disadvantage of an inherently limited capacity to divide. Therefore, current attention is focused on the production of functional liver cells from human pluripotent stem cells. The expectation is that in the future these cells will be a useful source of functional human liver cells for drug toxicity testing, although at present these stem cell derived liver cells are "immature" and do not develop the full range of characteristics of adult liver cells. This is a focus of much current research.

14.3 CHALLENGES AND OPPORTUNITIES IN DRUG DISCOVERY

Whether this process of drug discovery and lead optimization is successful depends to a large extent on the availability of the proper disease model systems to test all of the candidate chemical compounds. For this purpose, many different cell and animal model systems are being used. These are intended to represent or reflect the human disease of interest and can be used to test drugs. For some diseases, however, there are no appropriate animal or cell models that really reflect the human conditions.

Examples are inherited diseases that are typically human (human genes are different from those in animals) or chromosomal abnormalities where these genetic differences between animals and humans make it very difficult to design a model system. Down's syndrome, for example, is caused by a duplication of chromosome 21, resulting in three copies instead of two. Mice and rats, commonly used model animals, do not have chromosome 21 at all. Since the distribution of genes over the chromosomes is different for the various species, this does not actually matter for the animals (the same genes are on different chromosomes) but it makes Down's syndrome very difficult to model in animals. Another example is cleft palate or "hare lip" in humans. Until very recently there was no animal model that mimicked this defect: mutation of the genes strongly associated with this in humans, p63 or IRX6, in mice did not affect facial development. Actual elimination of the activity of both genes in mice does now, however, seem to have the expected effect and abnormal facial development does occur.

For pharmaceutical companies, there is growing interest in carrying out drug discovery on disease models based on human cells. Human cell lines currently in use are often isolated from malignant tumors so that they divide indefinitely, or are "immortalized" in the laboratory in specific ways. However, they often lose the specialized function and characteristics of normal (or "primary") cells once immortalized and they do not have the same genetic background associated with human inherited diseases. On the other hand, primary tissue cells, which could, in principle, be obtained from patients with a specific genetic disease, are often not able to divide and multiply in culture and, therefore, cannot be cultured continuously like an immortal cell line. As a consequence, primary human cells often cannot be produced in sufficient quantities over the long periods that would be necessary for developing a disease model and testing large numbers of chemical compounds.

Pluripotent stem cell lines, or for some organs, adult stem cells, may provide the much-needed solution as a renewable source of human cells that can also be used to create cell-based disease models. Among the pluripotent stem cells, human embryonic stem cells and induced pluripotent stem cells derived by reprogramming normal tissue cells (see Chapter 3, "What Are Stem Cells?" and Chapter 4, "Of Mice and Men: The History of Embryonic Stem Cells") can, in principle, be used for this purpose. One option for creating disease models is to introduce specific disease-causing mutations into the genome of the pluripotent cells by gene targeting. This is particularly appropriate for creating human models for diseases caused by one specific mutation; these are called monogenic diseases. The advantage here is that a whole series of these cell lines can be made, each with just one mutation, and by comparing them with each other, it is possible to say which mutation might cause the most severe form of the disease. Alternatively, some embryonic stem cell lines have been derived from early human embryos diagnosed as carrying a genetic disease (using preimplantation genetic diagnosis) and donated for stem cell isolation by their "parents." Such stem cell lines offer the opportunity to create models for diseases with a more complex genetic background. Embryonic stem cell lines like this already exist, for example, for Down's syndrome and Huntington's disease (a severe neurodegenerative disease that develops in later life). These complex diseases may be caused by a combination of unknown DNA changes in a number of different genes.

Induced pluripotent stem cell lines can, in principle, also be used to create disease models in a similar way, except that they can be derived from the patients themselves; they thus carry the genome of the patient. They have the advantage over embryonic stem cells, aside from ethical issues, that the severity of the disease in the cell donor will be known and perhaps also which drugs were effective or not in treating the disease. Taking genetic forms of depression or schizophrenia as examples, the response of the patient to a specific drug would be known and it would be possible to study the response to the same drugs in nerve cells derived from induced pluripotent stem cells of the same patient. Or taking Down's syndrome as an other example: sometimes the syndrome gives rise to very severe symptoms (severe heart defects, impaired intellect, physical defects), while in other cases symptoms might be quite mild. Induced pluripotent stem cells from these patients may provide clues on why disease severity varies between patients with the same genetic defect. Nongenetic factors that contribute to a disease could also be mimicked in these disease models with the help of advanced tissue engineering technologies and tailored variations in culture conditions. For example, cardiac hypertrophy, a medical condition that predisposes to the development of heart failure, can be induced in heart cells by the addition of certain chemicals to the cells.

Successful implementation of human stem cell technology in drug development will depend on the ability of the cells to differentiate properly to the functional cell type of choice. In general, differentiated cells derived from human stem cells are somewhat immature compared to identical cell types in the adult body. Heart muscle cells, for example, display both immature electrical activity as well as immature subcellular structures. It is thought that the use of new tissue engineering technologies to mimic an organ in three dimensions rather than two—as is the current standard in cell culture—might be an effective strategy to further enhance cell maturation (see Chapter 12, "Cancer Stem Cells: Where Do They Come from and Where Are They Going?).

New developments in this area include the surge of interest in adult stem cells, found in certain organs and tissues that may be accessible via biopsies. One example is stem cells of the large intestine, which appear to have the remarkable ability to divide over very long periods and under certain circumstances in culture, to form *organoids* or *mini-guts* (see Chapter 10, "Adult Stem Cells: Generation of Self-Organizing Mini-Organs in a Dish"). It is fairly easy to collect a biopsy from the large intestine and, in fact, for some diseases, such as cystic fibrosis, this is done routinely for diagnosis (Figure 14.3). Stem cells from the intestine of cystic fibrosis patients form extremely abnormal mini-guts in culture, and studies are already ongoing to see if any chemical in a compound library from a pharmaceutical company can restore these to normal. In that case, a lead compound will have been identified that could be developed as a drug for the patients. In much the same way, it turns out that colon adenomas—benign tumors or polyps of the intestine—also have a stem cell population and they can also be grown as mini-guts. Again, it is a very interesting model in which to investigate whether new drugs can be found in the pharmaceutical company libraries that prevent the formation of these adenomas, or maybe even the rare transition from such a benign tumor to a malignant colon carcinoma.

It is important to realize, however, that a stem cell based model will always remain a highly simplified version of just an organ, let alone a

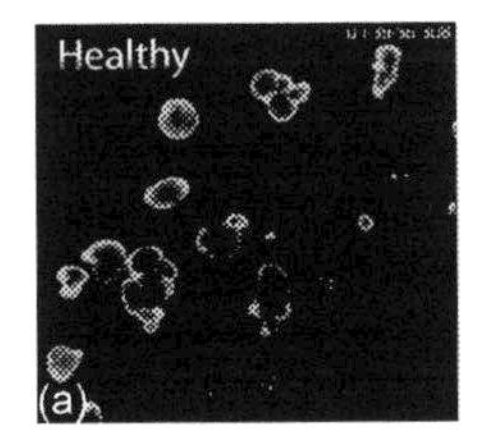

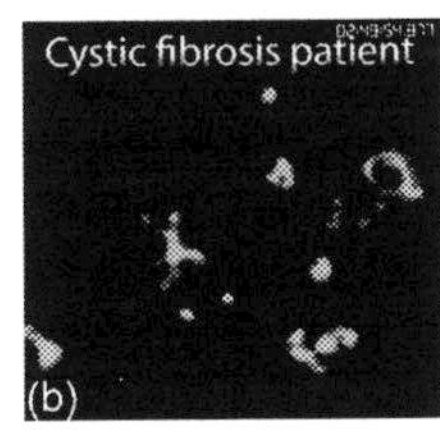

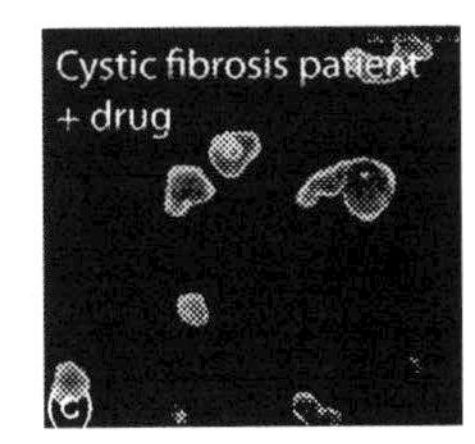

FIGURE 14.3 Organoids from intestinal stem cells are being used to develop drugs for curing cystic fibrosis. (a) Mini-guts made from stem cells of a healthy person, (b) mini-guts made from intestinal stem cells of a cystic fibrosis patient, and (c) the same structures after treatment with a candidate drug against cystic fibrosis. *Source: Rob de Vries and Hans Clevers, Hubrecht Institute, Netherlands.*

whole human being in which organs interact with each other, both in a healthy state as well as in disease. One of the most important features missing from *in vitro* models of normal and diseased cells designed to test drugs is the metabolism of the drug, which often takes place in the liver. Sometimes a drug first needs to be processed by the liver to become effective against the disease. Often, toxic effects of a drug are mediated by metabolites made in the liver. Some cell-based disease models include liver cells to enable investigation of this particular aspect of drug effects. At present, it is still difficult to create a completely cell-based model that can mimic the effects of more than one organ. Thus, it will be clear that not all diseases can be modeled using stem cell and tissue engineering technology and not all drugs will be testable in these systems even if they were near perfect. It is, nevertheless, valuable to identify those that can be useful models, and much research is presently being carried out in this area.

14.3

RESEARCH ON TREATING DEAFNESS WITH STEM CELLS AT THE CALIFORNIA INSTITUTE FOR REGENERATIVE MEDICINE

Hearing loss is one of the most common conditions affecting older adults. Approximately 17% of adults say that they have some degree of hearing loss. Roughly one-third of people between 65 and 74 years of age, and 47% of those 75 years of age and older experience hearing loss. Men are more likely to experience hearing loss than women.

People with hearing loss may find it difficult to have a conversation with friends and family. They may also have trouble understanding a doctor's advice, responding to warnings, and hearing doorbells and alarms (Figure B14.3.1).

Hearing loss comes in many forms. It can range from a mild loss, in which a person misses certain high-pitched sounds, such as the voices of women and children, certain bird songs, or various musical instruments, to a total loss of hearing. It can be hereditary or it can result from disease, trauma, certain medications, or long-term exposure to loud noises. It is anticipated that due to the frequent use of headphones attached to, for instance, MP3 players, particularly by young adults, the number of people with hearing loss will increase dramatically in the next few decades.

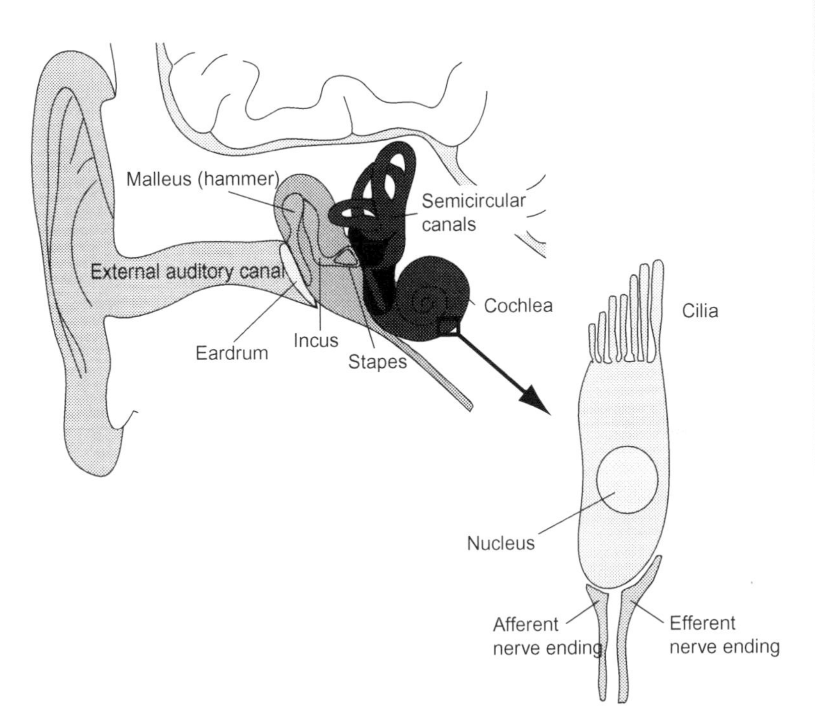

FIGURE B14.3.1 The location of sensory hair cells in the inner ear. Pluripotent stem cells can be induced to form sensory hair cells. The California Institute of Regenerative Medicine (CIRM) is one organization supporting research into their use curing some kinds of deafness.

A major cause of why acquired hearing loss is permanent in mammals lies in the incapacity of the sensory epithelia of the inner ear to replace damaged mechanoreceptor cells, or hair cells. Sensory hair cells are mechanoreceptors that turn fluid movements generated by sound into electrochemical signals interpretable by the brain. Degeneration and death of hair cells is the cause of hearing loss in >80% of individuals with progressive deafness.

The California Institute for Regenerative Medicine (CIRM) is one of the organizations that support research into the generation of hair cells from human embryonic stem cells. The projects aim to identify the particular cocktail of growth factors that can steer the primitive stem cells into the sensory cells that are crucial for hearing (Figure B14.3.2).

14.3 (*cont'd*)

FIGURE B14.3.2 The California Institute for Regenerative Medicine (CIRM) in San Francisco, U.S.A. *Source: CIRM, www.cirm.ca.gov.*

Very recently, this approach has been successful and cells with hair cell properties have indeed been derived in cultures of human embryonic stem cells. These hair cells are being used to study their interactions with the surrounding nerve tissue and to study the cause of hair cell malfunction. Malfunction can be caused by a genetic defect or predisposition, but may also be a side effect of certain drugs that affect hair cell survival. Some forms of chemotherapy, for example, can destroy hair cells, and hearing loss follows. Ultimately the aim of these projects is to produce cells for transplantation therapy or, alternatively, learn more about the biology of stem cell derived hair cells and how they might be stimulated to grow. Pharmaceutical companies could use these cell systems to develop novel drugs to stimulate residual hair cells in the ear of a deaf patient to regenerate the damages cells. Although still far away from clinical application, the promising results are music to the ears.

14.4 HOW THE SAFETY OF NEW DRUGS IS SECURED

All drugs have side effects. Depending on the severity of the disease for which they are used and the effectiveness of the drug, certain side effects are considered acceptable and seen as unavoidable toxicity, while other side effects are so serious that they preclude the drug from coming onto the market to be prescribed to patients (Figure 14.4).

Drugs can, in principle, exert toxic effects on all kinds of tissue cells, however, for some tissue types, such as liver or kidney, the effects are usually not immediately disastrous and can, in principle, be reversed, for example, by (temporarily) stopping the toxic treatment. In contrast, toxic side effects of a drug on the heart, for example, causing arrhythmia, can be immediately life-threatening and irreversible. Likewise, long-term heart failure due to drug-induced cardiomyopathy can be a major problem. Many patients with breast cancer now survive after chemotherapy, but 5% will develop this specific form of severe heart failure in the years that follow. This type of side effect is therefore, in principle, unacceptable. Indeed, patient deaths resulting from toxic effects on the heart have led to withdrawal of a number of drugs from the market. Aside from the tragedy for patients and their families, the costs to the pharmaceutical industry of drug withdrawal are enormous (development costs lost, damage claims, loss of a good name). Other drugs have been required by the regulatory authorities to carry a safety label to warn of potential cardiac risk. Overall, cardiotoxicity has been implicated in one third of drug withdrawals over the last 30 years.

Regulatory authorities, such as the U.S. Food and Drug Administration (FDA) in the United States and the European Medicines Agency (EMA) in Europe have an important task in controlling the entry of new drugs to the market, ascertaining their efficacy, and guarding the safety of patients to be treated with the drug. To prevent the introduction of new drugs into the market that may carry risks of

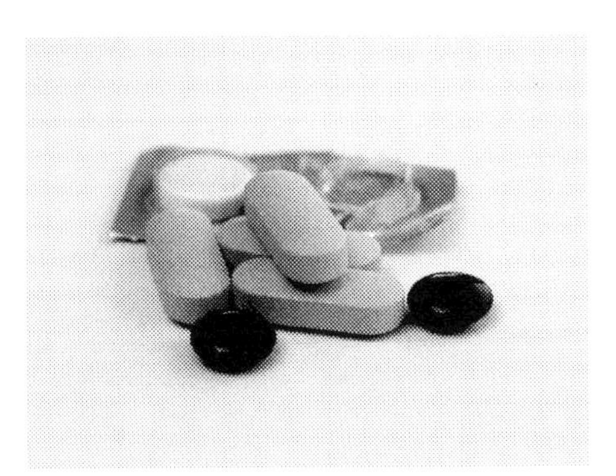

FIGURE 14.4 Pills that are on the market have to be thoroughly tested for efficacy (do they work?) and side effects (are they safe? do they adversely affect any body functions?). Stem cells will need to be similarly tested.

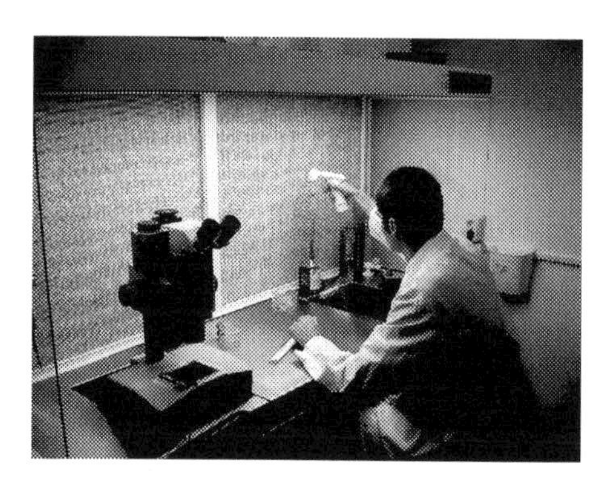

FIGURE 14.5 Much research is still needed on the efficacy and safety of stem cell therapy.

cardiotoxicity, testing guidelines to detect this type of side effect are being enforced by the FDA. In 2013, the FDA actually issued new guidelines in which it proposed that drug companies include tests for toxicity in their safety portfolio using cardiomyocytes from human pluripotent stem cells within two years (Figure 14.5).

Despite the awareness and regulations, it is still difficult to be sure whether a new drug will have cardiotoxic effects before the clinical stage of drug development, mainly because of shortcomings in the test systems presently used. Human embryonic and induced pluripotent stem cells could provide new opportunities in enabling screening systems to test drugs for cardiotoxic side effects in humans. As this is a high priority safety issue, and human pluripotent stem cells can be differentiated to cardiomyocytes relatively easily, there has been increasing interest by many different types of companies, not only those in the pharmaceutical industry, to invest in the development of human stem cell based model systems to screen for cardiotoxic side effects of drugs. This is not least because of the pending FDA guidelines: the pharmaceutical companies cannot afford to wait until it becomes obligatory to test potential drugs on human stem cells.

One way of measuring the effects of drugs on stem cell derived cardiomyocytes is to attach the cells to a multielectrode array (MEA) device, which enables measurement of the external field potential of the cells, rather in the same way that an electrocardiogram measures the heart's electrical activity through electrodes stuck to a patient's chest. After addition of the drug to be tested to the small chamber in which the heart cells are sitting next to the electrodes, any changes in the field potential can be measured rather accurately (Figure 14.6). These changes can be used to predict whether the drug is likely to induce dangerous arrhythmias at concentrations that would be present in the blood of patients. Model systems such as these are already showing their potential to predict cardiotoxic effects of some drugs

FIGURE 14.6 The external field potential of cells, such as stem cell-derived cardiomyocytes, can be measured with a multielectrode device (MEA). *Source: Caterina Martines Grandela, Leiden University Medical Center, Netherlands.*

fairly well. It is expected that model systems to screen for toxicity on other cell types as well for heart failure will also be developed in much the same way, provided that ways of quantifying the response can be found.

Thus, the use of human pluripotent stem cells will hopefully contribute to the development of more effective and less toxic drugs. In addition, it may be possible to find molecules using these test systems that actually prevent or reduce the toxic side effects of drugs without altering their therapeutic activity. As an important bonus, over time these developments may also support a reduction in the number of animals used for drug testing. Stem cell technology may well introduce the next revolution in drug development, with an impact not unlike those of the historical events described in the introduction to this chapter.

ACKNOWLEDGMENTS

The authors would like to acknowledge the contributions of Stefan Braam and Robert Passier in the writing of this chapter, and the contribution of Andre Hoekema and Onno van de Stolpe from Galapagos (*www.glpg.com*) to Box 14.4, "Discovery of Novel Drug Targets."

14.4

DISCOVERY OF NOVEL DRUG TARGETS

Target-based drug discovery starts with the availability of systems to identify genes that are associated with a disease. For the discovery and validation of disease-relevant genes, robust tools are needed to allow biology-driven target discovery. The discovery of RNA interference (RNAi) facilitated the knockdown of genes in human cells at a genome-wide level. In 2001, Thomas Tuschl and colleagues[1] identified short interfering (si) RNA as the active component of RNAi. This triggered the development of RNAi tools for drug discovery research, including target discovery and validation. RNA interference by siRNAs is caused by small, very specific RNA sequences that are artificially introduced into the cell of choice. They form a hairpin structure because they contain complementary RNA sequences on both ends that hybridize together to form a double-stranded RNA molecule. Through a few intermediate steps, one of the hybridized RNA parts is separated and can recognize and bind to a specific mRNA molecule present in the cell, which is subsequently degraded. In this way the production of the associated protein is stopped in a highly specific way. The effect of a rapid reduction in the concentration of that protein on the function of the cell can subsequently be studied. The siRNAs are designed in such a way that they recognize the mRNA of the protein of interest: target-specific siRNAs.

Target-specific siRNAs can be introduced into human cells by the transfection (introduction) into the cell of synthetic siRNAs or by delivery through viral vectors, which subsequently produce the siRNA sequence in the cell. The benefits of using viral delivery tools are that delivery is robust, efficient, and provides long-term expression.

Galapagos is a drug discovery company that uses a robust siRNA delivery technology based on adenoviral vectors (SilenceSelect™) to discover novel drug targets for major human diseases. These enable the development of new drugs directed toward the newly identified drug target (Figure B14.4.1).

Not all proteins are good drug targets. For example, some proteins are essential for the function of all cells in the body and, therefore, cannot be blocked by a drug without serious side effects. The interesting remaining drug targets, around 5000, are called "drugable." Several adenoviral libraries have been constructed that focus only on the drugable targets. Every mRNA that needs to be targeted with a specific siRNA is represented in the library by three adenoviral siRNAs.

Another essential element in setting up a disease-based target discovery is the design of *in vitro* cellular assays that model a disease, and

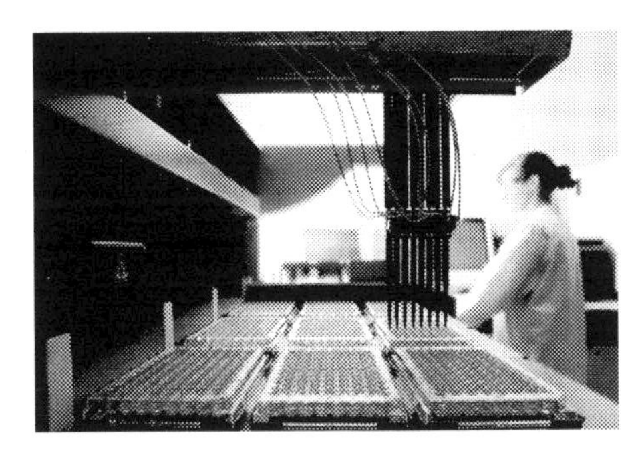

FIGURE B14.4.1 High throughput analysis used to search for novel drugs to treat disease. *Source: Galapagos-Biofocus BV, Leiden, Netherlands.*

FIGURE B14.4.2 Research in progress at Galapagos. *Source: Galapagos-Biofocus BV, Leiden, Netherlands.*

thus contain the right cell types to discover the drug targets in using the siRNA library. Cardiomyocytes derived from human embryonic stem cells provide an interesting platform to model various heart diseases (Figure B14.4.2).

These disease-model cells to which the siRNAs are delivered are present in multiple small culture wells in an array format, and the siRNA is delivered to each well. Subsequently, the sought-for change in the function of the cells is measured, for example, in a reduction in cell growth in case of a drug target for cancer. In those wells in which the wanted effect is seen, it is known which protein was "knocked down."

If the function of this identified protein could be blocked by a drug in a patient with the disease—without serious side effects—it might be expected to represent a new treatment. The next step, therefore, is to use libraries of small chemical molecules to select those compounds that can block the function of the protein, and mimic the effect on the cell seen with the siRNA. Such a result means not only a success, but also marks the beginning of a very costly and laborious—and high risk—drug development track.

[1] *S.M. Elbashir et al., Duplexes of 21-nucleotide RNAs mediate RNA interference in cultured mammalian cells,* Nature *411 (2001), pp. 494–498.*

CHAPTER

15

Patents, Opportunities, and Challenges: Legal and Intellectual Property Issues Associated with Stem Cells

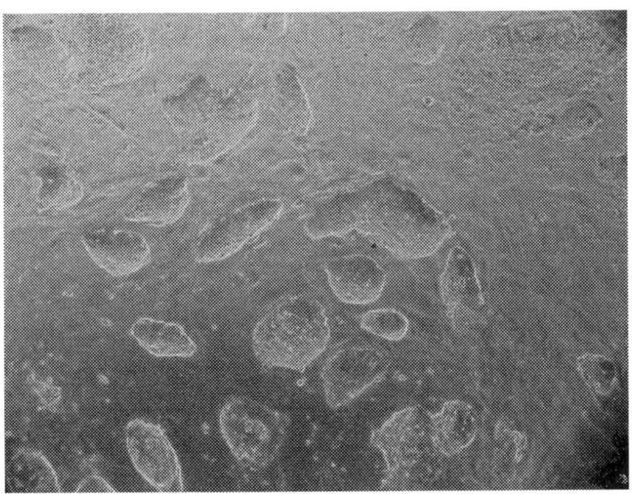

Stieneke van den Brink, Hubrecht Institute, Utrecht, The Netherlands

OUTLINE

Making stem cell products available to customers, whether for transplantation as therapy to patients or as specific derivatives of stem cells to pharmaceutical companies for drug testing and discovery, usually requires that there is some opportunity for the manufacturer to commercialize the product and make a profit. If there is no opportunity for profit, it is unlikely that the product will be made. Stem cells clearly have many different potential applications that could be commercialized. In fact some already are. These range from offering a unique service to a customer (e.g., producing a variety of "cell-based screening platforms" for neural, cardiac, liver, vascular, blood cells etc. for drug toxicity testing by a pharmaceutical company) to a stem cell-based "drug" that is sold as a therapeutic product (e.g., vials of cells to be used for treating a particular disease). Despite the large research and development investments already made by large and small biotechnology companies to generate stem cell products, the risks in this particular market are still very high. There are no guarantees that the products will meet regulatory requirements, be marketable, or have customers by the time they can be sold. Very many stem cell companies filed for bankruptcy or stopped producing stem cell products once the initial hype at the beginning of this century about their potential in regenerative medicine declined.

In the first few years after James Thomson's first isolation of human embryonic stem cells, many scientists across the globe filed patent applications for their inventions on the derivation, differentiation, or applications of stem cells, covering human embryonic and, later, induced pluripotent stem cell lines as well as adult progenitor stem cells. Associated with patent applications is often the intention to found a company that will make use of the protected knowledge. One early example of a scientist-initiated start-up stem cell company is that of James Thomson himself, who founded Cellular Dynamics International, which develops human model systems based on stem cells to test for drug toxicity. All of its current products are based on differentiated derivatives of human induced pluripotent stem cells. These include cardiomyocytes (iCM), neural cells (iN), and several other cell types that are sold as frozen vials of about one million cells costing about US$2000 for an academic group (the price is higher for the pharmaceutical industry). Many researchers find this very convenient for testing drugs or developing technology for stem cell derivatives, since they do not have to go to the trouble of establishing the culture and differentiation methodology in their own laboratories.

Several other stem cell companies (Geron, Advanced Cell Technologies, Novocell [now ViaCyte], Cellartis [now Cellectis], Stem Cell Sciences, Embryonic Cell International [now ESI Bio], to name just a few) were founded either to develop stem cell based therapies for a variety of diseases, or to exploit pluripotent stem cells as models for human diseases or use them for testing drug toxicity (known as *safety pharmacology*). Many of these companies were founded in the period 2002 to 2005, but because therapies, in particular, turned out to be further off than originally thought, several of these no longer exist. As the opportunities for safety pharmacology become clearer and some therapies in unexpected areas (e.g., to treat the eye disease macular degeneration) seemed to be feasible after all, a new wave of stem cell based biotech companies is now emerging.

The legal and patent issues, however, are still far from clear, in part because of a network of patent claims from the early companies that may or may not protect intellectual property from use by others, but also in part because individual countries have differing opinions on what is morally acceptable to patent. For example, patenting the production and use of adult stem cells has been much more straightforward than patenting the same for pluripotent stem cell lines derived from human embryos, as there are fewer perceived ethical issues. By contrast, patent protection of methods to derive and use induced pluripotent stem cells is an area of emerging legal complexity. This is partly because of differences in opinion on who discovered what first. For the general public, or even stem cell scientists and clinicians, such patent issues may not seem to be of relevance to the overall goal of using stem cells for therapy or drug discovery, but they do, in fact, have the potential to "make or break" the commercial viability of stem cell applications. Without patent protection, companies will not be eager to produce stem cells for therapy, for example, and most hospitals do not have access to the technology, infrastructure, and expertise themselves to produce a complex stem cell "product" to be given to a patient.

In this chapter, we consider some of these issues in order to provide a perspective on where the potential problems and solutions are likely to arise. We give some examples of companies and organizations currently working on stem cell applications, but this is neither exhaustive or set in stone because, as recent history has shown, companies in the stem cell area come and go, depending on their particular business model and its perceived feasibility by venture capitalists and corporate investors. It is, however, probably safe to say that interest from venture capitalists in stem cell products has declined over the last decade, and it has largely been left to publicly funded research to mediate the translation of stem cell technology to medical practice.

15.1 COMPANIES AND ALLIANCES

In addition to small and middle-sized biotech companies, several "big pharma" companies have started expressing interest in applications of stem cell technology, primarily for drug discovery and development. They have, as a result, invested large amounts of money in public–private collaborations to speed up stem cell research. One example is a not-for-profit initiative in the European Innovative Medicines Initiative (IMI), in which pharmaceutical companies including GlaxoSmithKline, AstraZeneca, and Roche have joined forces with academic research groups to establish platforms for drug discovery in diseased human induced pluripotent stem cells and testing toxicity of drugs on liver and heart cells derived from pluripotent stem cells. GlaxoSmithKline also invested US$25 million in the Harvard Stem Cell Institute in Cambridge, Massachusetts, while Roche recently initiated a collaboration with Massachusetts General Hospital and Harvard University to develop new disease models using human stem cell lines. In 2009, even the Healthcare division of the electronics company General Electric (GE) announced a partnership with one of the original regenerative stem cell companies, Geron. This has now been taken over entirely by GE, which markets cells with a machine for "high content imaging" (measuring many cell properties at the same time). Without doubt, the coming years will witness many more stem cell based initiatives, particularly in drug discovery and safety pharmacology (see also Chapter 13, "Human Stem Cells for Organs-on-Chips: Clinical Trials without Patients?"). This is in part because methods of producing human induced pluripotent stem cells are improving, and commercial suppliers of culture and differentiation media have very reliable products so that ethically more acceptable sources of pluripotent stem

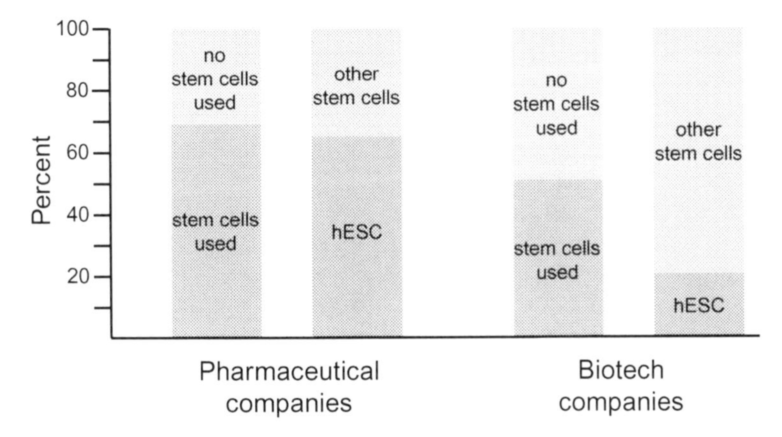

FIGURE 15.1 The use of stem cells by pharmaceutical and biotech companies (brown/blue bars) and the type of stem cells used (green/pink bars). *Source: Jensen J. et al., Human embryonic stem cell technologies and drug discovery. 2009, J. Cell. Physiol. 219, 513–519.*

cells and easier ways of growing them are becoming available for industry. In addition, the recent shift in U.S. presidential policy (many of the companies are U.S. based or owned) has increased public acceptability of the use of established human embryonic stem cell lines (Figure 15.1). The emerging alliances between academia and companies may be interpreted as a sign of pending implementation of some stem cell technologies. Nevertheless, return on investment is still relatively low, which means that many venture capitalists or large companies are waiting to see how the field develops and what it can actually offer them as a benefit above what they are using now. It is expected that they will only invest when the odds are high that something marketable will be produced in the short term. Patients, families, and physicians should, however, be encouraged that some applications may really be just around the corner.

15.2 PATENT ISSUES: CURRENT INTELLECTUAL PROPERTY LANDSCAPE

Legal disputes on stem cell patent rights are inevitable and more are likely to arise as more and more potential applications begin to emerge. Stakes are high, and very broad claims are being made in the most basic patents as a result. For example, with human embryonic stem cells, was it the first description of cells growing out from the inner cell mass of a five-day-old human embryo in 1994 that counts, or was it the first description of stable cell lines that were pluripotent in 1998? The 1998 finding was patented, but was it undermined by the 1994 publication, which was not? This is not yet clear. With respect to the method for derivation of induced pluripotent stem cells, the question arises whether derivation of the mouse cells was the actual invention, and obtaining the human cells was just a logical consequence, or was the derivation of the first human induced pluripotent stem cell line the invention that should be patented? To make things more complicated, the publications on human induced pluripotent stem cells from U.S. and Japanese authors appeared in the scientific literature simultaneously, but each described a different method. Are the two methods each eligible for separate patents, or was one first? And since these publications, new methods have been described using entirely different factors and methods; are these also patentable (Figure 15.2)?

15.2.1 Wisconsin Alumni Research Foundation

In the United States, there was a fierce dispute over the so-called Wisconsin Alumni Research Foundation (WARF) patents. These patents describe the production of first human embryonic stem cell lines by James Thomson at the University of Wisconsin. WARF is the organization set up to protect patents deriving from this discovery, and to distribute the original stem cell lines to other researchers. We give this as an example to show how complex patent claims in this field have become.

The claims in the WARF patents are directed to cell culture of primate (ape/monkey and human) embryonic stem cells that are capable of proliferation *in vivo*, for example, after transplantation into an embryo (ape) or adult organism (ape and human). WARF filed four

FIGURE 15.2 Many legal battles on stem cell patent rights are presently ongoing.

separate patents covering this research area. The Board of Patent Appeals and Interferences (BPAI) of the U.S. Patent and Trademark Office (USPTO) decided to allow three but disallow one of these four WARF patents (U.S. Patent Number 7,029,913). The BPAI considered that the claims of the patent were "obvious over the prior art" (meaning that there was already known and published information on the subject, so it was matter of simple deduction to use it). The BPAI considered that a skilled person would have had no problem to arrive at the essence of the claims from studying the "art" available at the time the patent was filed. This indicates that the "invention" is not actually novel and therefore not patentable. Specifically, at the time the patent was filed, pluripotent embryonic stem cells had already been derived from embryos from mice, rats, and hamsters, and the prior art explicitly demonstrated two ways to produce embryonic stem cells. In addition, in 1994, four years before James Thomson derived a proper human embryonic stem cell line, Ariff Bongso in Singapore had already published a paper showing that he could derive growing cells from a human embryo. He was just not able to get them to turn into a permanent cell line. The Patent Office decided that the path to deriving human embryonic stem cell lines had a definite starting point, making it obvious to try the techniques that were developed for animal embryos also for human embryos.

This WARF patent is one of four, three of which were upheld in earlier re-examination proceedings. The fourth member of this group of patents was granted in September of 2009. In the BPAI decision of April 2010, the USPTO seems to have a different view of the relevance of the prior art at the time of the original filing of this patent family. If this view is maintained in the future, the validity of the other WARF patents may be cast into doubt, as they all originate from the same first filing, made in 1995, and have more or less similar claims.

Even if all of the patents are valid and accepted by the Patent Office, they will probably expire before the relevant applications of the human embryonic stem cells can be commercialized. Depending on whether patent extension is obtained for specific products, the protection will expire in the years after 2015.

15.3 EUROPE VERSUS THE UNITED STATES

Europe has generally shown greater concern with ethical issues than the United States with respect to patent protection and filing in biotechnology, and has implemented means to address this within the European Patent Convention (EPC) system. From its inception, the EPC

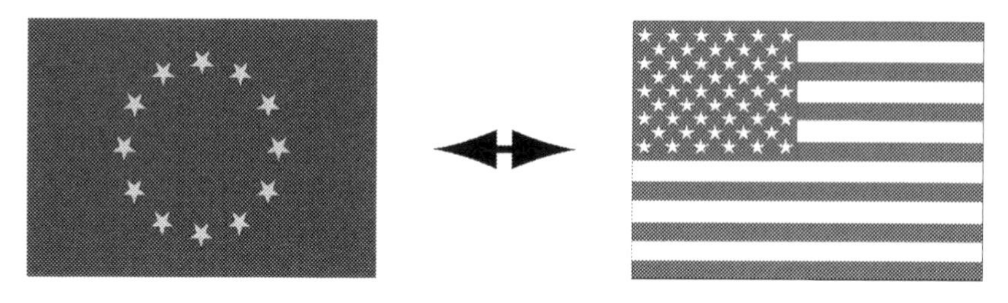

FIGURE 15.3 Patent filing and protection in biotechnology differs between Europe and the United States.

has excluded, for example, "morally offensive subject matter" from patent protection (Figure 15.3).

In 1999, the EPC was amended to incorporate provisions from European Community Directive 98/44/EC, the "biotechnology directive." These rules provide guidance on how to interpret "morality" in the field of biotechnology. Among the new rules was Rule 28(c) EPC, which states that European patents shall not be granted in the case of biotechnological inventions that involve "uses of human embryos for industrial or commercial purposes."

While the United States has granted a range of patents covering the generation and culture of human embryonic stem cells, the European Patent Office (EPO) had put granting of patents with such claims on hold, pending a decision of the Enlarged Board of Appeal on the issue. The Enlarged Board of Appeal was also asked to make a decision in appeal proceedings in the context of the WARF application.

The Enlarged Board decided that the European Patent Convention prohibits patenting of claims on human embryonic stem cell cultures that, as described in the application, at the filing date could only be prepared by a method that necessarily involved the destruction of the human embryos from which the cultures are derived. This was also true even when the destruction of the embryos themselves was not part of the claim. The Enlarged Board further considered that it was irrelevant whether at a later date a technique was developed through which a cell line could be produced without destroying a human embryo. The EPO subsequently rejected the WARF application.

Since the decision was issued, at least two other stem cell cases have had decisions made by the Boards of Appeal (T522/04 and T329/06). In both cases the claims were determined to violate Rule 28(c) EPC, as the applications only described the preparation of the embryonic stem cells through the use (involving their destruction) of embryos at a very early stage. However, a claim expressly disclaiming human embryonic stem cells was allowed. As one can imagine, the commercial applications of such claims is limited by such a disclaimer.

By focusing the decision on the destruction of embryos, the Enlarged Board of Appeal has provided no direction on whether stem cells prepared using methods in which the embryos remain viable would be allowable. Subsequently, for example, techniques were reported that appear to generate embryonic stem cells from human embryos without embryo destruction. This usually involves removing one of the cells at the eight-cell stage, much as would be done for prenatal genetic diagnosis. This cell is then used to derive a cell line instead of being used to search for disease-causing mutations. The remaining seven-cell embryo is, in principle, able to develop normally if transferred to the uterus of a recipient mother. Of course, the question could always be "which mother would let her embryo be used for this purpose?" but the issue is, at present, one of principle. In a first communication, however, the Examining Division has objected to these claims. It considered it irrelevant that the human embryo is not destroyed: Rule 28(c) EPC prohibits any use of human embryos for industrial or commercial purposes.

Another way to patent inventions with respect to human embryonic stem cells, without actually referring to the use of a human embryo, would be to refer only to established stem cell lines. No case law has yet been developed for these types of claims.

As described elsewhere in this book, technology that allows the generation of pluripotent stem cells (i.e., induced pluripotent stem cells) from nonembryonic (adult) cells could be a way around the issue. While the term *pluripotent stem cell* arguably encompasses embryonic stem cells, other methods that are not contrary to morality could also be used to produce such cells. Therefore, it would seem better in a patent application to claim pluripotent instead of embryonic stem cell subject matter, and to include in the patent various methods and sources for obtaining stem cells.

Although the Enlarged Board of Appeal has the final word in the interpretation of EPC law, practice before the EPO is generally harmonized with the national patent law of the member European Union (EU) states. In this respect, it is interesting that in January 2010 the Bundespatentgericht (German Federal Patent Court) referred a number of questions on embryonic stem cell patenting to the European Court of Justice. This is the case that became known as "Oliver Brüstle v Greenpeace eV" (see Box 15.1, "Patenting Inventions on Human Embryonic Stem Cells: The Case of Brüstle v Greenpeace"). One of the questions specifically addressed was whether under the "biotechnology directive" a technical procedure is considered nonpatentable if the destruction of human embryos is necessary for the application of that procedure, even though the use and destruction of the embryo does not actually form part of the procedure for which the patent is claimed. One could think of the scaffold (or carrier material) on which retinal epithelial cells, derived from human embryonic stem cells, are grown,

15.1

PATENTING INVENTIONS ON HUMAN EMBRYONIC STEM CELLS: THE CASE OF BRÜSTLE V GREENPEACE

Protecting intellectual property emerging from research in the biological sciences by filing a patent is, just as in many other areas, a very common way of ensuring that any commercial exploitation results in some kind of financial benefit to the inventor. It also offers protection to the company taking the financial risk of product development based on the patent from other companies who have invested less but by copycatting can bring a cheaper version of the product to the market because they do not require the same return on investment. However, it is not always considered morally acceptable to patent biological findings, and human embryonic stem cells are a case in point, on which views are strongly divided. In the United States, for example, a fundamental patent based on the first cell-lines isolated by James Thomson[1] was granted to the Wisconsin Alumni Research Foundation (WARF) and is still in force. This patent covers methods for the derivation, propagation, and use of human embryonic stem cells, thereby including many aspects of the research and medical applications of these cells. In Europe the situation is different: in 2008, a parallel European patent had been revoked in the so-called WARF decision of the European Patent Office (EPO). The EPO stated clearly that the WARF decision was not meant to ban the patentability of stem cell inventions in general. However, today stem cell inventions in Europe face an even stricter ban following the case of Brüstle v Greenpeace, which reached the European Court of Justice (ECJ) in October 2011.

The History of the Case

In December 1997, Oliver Brüstle filed a patent in Germany on isolating and growing large numbers of neural progenitor cells by differentiating embryonic stem cells (Figure B15.1.1). The patented inventions aim at the (future) use of these cells in treating neurological disorders such as Parkinson's disease. The invention does not actually relate to the derivation of stem cell lines from human embryos but only uses established embryonic stem cell lines.

In Europe, the patentability of biotechnological inventions is governed by the EU Biopatent Directive (Directive on the Legal Protection of Biotechnological Inventions 98/44/EC). The directive also contains exemptions from patentability including Art. 6(1), which states that patents contrary to public order and morality are excluded

FIGURE B15.1.1 Oliver Brüstle. His attempts to file a patent in Europe based on human embryonic stem cells have so far failed. *Source: Photograph courtesy of Oliver Brüstle.*

patentability. To provide national courts and patent offices with guidance on how to interpret this clause, an illustrative list of examples was incorporated in Art. 6(2) of the Biopatent Directive. Since the directive was issued in 1998 before human embryonic stem cells existed, it contains no specific language relating to the patentability of human embryonic stem cell based inventions. However, one of these examples has now proven to be key for the patentability of stem cell based inventions: Art. 6(2)(c), which states that, in particular, "uses of human embryos for industrial or commercial purposes" shall be excluded from patentability. However, there is no definition of any of the terms used in this provision to be found within the directive; neither the term *human embryo* nor what is to be understood by "uses for industrial or commercial purposes" are defined.

The patent was challenged in Germany by Greenpeace on the grounds of public order and morality. The German Federal Patent Court ruled that the patent was invalid in so far as it covers neural cells obtained from human embryonic stem cells and processes for the production of those neural cells. The defendant appealed against the judgment to the Bundesgerichtshof (German Federal Court of Justice, FCJ), which held that the matter depends on the interpretation of the Biopatent Directive.

15.1 *(cont'd)*

In 2010, the German FCJ asked the ECJ for clarification on how to interpret Article 6(2)(c), the Biopatent Directive, which it considered would need unified interpretation within Europe. The referring court sought, first, to ascertain whether the developmental stage from which human embryonic stem cells can be obtained already qualifies as "human embryo" within the meaning of that article. The court asked, secondly, for definition of the expression "uses of human embryos for industrial or commercial purposes." Third, and finally, the FCJ asked whether the provision includes cases in which the use of human embryos does not form part of the technical teaching claimed with the patent, but is a necessary precondition for its application.

The Court Ruling

The Court ruled that "any human ovum after fertilization, any non-fertilized human ovum into which the cell nucleus from a mature human cell has been transplanted, and any non-fertilized human ovum whose division and further development have been stimulated by parthenogenesis constitute a 'human embryo.'"

Whilst the language used is very specialized and complex, this definition of the embryo has far-reaching implications. It explicitly covers cloned embryos derived by somatic cell nuclear transfer and parthogenetic embryos. On the interpretation of "the exclusion from patentability concerning the use of human embryos for industrial or commercial purposes set out in the Directive 98/44 *also covers the use of human embryos for purposes of scientific research*, only use for therapeutic or diagnostic purposes which are applied to the human embryo and are useful to it being patentable", the court further held that research using human embryonic stem cells cannot be patented even if the use of embryos is not mentioned in the patent (although methods for diagnosis or therapy for the embryo—like the isolation of blastomeres for preimplantation genetic diagnosis—can be patented). The wording is as follows:

> "Directive 98/44 excludes an invention from patentability where the technical teaching which is the subject-matter of the patent application requires the prior destruction of human embryos or their use as base material, *whatever the stage at which that takes place and even if the description of the technical teaching claimed does not refer to the use of human embryos*".

Impact on the Stem Cell Industry in Europe

Inventions that involve destruction of human embryos are, by the ECJ ruling, deemed immoral and therefore not patentable. All patents already granted remain in force but are not enforceable as far as they cover human embryonic stem cells derived by embryo destruction. The value of those assets has, therefore, dropped significantly because all Member States that have ratified the Treaty of Europe are bound by rulings of the European Court of Justice, which overrides national law. Whereas Greenpeace and a few religious groups welcomed this decision, many scientific organizations now believe that this could be the end of the European pluripotent stem cell industry that was just beginning to emerge. The problem is that the business model of this industry is based on its ability to raise funds. Making stem cell products is very costly, and the best way to attract investors is to have intellectual property protection. If investors cannot obtain patent protection in Europe, are they still going to invest in European stem cell companies? These same inventions can still be patented in other countries such as Asia and the United States, making it likely that European stem cell companies would move outside the EU, mostly for cost reasons. Ironically, one goal of Directive 98/44 was to enhance and protect biotech investments in Europe. The directive states, "In the field of genetic engineering, research and development require a considerable amount of high-risk investment and therefore only adequate legal protection can make them profitable."

How this will affect the development of treatments for patients or applications of human embryonic stem cells to drug discovery and safety being examined by the pharmaceutical industry remains to be seen.

[1]*Publication in* Nature.

for example, with the aim of treating patients with macular degeneration. Is the method for producing such a scaffold and making the cells attach to it patentable or not? The best advice that patent lawyers are presently giving their biotech and research scientist clients is to try, if possible, to disclose inventions in a manner that includes noncontroversial uses or methods of preparation.

15.4 MORE LEGAL AND ETHICAL ISSUES

The creation and use of human embryonic stem cells still remains associated with ethical issues in some countries. Induced pluripotent stem cells that do not require destruction of embryos but a simple skin sample, for example, would be expected to be far less controversial with respect to ethical issues. However, this mainly refers to its derivation method and, in fact, the technology has given rise to many more legal and ethical issues. One rather far-fetched example could be that once a pluripotent stem cell line has been made, they could potentially be differentiated toward gamete cells (sperm and eggs), and human embryos could even be produced if the gametes were really functional. This would require strict regulation and adequate control with respect to these uses, much like those already in place in most countries on the use of human embryonic stem cells, which, of course, have the same potential to become gametes.

In the state of California, for example, guidelines relating to derivation and use of induced pluripotent stem cells are based on the earlier regulations set up for human embryonic stem cell lines. The EU seems to be taking much the same perspective through its Ethical Advisory Panels. Guidelines, however, differ between countries and, in the United States, even between states. The United Kingdom, Sweden, and Belgium are among those countries with the least restrictive regulations on working with all types of pluripotent stem cells, while in Germany, for example, there is a clear distinction between the derivation of pluripotent stem cell lines from human embryos (not permissible) versus adult cells (permissible). Because of the ethical discussion, larger companies with international visibility and shareholders with different cultural and religious backgrounds may still be reluctant to become associated with pluripotent stem cell technology, even when it does not concern embryonic stem cells, and thus be hesitant to invest in or partner with stem cell companies using this technology.

There are other ethical issues emerging as well. What happens, for example, if a serious disease is discovered as having the potential to develop in the individual who donated the cells for reprogramming? Should they be told or not? Is there a difference depending on whether the disease is treatable or not? Even more complex, what should be done in the case of diseases that may be preventable? And should donation of cells always be altruistic (as in much of Europe), or may there be compensation in the case of particularly valuable cells, as is the case for blood or even embryo donation in the United States?

15.4.1 Privacy and Ownership

Privacy is an issue to consider for all sources of human stem cells, unless used for an autologous transplantation where the donor of the cells is also the sole recipient. When embryos or adult cells (as for the derivation of induced pluripotent stem cells) are used to derive cell lines that will be cultured indefinitely by a commercial company, the genome inside each of the cells remains identical to the genome of the original donor(s). Informed consent from the cell donor will always be required. This consent will need to deal with issues such as anonymity of the cell line to be derived, the time period, as well as the range of applications the cells can be used for, and, finally, not-for-profit research applications versus commercial exploitation. Most informed consent documents result in the donor transferring the ownership of the cells to the organization that is carrying out further processing. However, it may be that the donor still has rights, such as the right at any time to withdraw consent with respect to a specific use of the cells, or restrict the consent to their noncommercial use. How would this work out if the donor dies?

As a possible scenario, it may turn out that to develop a therapeutic stem cell transplantation into an FDA-approved treatment, a commercial pharmaceutical company is in the end better positioned than an academic hospital, however, this might entail a switch toward commercial exploitation of the cells. In any case, this could only take place if permission to return to the donor is included in the informed consent, and if there is a "key" to decoding the donor identity of the cells or tissue. For this reason, various large organizations are trying to develop model informed consent forms so that there will be some international agreement on the wording, allowing for exchange of lines and commercial and noncommercial collaboration. These are complex issues with plenty of room for future debate, which illustrate the importance of a dedicated ethical and regulatory framework.

Within the context of this book, details of the current guidelines are not included since they are prone to frequent changes and adaptations as insight into these matters increases. The International Society for Stem Cell Research (ISSCR) has drafted guidelines that cover many of these points in the use of stem cells. More detailed information can be found at its website, *www.isscr.org*, which is often updated to reflect the current regulatory considerations.

ACKNOWLEDGMENTS

The authors would like to acknowledge the contribution of Mark Einerhand (*www.vo.eu*) to this chapter.

Further Reading

1. Caulfield T, et al. Nat Methods 2010;7:28–33.
2. Fitt R. New guidance on the patentability of embryonic stem cell patents in Europe. Nat Biotechnol 2009;27(4):338–9.

CHAPTER

16

Stem Cell Perspectives: A Vision of the Future

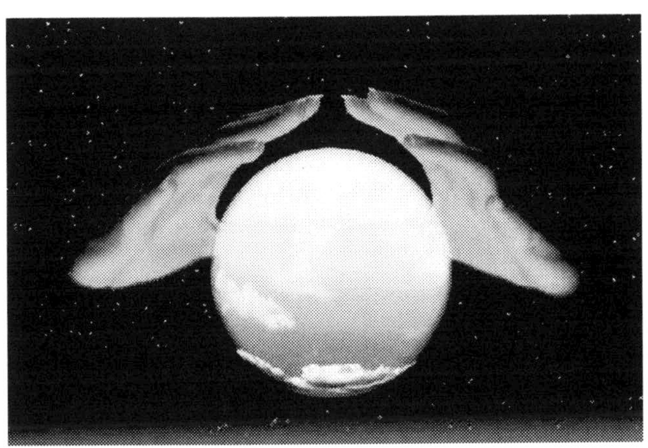

© 2008 Managed Objects Inc.

OUTLINE

In this closing chapter, we will take a bird's eye view of some of the interesting new ideas currently in development that illustrate the diversity of applications that will be enabled by stem cell technology in the coming few years.

The first decade or so of research on stem cells since the first isolation of embryonic stem cells in 1998 revolutionized the way we think about regenerative medicine: the restoration of tissue and organ function by replacing its cellular components, not simply looking for drugs that heal. For cell biology, this has been a little like the race to the moon was for engineering and computer science in the 1960s (Figure 16.1). It has brought us to the eve of being able to treat a number of chronic, debilitating, and even fatal diseases in a permanent way. As we have discussed in this book, treatments for diabetes, age-related blindness, intestinal ulcers caused by Crohn's disease, graft-versus-host disease, massive bone and cartilage damage, and some demyelinating diseases in children (although probably not diseases such as multiple sclerosis in adults) may well be on the immediate horizon. Within a decade we will know from first clinical trials whether these diseases actually benefit from stem cell therapy. And we will learn so much from the new human disease models based on stem cells, that drugs will be available for diseases now chronic and untreatable.

What else do we expect for the coming decade? In such a rapidly developing field, this is, of course, difficult to predict. Who would have expected in 2007 that therapeutic cloning to generate patient-matched embryonic stem cells would become obsolete, and within just one year be replaced by the concept of induced pluripotent stem cells? Or that adult stem cells would turn out not only to be identifiable directly in tissues but expandable in culture and on the verge of clinical application? With such unexpected, not just rapid, advances how can we make any predictions for even the next couple of years, let alone a decade?

FIGURE 16.1 Stem cell research has been a little like the race to the moon.

In the case of pluripotent cells, three years after their generation in humans, it is clear that human-induced pluripotent stem cells are very similar to human embryonic cells, not only in terms of gene and protein expression but also imprinting. However, they might not quite be there yet in terms of therapy: their long-term stability during *in vitro* culture still requires some study. This is an issue high on the scientific agenda, and which we might expect to be solved in the coming few years, particularly because of huge investments in this area in Japan.

Since being awarded the Nobel Prize, Japanese scientist Shinya Yamanaka has changed the scientific horizon in Japan: whole institutes are devoted to research on induced pluripotent stem cells and much of the life science research budget is dedicated to keeping Japan at the forefront in this area, particularly with respect to stem cell therapy. Once we have human induced pluripotent stem cells as true equivalents of human embryonic stem cells, it is likely that "banks" of these cells will be made as a resource for transplantation to patients with different human leukocyte antigen (HLA) gene combinations. Japan will need a different bank for matching cells to their patients than China, and China a different bank from Europe or the United States, and so on. In addition, human-induced pluripotent stem cells may be banked from individuals of mixed racial backgrounds for whom tissue matches are the most difficult to find. This is already the case for umbilical cord banks. An Asian father and African mother, for example, is a difficult combination to match. There may also be banking of cells from so-called universal donors who provide a match for a wider range of individuals.

How would these banks be used? For cell therapy, a clinician would have the HLA type of, say, a diabetic patient determined, but not know how to make the pancreas cells needed for treatment. This will remain a specialist activity. It is conceivable that specialist facilities will select the right HLA-matched cell line and produce properly differentiated cells in the required quantity under standardized, good medical practice (GMP) conditions, which could perhaps be even fully automated using robotics. These differentiated cells would be delivered live to the hospital for administration to the patient. This would keep costs to a minimum, one of the major obstacles that need to be overcome before introducing cell therapies as part of standard patient care. Costs are the major reason that most scientists do not expect human-induced pluripotent stem cell transplantation to be fully "individualized," meaning that it will probably not be possible to generate a personal cell line for each patient that needs cell transplantation within a reasonable budget, even if, in principle, such lines could be produced. Hundreds of thousands of dollars as expected costs for the production of just one cell line under GMP conditions and the same amount again to produce the differentiated cells required for transplantation would be prohibitive for any health insurance plan and most individual budgets.

Aside from use as a resource for transplantation, banks of induced pluripotent stem cells are likely to have their greatest potential in modeling diseases so that we learn more about what causes the disease and how to treat it. In addition, in combination with whole genome sequencing and knowledge of the medical history and drug responses of patients from whom they are derived, it may well be possible to predict who is likely to get the disease (disease *predisposition*) and whether there are any lifestyle factors that influence this probability. Prevention is always better than cure. Banks of induce pluripotent stem cells are already being established for this purpose in Japan, the United States (at the National Institutes of Health, the New York Stem Cell Foundation, and the California Institute for Regenerative Medicine), and the United Kingdom (by the Wellcome Trust). More will possibly follow. This will likely become an amazing source of information on health, disease, and longevity.

Transplantation with progenitor cells from postnatal tissues (adult stem cells) is likely to reach the clinic earlier than with pluripotent stem cells because the risk of forming tumors is lower. Questions of producing cells efficiently, at reasonable costs, and in sufficient numbers are issues for which there is partial overlap with pluripotent stem cells, although adult stem cells in general require less manipulation in culture. The two areas will, nevertheless, likely benefit from each other's expertise. Of the progenitor cells most likely to be in use for therapy soon, neural progenitors are high on the list. There are already a number of life-threatening neural diseases in children for which permits to test cell-based treatments have been given, although the disease in question (Batten's disease) is so rare that finding enough patients to include in the trial is a problem. It seems that neural stem cells have the ability to form derivatives that can form myelin and migrate throughout the brain and central nervous system in children, so that even demyelinating diseases in these young patients could be treated.

Parkinson's disease is also a neural degenerative disease that has seen real advances in the last few years in producing the right nerve cells for transplantation; the start of a trial is on track for 2017 in New York, and other sites may follow as they learn from the experience there. Also making great advances as treatments in mice and now looking very promising is the use of the new stem cell types recently identified in the intestine. Ulcers healed with these intestinal stem cells and first steps toward making this applicable to humans are underway.

Regarding stem cells in other tissue types—the mammary gland, skin, hair, sweat glands, sebaceous glands, pancreas, liver, and so on—we will know in the coming years whether these cells can be expanded from small numbers of cells in humans, either in the laboratory or *in situ* in the patients' own tissues and organs. And we will learn whether they have a clinical value in treating diseases of these organs and tissues. We will also learn whether the combination of gene therapy and stem or progenitor cells will work as a safe therapy: repairing the patient's defective gene and using his own repaired cells to treat the disease. We described one example of a serious skin disease treated by a combination of gene and stem cell therapy (Box 7.3, "New Skin for Claudio"). This approach is likely to expand in the coming years to be applied to more diseases, particularly as the methods for introducing genes into cells become safer and more precise so that other important genes that control cell behavior are not affected.

This sums up what we do expect from stem cell therapy in the coming decade, and at least some of the questions for which we expect an answer. It is important to realize that most new therapies of any sort (e.g., bone marrow transplantation to treat blood diseases, or monoclonal antibodies against cancer), in general, take 30 years or more to become clinically useful and routine practice. Stem cell therapy in its recent form is, in this respect, still very young.

Another question could be: what do we *not* expect of stem cells in therapy in the coming decade? This type of question is always most difficult to speculate on, because scientific research often surprises us, but it is worth mentioning a few of the more challenging diseases where expectations for stem cell therapy are lower: Alzheimer's disease, multiple sclerosis in adults, atherosclerotic heart disease, and stroke. Reasons include challenges of cell delivery to the appropriate location in the body, inability to produce the required cell types, and the nature of the tissue destroyed.

Beyond direct cell therapy, we do have the exciting prospect of using stem cells (both from diseased and healthy individuals) for drug discovery: if we can model disease in a culture dish in the laboratory we may be able to discover novel drugs, as necessary in combination, to slow down the rate at which a disease develops or ameliorate the symptoms. This would ultimately also be of huge benefit to patients, particularly those with chronic progressive disease, since it may extend the period in which they live in relatively good health, prevent or slow down development of their most debilitating symptoms, or even reverse them after their onset.

16.1

WILL HUMAN STEM CELL BASED MODEL SYSTEMS REPLACE ANIMAL EXPERIMENTS?

Society increasingly voices reluctance to experiment on animals if there are alternatives, such as those based on cultured cells. Pluripotent and multipotent human stem cells offer opportunities to develop model systems for tissues very similar to those in the human body. The more these human tissues "in a dish" resemble bona fide human tissue, the greater the chance that they will be able to contribute to the "3Rs" in animal experiments (replacement, reduction, and refinement) and become highly predictive models for drug discovery and safety pharmacology.

As an example, tissue engineers in France and the United States synthesized model systems based on stem cells for skin which can be used in combination with inflammatory cells as alternatives to rabbit experiments to test chemical substances for skin irritation.

16.1 COMBINING TECHNOLOGIES: NEW HUMAN DISEASE MODELS FOR DRUG DISCOVERY

Many diseases are typically human, so there may be no appropriate cell or animal models the pharmaceutical industry could use for developing and testing new drugs. With the rapidly developing technologies to culture and differentiate human pluripotent stem cells in a controlled manner, human disease models come within reach of pharma companies, eager to incorporate them into their drug development program for discovery and development of new drugs. Often, however, cultured cells grown on standard tissue culture plastic, even if they are human, show very different responses to drugs than cells in a three-dimensional tissue composed of multiple cell types attached to a matrix protein scaffold. Tissue engineering, physics, microfluidics (technology to manipulate fluid flow in microchannels), nanotechnology, advanced microscopy, and microfabrication (organs-on-chips, the technology used to make "chips," electronic integrated circuits) are all techniques that will contribute to creating human tissue "look-alikes" in the laboratory in the coming decade. This creates unprecedented opportunities for development of unique human disease models.

Tissue engineers in Germany, the United States, Canada, and the Netherlands all have research groups which are very actively pursuing solutions to manufacturing human heart and vascular tissue. This includes not simply human heart cells derived from stem cells attached to the commonly used plastic surface, but mixtures of different heart cells layered and precisely aligned in patterned strips. In some cases they are attached to flexible substrates that can undergo cyclic stretch and strain to mimic the mechanical environment of the beating heart, and even the way that it undergoes stress during physical exercise.

This assay system, and others like it for liver, kidney, blood vessels, bone, and cartilage, for example, will likely be the new drug-testing systems in the coming years. The Wyss Institute at Harvard University in Boston and researchers at Stanford University in the United States and in Toronto, Canada are also making major investments and advances in these areas. In addition, model systems based on stem cells to investigate how drugs pass the blood-brain barrier so that they can access the brain as their therapeutic site of action, are already under development. These exciting approaches could in the future be combined with second generation model systems based on human-induced pluripotent stem cells, for example, from patients with relatively common and typically human diseases with a complex genetic background such as atherosclerosis and diabetes type 2, but also, for example, amyotrophic lateral sclerosis (ALS) and some forms of Parkinson's disease. Induced pluripotent stem cell based model systems are expected to provide an entirely new range of opportunities to understand how these human diseases develop and progress, and open a new era for drug target discovery and drug development.

16.2

WHAT COULD STEM CELLS MEAN FOR FUTURE HEALTH?

A question often asked of stem cell researchers is will all of the research on stem cells lead to immortality? Will we live forever? The consensus is probably not. If stem cell research is as successful as we hope, it may be possible to repair vital organs repeatedly throughout a lifetime and, through knowledge of stem cell behavior, prevent damage from disease happening so quickly. The aim would be to prolong the period of health so that "healthy aging" allows people to spend more of their lives contributing to society rather than depending on it.

One can make an analogy with cars in Cuba. There were many huge American Chevrolets, Plymouths, and Chryslers imported during

16.2 *(cont'd)*

the heydays of the 1940s and 1950s. These cars still drive, the chassis is often in top condition, but what about the engine? The original engine is long gone and usually replaced by another from Toyota, Kia, or Lada. Brilliant local engineers replace the original engine with a completely different brand, but it works fine. That is how we see stem cell and regenerative medicine at its best: the same old chassis, just new parts and a new motor! (See Figure B16.2.1).

FIGURE B16.2.1 Regenerative medicine and the use of stem cells can be likened to a car in Cuba: old chassis, new engine. This classic model from 1949 is running perfectly well on a new Korean engine. *Source: Kelly Hosman.*

16.2 PERSONALIZED MEDICINE AND SAFER DRUGS

Many drugs have more or less serious side effects. However, not all patients receiving a drug for a specific disease will experience these side effects; in fact they may be very rare. Likewise, drugs designed to treat specific diseases do not always benefit all patients. The reasons for this are often unclear, but there may be a genetic contribution, determined by the patient's own genetic makeup and predisposition. Drug prescription as a result is often necessarily empirical: "Try this drug first and if it does not work, we'll prescribe another." Likewise for side effects: "Let's try another one and see whether that one is tolerated better."

It is currently not possible for pharmaceutical companies to determine individual risk of a side effect prior to starting clinical trials, since no model systems for screening toxicity of drugs are available that enable investigation of the relationship between side effects and the particular variations in the genome of the patient. However, the current

FIGURE 16.2 A lot of fundamental and applied research is still necessary to make full use of the potential of stem cells in health.

trend in development of novel drugs is based on the concept of "personalized treatment." This means that in the future, the decision to treat a patient with a specific drug will become more dependent on whether the individual patient is expected to respond favorably to that drug. Toward this aim, a patient may undergo a special diagnostic test, called a companion diagnostic assay, to predict their response to the drug prior to starting therapy. Tissue and disease model systems based on human stem cells may help pharmaceutical companies to develop these companion diagnostic assays prior to starting clinical trials involving real patients. One could imagine building a collection (a library or bank) of genetically different human stem cell lines, which, as a whole, could be representative for the entire human population, and use this as an *in vitro* laboratory-based clinical trial. Cardiotoxicity (the side effects of a drug on the heart) can be used as an example to illustrate this concept. A side effect of many drug compounds is that they can block ion channels in heart cells, which increases the risk of life-threatening arrhythmias. Arrhythmias usually only occur in a very few patients and may be due to their specific genetic predisposition. In its simplest form, such a cell-line bank could be composed of genetically identical pluripotent cell lines in which only the genes coding for the proteins of the ion channels have been changed to represent the genetic variants found in the general population. The modified cell lines could be differentiated to heart cells and used to screen for toxic side effects of a drug compound on the heart. If an association with a specific ion channel gene variant were found, a simple companion diagnostic test could be developed to identify patients at risk, prior to administering the drug. This is already becoming a reality. Similar banks could be developed for a variety of cells and tissues to predict predisposition to toxic effects of drugs on liver, brain, blood, nerve; in fact any cell type that can be made from a pluripotent cell (Figure 16.2).

16.3 FINAL NOTE

While regretting much of the hype and false hope surrounding stem cells and acknowledging the challenges ahead, many scientists share the conviction that human stem cell technology and greater understanding of stem cell biology will, in the coming decades, fulfill much of its promise and revolutionize medicine as we know it today, improving patient care without past precedent.

16.3

HUMAN ORGANS IN ANIMALS?

One possible new application of stem cells has arisen from research published in . . .which demonstrated that stem cells could be turned into whole organs if grown inside the body of an animal. If this worked with human stem cells, these cultured organs could in principle be collected from the animals and transplanted into human patients.

For tissues that consist of one or only a few cell types it may be feasible to develop these outside the body from stem cells just simply in laboratory culture and use them for organ repair, but organs have much more complex structures and are composed of various cell types that need to be at the correct position for the organ to function. To create these outside the body would be extremely challenging. What scientists find virtually impossible in the laboratory, however, takes place routinely in developing embryos: embryos make organized tissues and organs all the time. For this reason, scientists are now following up the first published study using animal stem cells and exploring the possibilities of using animals for the development of human organs from human stem cells.

This is how it works: when pluripotent mouse stem cells are injected into a blastocyst stage mouse embryo, the cells can contribute to the formation of the developing fetus. If the embryo is transferred to the uterus, or womb, of a female mouse, a baby mouse will be born that is composed of two different cell types: cells derived from the original embryo and cells derived from the pluripotent stem cells. Such an animal is called a chimera, a word that originates from Greek mythology were the Chimera refers to a ferocious animal that is partly lion, partly goat and with a serpent as tail (see also Chapter 3).

Japanese researchers then took this a step further. They injected rat pluripotent stem cells into a mouse blastocyst that was genetically modified in such a way that it could not produce a pancreas.

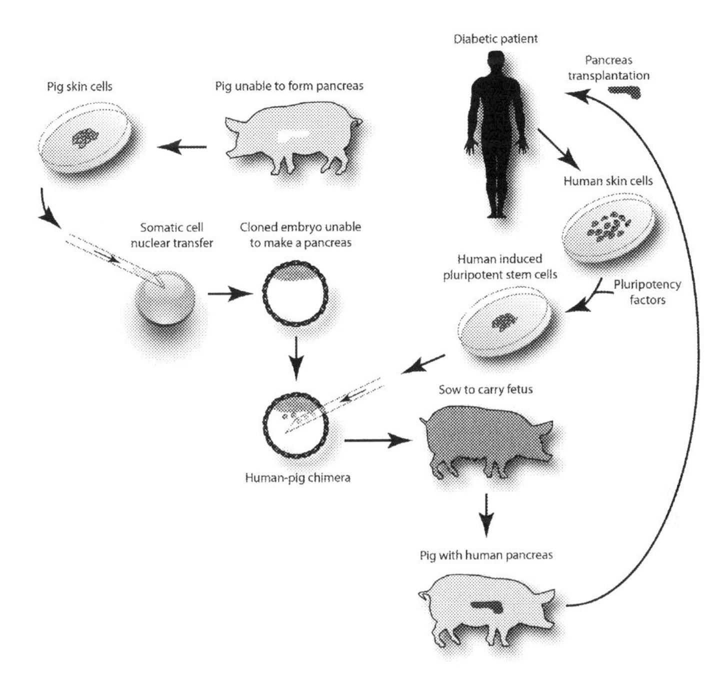

FIGURE B16.3.1 Skin cells from a pig genetically unable to form a pancreas are used for cloning by somatic cell nuclear transfer to make an embryo that is therefore unable to form a pancreas. Skin cells from a diabetic patient are used to make patient-specific induced pluripotent stem cells. These cells are injected into the pig embryo, where they are able to contribute to the formation of the fetus. The embryo is transferred into a carrier sow, in which the fetus will develop. Since the fetus cannot form a pancreas, the human cells will form this organ: a human organ in a pig. When the pig is born, it will have a human pancreas that is genetically identical to the patient and can be used for transplantation.

The resulting embryos were transferred to the womb of a recipient female mouse, which carried the embryos and fetuses to term. The resulting animal was predominantly of mouse origin, with some rat cells in various organs, but the pancreas was entirely of rat origin, so completely formed from the injected rat stem cells. Similar experiments have resulted in mice with a rat thymus and pigs from one strain have been generated with a pancreas completely formed from a different pig strain. These experiments open the possibilities of more extreme engineering: could it be possible to develop and grow a human organ, say a pancreas, in a large animal, say a pig? For this one would need an embryo of a pig that cannot form a pancreas from its own cells, and

16.3 (*cont'd*)

human pluripotent cells. Cells of a pig that is genetically unable to form a pancreas could be used for cloning by somatic cell nuclear transfer (see Chapter 6), resulting in pig embryos that because of the genetic engineering cannot form a pancreas. Simultaneously, skin cells could be derived from, say, a diabetic patient, with defective pancreas function, and turned into induced pluripotent stem cells. These human stem cells could be injected into the pig embryo, and then the embryo transferred to a sow to that it could develop to term. The resulting piglet would have a human pancreas that could be used for transplantation to the specific diabetic patient from which the stem cells were derived. Since the pancreas is made of cells that are genetically identical to the patient, the pancreas would be recognized as self by the immune system and not be rejected. One of the issues that would need to be solved if the whole pancreas were transplanted would be how to make the blood vessels be of human origin. If only the islets containing the insulin secreting beta-cells were transplanted to the patient though, this would not be a problem. For other organs, like kidneys, it would of course. Of note though, chimeras can actually occur in humans, albeit rarely, but in this case one individual originates from two individually fertilized eggs which fuse early in development, before implantation into the womb. These two embryos would normally have resulted in fraternal twins, but because they stick together, just one baby is born. When two "brothers" or two "sister" embryos fuse, this is usually not noticed throughout the entire life of the person and might only be discovered when they need medical treatment like a bone marrow transplant.

Since early embryos are sticky, embryos of two different species can even be fused artificially, something that would obviously never happen in nature. This has for instance been done with sheep and goat embryos. The resulting animals had characteristics of both sheep and goat.

Currently scientists, policy makers, and ethicists are discussing whether such an approach would be beneficial for the community and whether it is ethically and morally justifiable. Should we allow the breeding of animals with human organs? How much can this be controlled, or would this result in animals that are "humanized"? What happens when for instance a pig develops with human brain cells? Or how can we exclude that such tissues are formed? Is it acceptable to "instrumentalise" the use of animals?

All questions that need to be addressed before human organs in animals become reality even if the procedure is feasible.

Glossary

Adult Stem Cells Stem cells present in a tissue or organ after birth; for example, blood stem cells from the bone marrow or intestinal stem cells in the crypts of the gut. This type of stem cell is usually multipotent, not pluripotent.

Allogenic Foreign to the own body; in the case of an allogenic transplantation, the transplanted tissues or cells are genetically different from the recipient.

Autologous In the case of autologous transplantation, the transplanted tissues or cells are derived from the recipient.

Bioreactor A device used to scale up biological processes. It can be applied to stem cell culture, for example, to obtain sufficient cell numbers for transplantation or use in biotechnology applications.

Blastema A population of (mesenchymal) stem cells that form, for example, at the amputation site of a salamander limb. A healthy new limb can regenerate from the cut surface.

Blastocyst The early embryo of a placental mammal, usually at the stage just before it nestles in the uterus for further development. It consists of an outer trophectoderm layer and an inner population of cells from which the fetus derives, called the inner cell mass.

Blastomere Cells from an early cleavage stage embryo. In some species, they may be pluripotent.

Blood Stem Cells Multipotent stem cells that are present in the bone marrow, umbilical cord blood, or placenta that give rise to all the different types of blood cells. They are usually not present in the blood stream unless "mobilized" from the bone marrow by specific growth factors, which may be induced by tissue injury (as in a heart attack) or administered as part of therapy.

Bone Marrow Transplantation Transplantation with cells obtained from the bone marrow. Under anesthesia, healthy bone marrow is aspirated from the pelvic bone of the donor and administered intravenously to a recipient patient as therapy. The bone marrow cells find their own way (or "home") to the recipient's bone marrow.

Cardiomyocyte Heart muscle cell that contracts synchronously with other heart muscle cells in the beating heart.

Cell Line A population of cells maintained in a laboratory that can divide indefinitely without changing. For example, some cancer cells, obtained from tumors, can be cultured as a cell line in the laboratory and give rise to very large cell numbers. Pluripotent stem cells that can be cultured as a cell line are unique in that they are "normal" cells and not cancer cells or "transformed."

Chimera Organism made up of cells from more than one individual. For example, a chimeric mouse can result from insertion of embryonic stem cells into a blastocyst-stage embryo that is placed into the uterus of a surrogate mother mouse for further development. The mouse that is born contains cells from the blastocyst and from the embryonic stem cells. The definition is sometimes used broadly to include organisms in which cells from another individual have been introduced at later stages of fetal development or in an adult.

Chimerism in the Bone Marrow Usually refers to blood stem cells from a transplant donor being introduced into the bone marrow recipient prior to an allogenic cell or organ transplantation. This form of chimerism may make recipients "immunotolerant" to other cells and tissues from the same donor and prevent transplant rejection.

Cleavage Division Cell division that is not associated with cell growth. Each cell divides in two and becomes half of its original size. The first cell divisions in a mammalian embryo are of this type.

Clinical Trial Clinical research in which a new treatment is tested in a group of patients. If the clinical trial is "randomized and double-blind," the patients are divided randomly into two groups, one group receiving treatment, the other not. The trial is "blind" if neither patient nor clinician administering the treatment knows if the patient is in the control or experimental group.

Clone An individual genetically identical to another in which there is a common ancestor or one is derived from the other. An example is twins originating from one egg.

Cloning *See* Reproductive Cloning and Therapeutic Cloning.

Collagen An extracellular matrix protein in connective tissue to which cells attach easily. It can be used in the laboratory as a surface on which to grow cells, often in its denatured form called gelatin.

Confocal Laser Scanning Microscope (CLSM) A microscope able to make "optical sections" of a three-dimensional specimen using a laser to focus on one optical plane.

Cord Blood Also known as *umbilical cord blood,* the blood in the umbilical cord and placenta that supplies nutrients from the mother to the developing fetus. It contains a relatively large number of blood stem cells and can be stored after birth for research or use in transplantation.

Cultured Meat Also known as *in vitro meat.* Edible muscle tissue made from *in vitro* cultured stem cells from cow or pig that have been differentiated to skeletal muscle tissue.

Culture Medium Fluid used to culture cells in the laboratory containing the nutrients they require to divide.

Cytoplasm The fluid in a cell, surrounded by the outer cell membrane (*plasma membrane*). The nucleus and cell organelles are located in the cytoplasm.

Differentiation A process by which a cell changes from one type to another, losing its precursor properties and acquiring new characteristics of a specialized (tissue-specific) derivative. A stem cell is said, for example, to differentiate into a cardiomyocyte, nerve cell, or pancreas cell.

Diploid Having two sets of chromosomes, one from each parent.

Ectoderm One of the three cell lineages in the early embryo. The ectoderm gives rise to the outer cell layers of the embryo, forming, for example, the skin and nerve cells.

Electrophysiology Technique for measuring electrical currents, voltages, and resistance over the cell plasma membrane that results from ion fluxes.

Embryo Early stage of development, after fusion of egg and sperm.

Embryoid Body Structure formed after aggregation of pluripotent stem cells. Tight contact between the cells results in outer and inner cell layers that differentiate differently. The outer layer becomes epithelial endoderm, while the inner cells differentiate to cells such as ectoderm, rather like the early differentiation steps that take place in an embryo during *gastrulation*. Early mesoderm differentiation also takes place within an embryoid body.

Embryonic Germ (EG) Cells Stem cells derived from primordial germ cells. These cells are pluripotent and behave rather similarly in culture to embryonic stem cells, although they do not form germ cells in chimeric mice.

Embryonic Stem (ES) Cells Pluripotent stem cells derived from the inner cell mass of a blastocyst-stage embryo and grown in culture. Under the right conditions, ES cells can be cultured as a cell line (an embryonic stem cell line), which is immortal.

Endoderm One of the three cell lineages in the early embryo. Endoderm gives rise to the innermost cells of the fetus: the gastrointestinal tract (e.g., intestine, liver, and pancreas) and the lungs.

Epigenetic Regulation A change in gene function in the absence of a nucleotide change (*mutation*). During embryonic development and differentiation, genes are increasingly blocked for use in the cell by epigenetic changes, leaving only the genes that are required for the specialized functions of a differentiated cell available for use. This is the reason that cells cannot easily "de-differentiate" and change into another cell type. The epigenetic regulation may be disturbed after artificial reprogramming, for example, during the formation of induced pluripotent stem cells and cloned animals.

Epithelium, Epithelial Cells Epithelial cell layers cover the body (skin) and line the body cavities, such as the intestinal lumen and lungs.

Ex Vivo Outside the body, for example, in the laboratory

Extra-Embryonic Tissue Part of the conceptus that will not contribute to the embryo proper but, for instance, gives rise to the embryonic part of the placenta, the umbilical cord, and the membranes surrounding the fetus.

Feeder Cell Layer A cell layer that "feeds" other cells. Often used in the context of stem cells that may be dependent on feeder cells to remain undifferentiated. They are sometimes treated to stop cell division so that they do not overgrow stem cells but can still deliver the right factors for their growth.

Fetus Unborn offspring of certain mammals, usually in the later stages of development.

Flow Cytometer Also known as *fluorescence activated cell sorter (FACS)*, a device that can count (and sometimes also sort or select) fluorescently labeled cells in a cell suspension.

Gamete Sex cell, either sperm or egg. After full maturation these cells contain half the number of chromosomes for a new individual and are therefore termed *haploid*.

Gene A DNA sequence in a chromosome that contains the genetic code for a protein and that functions as a template to produce mRNA.

Gene Therapy Introducing a genetic change in the genome in cells to treat or cure hereditary disease.

Genome The total amount of DNA in a living organism. In humans, this is contained in 23 chromosomes and is made up of around 3 billion base pairs encoding ~23,000 genes.

Genotype The genetic makeup of an individual, with all the individual variations in the genome.

Germ Cell Gamete or sex cell.

Germ Layers The three cell lineages that originate from the blastocyst: ectoderm, endoderm, mesoderm.

Graft-versus-Host Reaction/Graft-versus-Host Disease (GvHD) Immune-mediated rejection reaction of white blood cells in a bone marrow transplant from a not HLA-identical donor, directed against the host cells. This can lead to serious organ damage and even death if it develops into GvHD.

Granulocyte-Macrophage-Colony Stimulating Factor (GM-CSF) A naturally occurring growth factor that is secreted as a result of stress or tissue damage or is administered intravenously to mobilize blood stem cells from the bone marrow into the blood.

Haploid Containing one set of chromosomes.

Heterozygous Containing different copies of the chromosome pair. A person is usually referred to as heterozygous for a mutation or DNA variant if the copy (allele) on the other chromosome is normal.

Homozygous Containing the same copy of the chromosome pair. A person is homozygous for a mutation or DNA variant if it is present on both copies (alleles) of a gene.

Immune Response The immune reaction against an invader (e.g., a pathogen or foreign (transplanted) cells) that results in its removal or destruction.

Immunofluorescence Usually used to demonstrate the presence of a protein in or on a cell; a fluorescently labeled antibody binds to another antibody that in turn recognizes a specific protein, usually in a tissue or cell. In the dark, the fluorescent label can be excited by light of a particular wavelength and the resulting fluorescent signal can be seen in a microscope. This shows where the labeled protein is and how much of it is present.

Immunoprivileged Sites Locations in the body that are relatively well protected against attack by the immune system, such as brain, central nervous system, eye, and testicle.

Imprinting A form of epigenetic regulation where one of the two gene copies (alleles) are permanently inactivated. An example is the genes on the inactivated X chromosome in women.

Induced Pluripotent Stem Cells A method of "reprogramming" adult cells to a pluripotent state so that they can divide indefinitely and form all cells of the body. They are often referred to as *iPS cells*. The discovery of the reprogramming method resulted in a shared Nobel Prize for Shinya Yamanaka in 2012.

Inner Cell Mass The inner group of pluripotent cells in a blastocyst-stage embryo.

In Vitro In the laboratory, outside the body.

***In Vitro* Fertilization** Artificial fertilization of an egg in the laboratory.

Karyotype The chromosomes in a nucleus, visualized by a specific staining.

Knock-out Mouse A mouse in which a specific gene is inactivated by deletion of part of its DNA usually using homologous recombination in embryonic stem cells.

Leukemia Inhibitory Factor (LIF) A blood cell protein originally identified as essential for keeping embryonic stem cells undifferentiated.

Mass Spectrometry A protein analysis technology where proteins are "cut" into peptides and their amino acid sequence determined. By comparing this information with a protein database, the original protein can be identified.

Maternal Originating from the mother. A normal human nucleus contains 2×23 chromosomes, of which one set comes from the mother and one from the father (paternal).

Matrix A structure or scaffold on which cells can grow in two or three dimensions.

Meiosis Reduction division, a type of cell division that results in a reduction (usually by half) in chromosome number.

Mesenchymal Stem (or Stromal) Cells (MSC) Multipotent stem cells derived by culturing various tissue of the body in the laboratory on plastic. MSCs can easily be derived from bone marrow and fat tissue.

Mesoderm One of the three germ layers in the developing embryo, forming tissues such as bone, muscle, and fat.

Methylation of DNA Cytosine nucleotides (when followed by a guanine) can be methylated by adding a methyl group. This pattern is transferred to daughter cells after cell division.

Microarray Technology Method to analyze expression of many genes simultaneously in the same cell sample. Short single-stranded DNA molecules with known sequences are attached to a "chip" or glass slide in an array (or matrix) format. Fluorescently labeled DNA fragments or RNA molecules can be added and will bind specifically to the complementary sequence. The fluorescent signal of the bound DNA or RNA is detected and quantified with specialized software.

Migration of Cells Cells can move to different locations individually or as group, actively in response to attracting or propelling signals, or passively during growth. Considerable cell migration takes place during embryonic development and during formation of metastases in cancer.

Mitochondrial DNA Circular DNA present in mitochondria that codes for about 10 genes and plays an important role in cell metabolism. The "powerhouse" of a cell.

Monogenetic Disease A disease caused by a DNA mutation in a single gene.

Morula A very early mammalian embryo ($\sim$2.5 days after fertilization in mice, 3–4 days in humans) just before the blastocyst forms; a small clump of cells, not bigger than the period at the end of this sentence.

mRNA Messenger RNA, the molecule carrying the code for the series of amino that make up a protein; transfers the information to the ribosomes in the cytoplasm.

Multipotent Capacity of a cell to differentiate to multiple cell types.

Mutation A change in one or more nucleotides in the DNA is called a variation or polymorphism if it is relatively frequent (and therefore not associated with serious disease) or a mutation if it is very rare; this can be associated with serious disease.

Myocardial Infarction or Heart Attack Lack of oxygen to parts of the heart cause heart cells to die. This may be life-threatening due to acute loss of heart function, and caused by block of a cardiac blood vessel (e.g., a blood clot).

Niche A small space. Undifferentiated stem cell populations (progenitor cells) reside in many organs in a stem cell niche, where specialized factors are locally active in maintaining the cells in a healthy but quiescent state until needed.

Nuclear Transfer Transfer of a nucleus, or complete cell, from one cell to the other. This technique is used for cloning and the recipient cell is then an egg cell.

Omnipotent *See* Totipotent.

Organ or Tissue Stem Cells In an adult, most organs and tissues contain a small number of stem cells, which are often quiescent but are capable of dividing when needed (for repair or replacement) and developing into the cell types that are normally present in the organ of tissue they are multipotent or unipotent.

Organoid Small, three-dimensional organ-like structure formed ("self-assembled") by cultured adult stem cells that can give rise to all different cell types of that organ; organoids can mimic some of the organ's function

Organ-on-a-Chip Models Technology to microengineer the smallest functional organ or tissue modules in the laboratory on a "chip," where the chip enables recreation of the cellular/tissue environment, very much like *in vivo*. Their envisioned use is to develop organ and disease models for drug development.

Osteoarthritis Breakdown of cartilage particularly in joints such as knees and hips. It can be caused by aging ("wear and tear") but can also be hereditary or arise from metabolic disorders. The main symptoms are pain and stiffness of the joints.

Paneth Cell A highly specialized cell type that is part of the epithelial layer of the intestine.

Parthenogenesis An unfertilized egg is artificially stimulated to start dividing (by an electrical or chemical stimulus) just like a normal embryo. It can often develop to the blastocyst stage but not further.

Passage Refers to cell culture. When cells have filled a culture flask they need to be transferred (passaged) to new culture flasks or dishes. This usually happens twice to three times per week for tumor cells and cell lines, less for primary (nontransformed) cells.

Paternal From the father. (*See also* Maternal.)

Placebo Pharmaceutical drug or treatment that by itself is ineffective. It is used as control drug or treatment to test efficacy of real drugs and treatments. Because of the subjective perception of treatment effects, the condition of the patient may improve. This is known as a *placebo effect*.

Pluripotent Capacity to form all cell types present in an organism.

Pluritest A microarray-based method for gene expression profiling, used to assess whether a stem cell is pluripotent. It is expected that pluritest will replace teratoma formation in mice as an assay of pluripotency.

Polymorphism *See* Mutation.

Prenatal Diagnostics Genetic analysis of an early embryo. In one variant, one cell is removed from an 8-cell stage embryo to perform a diagnostic test to exclude a genetic disease or determine the gender of an embryo. The embryo with seven remaining cells can continue to develop in the laboratory while the diagnostic test is done and, if healthy, be transferred to the womb of the mother to develop into a baby. In a second variant, a small sample of extra-embryonic tissue is collected from the conceptus (from the chorion) or from the amniotic fluid surround the fetus (amniocentesis) and also used for genetic diagnosis.

Primordial Germ Cells Progenitor cells of the gametes (sex cells). These cells are around the first to be formed in the developing embryo. They later travel to the gonads where they differentiate into either sperm or egg.

Progenitor Cells Organ stem cells that still have the capacity to divide but have restricted differentiation capacity.

Protein Polymer of amino acids.

Recessive A gene or allele is inherited in a recessive manner when it does not cause a phenotype (e.g., a disease) when it is present only once: heterozygous.

Reproductive Cloning Creating an individual that is genetically identical to an individual already existing.

Reprogramming Modifying the structure in the nucleus in such a way that it is capable of choreographing the development of a new individual. The cytoplasm of an egg cell contains several factors that can induce reprogramming of the nucleus during cloning. Some *pluripotent transcription factors* also induce reprogramming of adult cells during the formation of induced pluripotent stem cells.

Satellite Cell Stem cells present in adult muscle tissue that aid in regeneration of muscle after injury.

Self-Renewal Stem cells divide without differentiating thus maintaining and expanding the stem cell population.

Signal Transduction Cells communicate with each other via direct contact or through signal molecules (e.g., hormones) that recognize specific receptors on the cell membrane or in the cell itself to convey a message to the cell to change a specific function: signal transduction.

Somatic Cell Nuclear Transfer Technique with which the nucleus of an egg cell is replaced by the nucleus from a donor cell to produce a cloned embryo.

Spermatogonial Stem Cells Unipotent progenitor cells that can still divide and differentiate into sperm cells.

Stem Cell Cell that can self-renew and differentiate into one or more cell types.

Stem Cell Line Stem cells that can divide indefinitely in the laboratory.

Stem Cell Tourism Term used to refer to the growing practice of searching for stem cell treatments in foreign countries. There is no evidence that any of these treatments work.

Superficial Digital Flexor Tendon (SDFT) A band of fibrous connective tissue that connects muscle and bone (tendon) to the back of a horse's leg. This particular tendon flexes the joint.

Tendon Fibrous tissue that connects muscles to bones.

Teratocarcinoma A (benign) teratoma that also contains undifferentiated stem cells and is therefore malignant.

Teratoma A benign (nonmalignant) tumor usually of the testis in humans characterized by the presence of many different cell and tissue types (hair, teeth, muscle). It is thought to develop from incorrectly programmed or mislocated primordial germ cells. Teratomas can also be artificially induced in mice by introducing pluripotent cells or early embryos at nonuterine sites.

Therapeutic Cloning Technology whereby somatic cell nuclear transfer is used to create a cloned blastocyst from which embryonic stem cells can be derived. If the donor nucleus is from a patient, the embryonic stem cells derived from the cloned embryo would be identical to the donor and not rejected by the immune system after transplantation. It has not been successful in humans and was effectively replaced by direct reprogramming; for example, through the derivation of induced pluripotent stem cells.

Tissue Engineering A discipline to construct or engineer organs or parts of organs from cell and connective tissue components to treat disease or trauma or generate disease model systems.

Totipotent Capacity to form all cell types in the adult body as well as the extraembryonic tissues, such as placenta, umbilical cord, and fetal membranes.

Transdifferentiation Concept that a cell, originating from one germ layer, can change into a cell type belonging to another germ layer.

Trophectoderm The outer layer of cells of the blastocyst-stage embryo. These cells do not become part of the developing fetus but contribute to the extraembryonic tissues, such as the embryonic part of the placenta.

Unipotent Capacity to divide and form one cell type.

United States Department of Agriculture (USDA) Ministry of agriculture in the United States that develops and organizes policy on farming, agriculture, forestry, natural resources, and food.

United States Food and Drug Administration (FDA) The federal agency that monitors the safety and quality of food, food additives, and pharmaceutical drugs. In addition, it regulates and supervises, for instance, blood transfusions, vaccines, medical devices, electromagnetic radiation emitting devices, and cosmetics.

Wnt Pronounced "wint." A group of proteins that can be secreted by cells and bind to receptors on the cell surface to instruct them what to do. (*See also* Signal Transduction.) Wnt proteins have been conserved during evolution, indicating their importance, and are present in animals ranging from fruit flies to man.

Zona Pellucida A noncellular protecting protein layer surrounding and protecting the egg and early embryo.

Index

Note: Page numbers followed by "*b*" and "*f*" refer to boxes and figures, respectively.

D

W

X

Y

Z

Photo Credits

Chapter 1

Figure 1.1 Stamcellen Veen Magazines.

Figure 1.2 Stamcellen Veen Magazines.

Figure 1.3 Stamcellen Veen Magazines.

Figure 1.4 Stamcellen Veen Magazines.

Figure 1.6 Anja van de Stolpe.

Figure 1.8 Hans Kristian Ploos van Amstel, University Medical Center Utrecht, Netherlands/Stamcellen Veen Magazines.

Figure 1.10 Stamcellen Veen Magazines.

Figure 1.11 Stamcellen Veen Magazines.

Figure 1.14 Bernard Roelen/Stamcellen Veen Magazines.

Figure 1.15 Stamcellen Veen Magazines.

Figure 1.17 Stamcellen Veen Magazines.

Figure B1.1.1 Stamcellen Veen Magazines.

Figure B1.1.2 Stamcellen Veen Magazines.

Figure B1.2.1 Stamcellen Veen Magazines.

Figure B1.2.2 Alfred Roelen/Stamcellen Veen Magazines.

Figure B1.3.1 Stamcellen Veen Magazines.

Chapter 2: *Stamcellen* Veen Magazines

Figure 2.1 (a) Hubrecht Institute, Netherlands/Stamcellen Veen Magazines; (b) Archive NWT, Netherlands/Stamcellen Veen Magazines.

Figure 2.2 Stamcellen Veen Magazines.

Figure 2.3 Henri van Luenen, Netherlands Cancer Institute, Amsterdam, Netherlands/Stamcellen Veen Magazines.

Figure 2.4 Stamcellen Veen Magazines.

Figure 2.5 Astrid van der Sar/Jeroen den Hertog, Hubrecht Institute, Netherlands/Stamcellen Veen Magazines.

Figure 2.6 Susana Chuva de Sousa Lopes/Stamcellen Veen Magazines.

Figure 2.7	Hubrecht Institute, Netherlands/Stamcellen Veen Magazines.
Figure 2.9	Jurriaan Hölzenspies.
Figure 2.10	Stamcellen Veen Magazines.
Figure 2.12	Bernard Roelen/Stamcellen Veen Magazines.
Figure 2.13	Bernard Roelen.
Figure 2.14	Susana Chuva de Sousa Lopes. Reprinted with permission from Lanza (Ed.), Handbook of Stem Cells, Elsevier, 2005/Stamcellen Veen Magazines.
Figure 2.15	Bernard Roelen/Stamcellen Veen Magazines.
Figure 2.16	Stamcellen Veen Magazines.
Figure 2.17	Ewart Kuijk/Leonie du Puy.
Figure 2.18	Susana Chuva de Sousa Lopes, Gurdon Institute, U.K./Stamcellen Veen Magazines.
Figure 2.19	Wolter Oosterhuis, Josephine Nefkens Institute Erasmus MC, Netherlands/Stamcellen Veen Magazines.
Figure B2.1.2	Sebastiaan Blankevoort, Leiden University Medical Center, Netherlands.

Chapter 3: Jeffrey de Gier/*Stamcellen* Veen Magazines

Figure 3.1	Joep Geraedts, Academic Hospital Maastricht, Netherlands/ Stamcellen Veen Magazines.
Figure 3.2	Stamcellen Veen Magazines.
Figure 3.3	Ester Tjin, Utrecht University, Netherlands/Stamcellen Veen Magazines.
Figure 3.5	Stamcellen Veen Magazines.
Figure B3.1.1	Susana Chuva de Sousa Lopes, Leiden University Medical Center, Netherlands.

Chapter 4: Bernard Roelen/*Stamcellen* Veen Magazines

Figure 4.1	Stamcellen Veen Magazines.
Figure 4.2	Leonie du Puy/Stamcellen Veen Magazines.
Figure 4.3	Stamcellen Veen Magazines.
Figure 4.4	Thorold Theunisen, Hubrecht Institute, Netherlands/Stamcellen Veen Magazines.
Figure 4.5	Stieneke van den Brink, Hubrecht Institute, Netherlands.
Figure 4.6	Stamcellen Veen Magazines.
Figure 4.7	Susana Chuva de Sousa Lopes/Stamcellen Veen Magazines.
Figure 4.8	Stamcellen Veen Magazines.

Figure 4.10 Dagmar Gutknecht, University Medical Center Utrecht, Netherlands.

Figure B4.2.1 Stamcellen Veen Magazines.

Figure B4.4.1 Stamcellen Veen Magazines.

Figure B4.5.1 © Martin Evans.

Figure B4.5.2 © Martin Evans.

Figure B4.6.1 Photograph courtesy of James Thomson.

Figure B4.7.1 Center for iPS Research and Application (CIRA), Kyoto, Japan.

Figure B4.7.2 Center for iPS Research and Application (CIRA), Kyoto, Japan.

Chapter 5: *Stamcellen* Veen Magazines

Figure 5.1 Rui Monteiro, Hubrecht Institute, Netherlands.

Figure 5.2 Stamcellen Veen Magazines.

Figure 5.3 Stamcellen Veen Magazines.

Figure 5.4 Stamcellen Veen Magazines.

Figure 5.5 Stamcellen Veen Magazines.

Figure 5.6 Stieneke van den Brink, Hubrecht Institute, Netherlands/Stamcellen Veen Magazines.

Figure 5.7 Stamcellen Veen Magazines.

Figure 5.8 Susana Chuva de Sousa Lopes, Gurdon Institute, U.K./Stamcellen Veen Magazines.

Figure 5.9 Susana Chuva de Sousa Lopes, Gurdon Institute, U.K./Stamcellen Veen Magazines.

Figure 5.10 Stamcellen Veen Magazines.

Figure 5.11 Karlijn Wilschut, Utrecht University, Netherlands.

Figure 5.13 Stamcellen Veen Magazines.

Figure 5.14 Reinier Raymakers, University Medical Center Nijmegen, Netherlands/Stamcellen Veen Magazines.

Figure 5.16 Susana Chuva de Sousa Lopes.

Figure B5.1.1 Alexander Damaschun, hESCreg, June 2010.

Figure B5.2.1 Katsuhiko Hayashi.

Chapter 6: Bernard Roelen

Figure 6.1 Stamcellen Veen Magazines.

Figure 6.6 Stamcellen Veen Magazines.

Figure 6.7 Marga van Rooijen, Hubrecht Institute, Netherlands/Stamcellen Veen Magazines.

Figure 6.8 Chris Goodfellow/Gladstone Institutes.

Figure B6.4.1 Roslin Institute, Royal (Dick) School of Veterinary Studies, University of Edinburgh, Edinburgh.

Figure B6.5.1 András Dinnyés.

Chapter 7: Bernard Roelen/*Stamcellen* Veen Magazines

Figure 7.1 Stamcellen Veen Magazines.

Figure 7.2 Stamcellen Veen Magazines.

Figure 7.8 Stamcellen Veen Magazines.

Figure 7.9 Leonie du Puy, Utrecht University.

Figure 7.10 Leonie du Puy, Utrecht University.

Figure 7.11 Stamcellen Veen Magazines.

Figure 7.12 Stamcellen Veen Magazines.

Figure 7.13 Lynne M. Ball, Departments of Pediatrics, Immunology and Hematology (Professor W. Fibbe), Leiden University Medical Center, Netherlands.

Figure 7.14 Adapted with permission from Moroni L., et al., J Biomater Sci Polym Ed 2008; 19(5):543–72.

Figure 7.15 Lorenzo Moroni and Clemens van Blitterswijk, Tissue Regeneration Department, University of Twente, Netherlands.

Figure 7.16 Stamcellen Veen Magazines.

Figure 7.17 Petra Dijkman, Anita Driessen-Mol, Carlijn Bouten, Frank Baaijens. Soft Tissue Biomechanics and Tissue Engineering, Department of Biomedical Engineering, Eindhoven University of Technology, Netherlands.

Figure B7.1.1 Melissa Temple-Smith.

Figure B7.1.2 Hospital Clinic of Barcelona, Spain.

Figure B7.2.1 Christine Baldeschi, I-Stem, Evry, France and Marcelo Del Rio, CIEMAT, Madrid, Spain. With thanks to Marc Peschanski, I-Stem, for input.

Figure B7.3.1 Michele De Luca, Centre for Regenerative Medicine "Stefano Ferrari", Modena, Italy.

Figure B7.4.1 Graziella Pelligrini, University of Modena and Reggio Emilia, Modena, Italy.

Figure B7.5.1 Stamcellen Veen Magazines.

Figure B7.9.1 Stamcellen Veen Magazines.

Figure B7.10.1 Stamcellen Veen Magazines.

Figure B7.10.2 Jan Roelen/Stamcellen Veen Magazines.

Figure B7.16.1 David McKeon, NYSCF, New York, NY, U.S.A.

Figure B7.16.2 David McKeon, NYSCF, New York, NY, U.S.A.

Chapter 9: Susana Chuva de Sousa Lopes

Figure 9.1	Linda van Laake, University Medical Center, Utrecht, Netherlands.
Figure 9.3	Stieneke van den Brink, Hubrecht Institute, Netherlands/Stamcellen Veen Magazines.
Figure 9.4	Stieneke van den Brink, Hubrecht Institute, Netherlands/Stamcellen Veen Magazines.
Figure 9.5	Dorien Ward, Hubrecht Institute, Netherlands/Stamcellen Veen magazines.
Figure 9.6	Stieneke van den Brink, Hubrecht Institute, Netherlands/Stamcellen Veen Magazines.
Figure 9.7	Stieneke van den Brink, Hubrecht Institute, Netherlands/Stamcellen Veen Magazines.
Figure 9.8	Anna de Lisio/Susana Chuva de Sousa Lopes, Hubrecht Institute, Netherlands/Stamcellen Veen Magazines.
Figure B9.1.1	Cees van Echteld, University Medical Center, Utrecht, Netherlands/ Stamcellen Veen Magazines.
Figure B9.1.2	Cees van Echteld, University Medical Center, Utrecht, Netherlands/ Stamcellen Veen Magazines.
Figure B9.1.3	Cees van Echteld, University Medical Center, Utrecht, Netherlands/ Stamcellen Veen Magazines.
Figure B9.2.2	Hubrecht Institute, Netherlands/Stamcellen Veen Magazines.
Figure B9.2.3	Ewart Kuijk.
Figure B9.3.1	Stefan Braam, Hubrecht Institute, Netherlands/Stamcellen Veen Magazines.
Figure B9.3.2	Stefan Braam, Hubrecht Institute, Netherlands/Stamcellen Veen Magazines.
Figure B9.4.1	Stamcellen Veen Magazines.
Figure B9.5.1	Dennis van Hoof/Jeroen Krijgsveld, Hubrecht Institute, Netherlands/ Stamcellen Veen Magazines.
Figure B9.6.1	Adapted from "Germline Genomics" by Valerie Reinke, Department of Genetics, Yale University, New Haven, Conn., U.S.A.; NCBI/NLM, open domain.
Figure B9.10.1	Stamcellen Veen Magazines.
Figure B9.12.1	Leon Tertoolen, Leiden University Medical Center, Netherlands.

Chapter 10: Anja van de Stolpe, Philips, Eindhoven, The Netherlands

Figure 10.1	Reproduced from Hans Clevers, The intestinal crypt, a prototype stem cell compartment, Cell, 2013, 154, pp. 274–284.

Figure 10.2 Reproduced from Hans Clevers, The intestinal crypt, a prototype stem cell compartment, Cell, 2013, 154, pp. 274–284.

Figure 10.3 Reproduced with permission from Gilbert, P.M. and Blau, H.M., Engineering a stem cell house into a home, Stem Cell Res. Ther., 2011, 2, p. 3.

Figure B10.1.1 Reproduced from Hans Clevers, The intestinal crypt, a prototype stem cell compartment, Cell, 2013, 154, pp. 274–284.

Figure B10.2.1 Reproduced from Hans Clevers, The intestinal crypt, a prototype stem cell compartment, Cell, 2013, 154, pp. 274–284.

Figure B10.3.1 Reproduced with permission from Lancaster, M.A. et al., Cerebral organoids model human brain development and microcephaly, Nature, 2013, 501, pp. 373–379.

Chapter 11

Figure 11.1 www.lougehrig.com.

Figure 11.5 Susana Chuva de Sousa Lopes/Stamcellen Veen Magazines.

Figure B11.1.1 Photograph courtesy of Joke Kniest.

Figure B11.2.1 Photograph courtesy of the ISSCR and Jim Ezell, photographer.

Figure B11.2.2 Photograph courtesy of Nancy Witty.

Chapter 12: Rui Monteiro, Hubrecht Institute, Utrecht, The Netherlands

Figure 12.3 Reproduced with permission from Vermeulen L. et al., Wnt activity defines colon cancer stem cells and is regulated by the microenvironment, 2010, Nature Cell Biology 12, 468–476.

Figure 12.4 Hans Clevers, Hubrecht Institute, Netherlands.

Figure 12.6 Reproduced with permission from Vermeulen L. et al., Wnt activity defines colon cancer stem cells and is regulated by the microenvironment, 2010, Nature Cell Biology 12, 468–476.

Figure 12.7 Reprinted with permission from Anja van de Stolpe: On the origin and destination of cancer stem cells: a conceptual evaluation, 2013, Am. J. Cancer Res. 3, 107–116.

Figure B12.1.1 Susana Chuva de Sousa Lopes.

Figure B12.1.2 Reprinted with permission from Barker N., et al., Identification of stem cells in small intestine and colon by marker gene Lgr5, 2007, Nature 449, 1003–1007.

Figure B12.1.3 Hans Clevers, Hubrecht Institute, Netherlands.

Figure B12.1.4 Hans Clevers, Hubrecht Institute, Netherlands.

Figure B12.1.5 Reprinted with permission from Sato T., et al., Single Lgr5 stem cells build crypt-villus structures in vitro without a mesenchymal niche, 2009, Nature 459, 262–265.

Chapter 13: Ronald Dekker, Philips, Eindhoven, The Netherlands

Figure 13.1 Photographs courtesy of Ronald Dekker, Philips, Eindhoven, Netherlands.

Figure 13.2 Reproduced with permission from van de Stolpe A. and den Toonder J, Workshop meeting report Organs-on-Chips: human disease models. 2013, Lab Chip 2013, 13, 3449–3470.

Figure 13.3 Photograph courtesy of Robert Passier and Ronald Dekker.

Figure 13.4 Reproduced with permission from van de Stolpe A. and den Toonder J, Workshop meeting report Organs-on-Chips: human disease models. Lab Chip 2013, 13, 3449–3470.

Figure 13.5 Reproduced with permission from van de Stolpe A. and den Toonder J, Workshop meeting report Organs-on-Chips: human disease models. Lab Chip 2013, 13, 3449–3470.

Figure B13.2.1 Drawing adapted from Kim H.J., Human gut-on-a-chip inhabited by microbial flora that experiences intestinal peristalsis-like motions and flow. Lab Chip 2012, 12, 2165–2174.

Figure B13.2.2 Uwe Marx.

Chapter 14: Bernard Roelen

Figure 14.3 Rob de Vries and Hans Clevers, Hubrecht Institute, Netherlands.

Figure 14.6 Caterina Martines Grandela, Leiden University Medical Center, Netherlands.

Figure B14.3.2 CIRM, www.cirm.ca.gov.

Figure B14.4.1 Galapagos-Biofocus BV, Leiden, Netherlands.

Figure B14.4.2 Galapagos-Biofocus BV, Leiden, Netherlands.

Chapter 15: Stieneke van den Brink, Hubrecht Institute, Utrecht, The Netherlands

Figure 15.1 Jensen J. et al., Human embryonic stem cell technologies and drug discovery. 2009, J. Cell. Physiol. 219, 513–519.

Figure B15.1.1 Photograph courtesy of Oliver Brüstle.

Chapter 16: © 2008 Managed Objects Inc.

Figure B16.2.1 Kelly Hosman.